GYMNOCALYCIUM

IN HABITAT AND CULTURE

Fig.1 *G. berchtii,* GC939.01 in habitat near to Los Chañares, Prov. San Luis, Argentina

GYMNOCALYCIUM

IN HABITAT AND CULTURE

Graham Charles

Fellow of the Cactus & Succulent Society of America
Member IOS

Dedication
To my wonderful wife Elisabeth for her love and inspiration. Sharing our plant hobby has been a joy for me, and I have indeed been fortunate to have her blessing for my many trips to South America in search of cacti.

Published by:
Graham Charles, Briars Bank, Fosters Bridge,
Ketton, Stamford, PE9 3BF England
E-mail: graham.charles@btinternet.com

ISBN 978-0-9562206-0-8

Printed and bound in the U.K. by
Butler Tanner & Dennis, Frome, BA11 1NF

Contents

There are two new combinations in this book:
Gymnocalycium horridispinum subsp. *achirasense* on page 154.
Gymnocalycium marsoneri subsp. *megatae* on page 241.

Foreword

It seems a rather long time since John Pilbeam's *Gymnocalycium, A Collector's Guide* came out in 1995 and our knowledge of the genus has surely moved on a great deal. Surprising, then, that it has taken 14 years before anyone has managed to bring out another tome on the genus, or is it? Perhaps not, because this is arguably one of the most complicated genera of cacti to get to grips with, both in terms of its diversity and the huge burden of modern literature to be analysed.

Before getting involved in the *New Cactus Lexicon* project I must confess to having felt severely challenged by *Gymnocalycium* in terms of being able to recognize its myriad species. Even after the Lexicon process, to which Graham Charles was a major contributor, I'm not sure I knew much more and regarded the genus as too complex to understand. Graham's masterful treatment, now laid before us, has put these Gymno demons to bed for me and with this new book we can all begin to appreciate these South American discoid cacti a lot more, whether from the viewpoint of botanist or grower. It seems clear to me that *Gymnocalycium* can expect a huge increase in popularity over the next few years and this will be a direct consequence of Graham's book and well deserved.

The book has two great strengths. First, it documents what is known about the plants very thoroughly and provides rather more in the way of objective information than author's opinion, which is a welcome change from the current literature, full as it is with subjective polemics on the pros and cons of this or that (micro-) species. Secondly, it is extremely well illustrated, with modern photographs, historic drawings and clear distribution maps, the images all carefully documented so that they have the value of a pictorial herbarium. All this also says a lot about the author's ability to liaise with other Gymno folk and the book above all is a great example of collaborative effort brought together with much drive and determination.

I for one will look on Gymnos with different eyes in future and may even be persuaded to grow a few more!

Nigel Taylor

Curator, Royal Botanic Gardens, Kew
March 2009

Preface

Fashions in cactus genera are difficult to predict and, in my 45 years of growing them, old favourites have taken a back seat and new 'stars' have become the rage. Gymnocalyciums appear to have remained popular during all of this time and remain much in demand today. So why is this? While *Notocactus* species, so sought after in the 1970's, are now on nobody's wants list. Every minute variant of *Sulcorebutia* is eagerly sought, even though there is more diversity to be enjoyed in the currently unpopular genus *Rebutia*.

I must admit that Gymnocalyciums have fascinated me ever since I first bought one all those years ago. I think one of the reasons was my difficulty at telling them apart. So many plants supplied with different names looked just the same! My interest was reinforced by my first encounter in the wild, a trip to Argentina in 1992. I was impressed by the diversity of habitats where Gymnocalyciums grow, from the most arid high altitude plains to grassland and even peaty moorland. Their ability to thrive in my glasshouse also endeared them to me, and in the 1970's, nurseries in Britain were offering habitat-collected plants from the activities of Alfred Lau, Frau Muhr and others. Although said at the time to be difficult to establish, I still have the plants today and a few are illustrated in this book.

Wanting to learn more about these plants, I searched the English language journals for information. Articles by John Donald, the then champion of South American cacti, further stimulated my interest and discussions in 'The Chileans', often translated from foreign periodicals, made me realise how much there was to know. The only dedicated work in English was the small booklet *Gymnocalycium* by Bill Putman but there was little information based on habitat observations which, for me, was the key to understanding relationships. An important book written in Czech by Bohumil Schütz appeared in 1986 *Rod Gymnocalycium* where he introduced his classification based on seeds, and gave so much interesting information about the first introduction to cultivation of the species. Then in 1995, John Pilbeam wrote his book *Gymnocalycium A Collector's Guide*, the first comprehensive account of the genus in English with colour pictures and a useful checklist for collectors. It proved very popular and since going out of print, it has become scarce and sought after.

For real enthusiasts, detailed information about the plants can only be found in journals published by specialist groups and the genus *Gymnocalycium* has benefited from two. The German A5 publication "GYMNOS" appeared in 1984 and continued for 26 issues until 1997. A group of Austrian enthusiasts has been publishing a well-illustrated A4 loose-leaf periodical called "GYMNOCALYCIUM" since 1988. It contains much valuable information about history, habitats and taxonomy, illustrated with plants in the wild and cultivation. Many of the most recently described *Gymnocalycium* taxa have been published in its pages.

The internet has recently become an important medium for finding information about the cactus hobby. There is a wealth of specialist detail that is only available on the web, for instance, there are pages about *Gymnocalycium* with pictures and text not available elsewhere. The gateway to all this knowledge is The Cactus and Succulent Plant Mall which can be accessed at www.cactus-mall.com.

My decision to embark on the writing of this book was based on two primary objectives. The first was to make available in English some of the information about Gymnocalycium that has only been published in other languages. Secondly, to illustrate the text with good quality colour pictures of plants of documented origin both in habitat and cultivation. Because it is so easy to inadvertently create hybrids in cultivation, many plants in collections today are not pure species, so without a documented wild origin, they are unreliable for serious study.

When embarking on a book project like this, one of the decisions one has to make is whether to simply list all the described taxa, or to endeavour to rationalise the list by identifying synonyms. Like most genera of the family Cactaceae, *Gymnocalycium* has many examples of new names that are just forms of existing species. It is a matter of how much variation one is willing to accept within a species. Specialists tend to look for differences rather than similarities, so when a new population of plants is discovered, the temptation is to give it a new name. It is much easier to do that than to determine to which existing species the plants might belong.

Many fellow enthusiasts have generously given me information and advice, for instance about places I have not visited and plants I have not seen in the wild. The result is that I have endeavoured to decide whether a newly described taxon is different enough to be recognised, or whether it is best treated as a synonym of an existing species. The extent to which this is done is a matter of individual judgement, so it may be that when more information becomes available, my treatment will need to be revised. This activity, commonly refered to as lumping, has been the approach recently taken in books produced by professional botanists. In contrast, amateurs continue to describe new taxa for minor variants and split established species into multiple segregants as subspecies, varieties and forms (see the table of accepted species numbers on page 12). Accepting so many names at so many ranks becomes cumbersome so, for simplicity, this book only deals with taxa at the rank of subspecies and above. The infraspecific names in the picture captions are of subspecies rank. Other names of lower rank, synonyms or names that are not validly published are shown in brackets where appropriate. Of course, you are not obliged to follow any proposed treatment. It is your choice as to what name you put on your plant label. I hope this book will help you decide.

There are a number of examples of groups of related species which are incompletely understood, so for these, my treatment has to be regarded as provisional. Firstly, the named species of subgenus *Gymnocalycium* from Prov. San Luis, Argentina, are in need of revision. There are likely to be less recognizable species than accepted in this book. A similar situation exists with subgenus *Gymnocalycium* in northern Prov. Córdoba, Argentina, and the neighbouring provinces, where rationalisation of the names is likely to be possible. Finally, more research is needed into the confusing taxa belonging to subgenus *Muscosemineum* in Paraguay and Bolivia.

With writing this book, I am aware that I shall upset some people who will not agree with my treatment of the published names. I have tried to reach a reasonable view, but as more information becomes available, my conclusions will probably be challenged. All I can do is record what I have learnt about the genus up to the end of 2008, and share it with you. Even as I write these words, yet more *Gymnocalycium* names have been published and so the assessment of new discoveries will continue.

Graham Charles

Stamford, England
February 2009

Acknowledgments

People who have an interest in cacti are friendly and willing to encourage others in all aspects of the hobby. Many such people have become my friends over the 45 years since I first took an interest in these remarkable plants. Without their help, I would not have been able to write this book. Many have encouraged me, whilst others have provided practical help, for instance, by generously sharing their knowledge, or by providing pictures. At the risk of omitting someone, I would like to acknowledge those who have been particularly helpful in the preparation of this book, starting with John Pilbeam, who put the idea in my head while he was considering whether to update his much sought-after book on the subject. His meticulous checking of my text has enabled me to correct many errors I made during its creation. I must also thank Dr. Nigel Taylor for reading the treatment of the recognized taxa and pointing out where I had exhibited my lack of botanical training in the text. I am also grateful to him for contributing the foreword to the book.

My interest in the genus was greatly stimulated by visiting their natural habitats in South America. I have been fortunate to have been able to go there with many travelling companions, notably Chris Pugh who has accompanied me on most occasions and whose ability to spot plants in the wild is unsurpassed. His calm approach to the difficulties which inevitably arise during such adventures has been a great source of comfort. My most recent 'Gymno' trip was undertaken at the end of 2007 with an objective of seeing some of the species in habitat which had eluded me in the past. I had the pleasure of the company of Roberto Ariu from Italy and Guillermo Rivera who lives in Córdoba, Argentina. Guillermo shares my interest in the genus and was very helpful with showing me a number of localities which I am sure I would have struggled to find.

The majority of literature about *Gymnocalycium* has been published in languages other than English, for instance German and Czech. I am indebted to Francis Fuschillo, another 'Gymno' enthusiast, for sharing the translations he has made over many years. Francis is probably best known for his photographs of seeds that required considerable patience to produce, and have featured in many articles over the years.

I really enjoy reading about the history of cactus exploration and the naming of new discoveries. Some of the early literature is rarely, if ever, offered for sale, so I don't own some of the important publications in the history of *Gymnocalycium*. My thanks go to Gordon Rowley for his generous help in providing photocopies of rare references, and allowing me to photograph some beautiful early illustrations in his library, so enabling me to reproduce them here. I have also been privileged to be able to use the splendid library at the Royal Botanic Gardens, Kew, and I would like to thank the library staff for their assistance in finding the items I sought.

Many students of *Gymnocalycium* have helpfully shared their knowledge about the genus and I have tried to represent their views here, even if I do not agree with them. I am especially indebted to Massimo Meregalli who made his researches available to me, especially about plants from Uruguay. My travels have yet to take me to Paraguay, so I am very thankful for the advice and help given to me by Volker Schädlich and Ludwig Bercht concerning the plants from there, and other taxa about which they have so much knowledge. I have also had many hours of fruitful discussion with Graham Hole who lives near to me and has an excellent documented collection of plants.

It is very difficult for one person to assemble a reasonably complete set of photographs to illustrate a book like this, particularly pictures of plants in habitat. So, my grateful thanks go to the many people who have allowed me to use their pictures including Geoff Bailey, Ludwig Bercht, Pat and Sally Craven, Norbert Gerloff, Franz Kühhas, Alfredo Lorenzini, Marlon Machado, Massimo Meregalli, Detlev Metzing, Roy Mottram, Guillermo Rivera, and Volker Schädlich.

The Maps

Most of the distribution maps are based on the series of maps published by Nelles Verlag GmbH, Munich. Their mapping clearly shows physical features, giving a good impression of the mountains that often influence where the plants can be found.

I am very grateful for permission from the publisher to reproduce parts of their maps in this book.

Details of their extensive range can be found at:
www.nelles-verlag.de

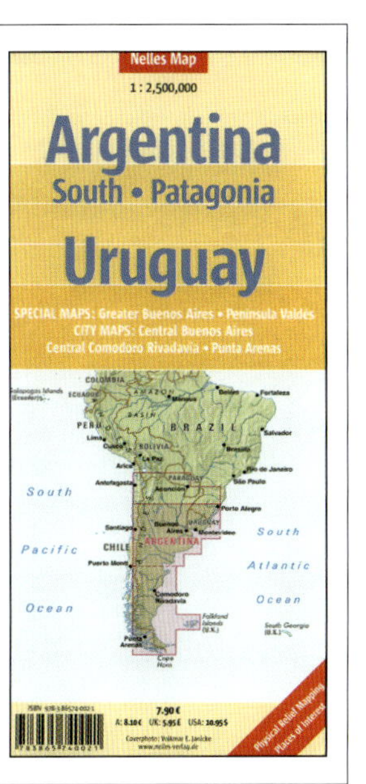

History of the Genus

The first description of a plant which would later be classified as a *Gymnocalycium* was in 1812 when Haworth named *Cactus gibbosus*. This was followed by *Cactus reductus* Link in 1822 and *Echinocactus denudatus* Link & Otto in 1828. These were the three species listed at the time of the creation of the genus *Gymnocalycium*.

Until recently, the author of the generic name *Gymnocalycium* had been attributed to Pfeiffer (1845). He was a doctor in Kassel, Germany, and a keen cactophile. From what he wrote, we know that the name *Gymnocalycium* first appeared in a catalogue of the collection of Schelhase in 1843. It was probably just a list without descriptions, but no copy is known to have survived. The type species was taken to be *G. denudatum* by Britton & Rose (1922) because it was the first of the three species listed by Pfeiffer.

Gymnocalicium Pfr.

Nacktdistel.

Diese neue Gattung bildete Herr Dr. Pfeiffer aus den schon länger bekannten drei Species, welche früher theils den Cereen, theils den Echinocacten zugetheilt waren. Blüthen wie bei Echinopsis-Arten aus den ausgebildeten Knoten, röhrenförmig; Kelch nackt.

Gymnocalicium denudatum Pfr. (Echinocactus denudatus Lk. et O.)
— gibbosum Pfr. (Echinocactus gibbosus H. Berol.)
— reductum Pfr. (Cereus reductus DC.)

Fig.2 Mittler's description of *Gymnocalicium* from his book *Taschenbuch für Cactusliebhaber* (1844)

It was Metzing (1988) who first pointed out an earlier description of the genus by Mittler (1844) in his *Taschenbuch für Cactusliebhaber*, a general book for cactus enthusiasts. He used the spelling 'Gymnocalicium' and acknowledged Pfeiffer as having proposed the name (Fig.2).

The correct designation of the genus was formally established by Eggli and Metzing (1992) when they published a proposal to conserve the orthography of *Gymnocalycium* Pfeiffer ex Mittler in Taxon 41(1); p.141-142. This interesting example of taxonomic detective work settled the authorship of the genus as 'Pfeiffer ex Mittler', which means that although Mittler published the validating description, he attributed the name to Pfeiffer. It also fixes the usual, familiar spelling of *Gymnocalycium* and corrects the type species to *G. gibbosum* since this species was chosen by Pfeiffer himself in a later work 'Nomenclator botanicus' of 1874.

Fig.3 Fric catalogues from the 1930s

The genus was not universally accepted: Prince Salm-Dyck refused to use it, leaving the plants in *Echinocactus,* as did Förster (1846) and Labouret (1853); Schumann (1899) erected a subgenus of *Echinocactus* which he called *Hybocactus* to accommodate those taxa which would later be classified as *Gymnocalycium,* but he also included species which were eventually placed in other genera.

Following Schumann, two botanists in South America then described more new species using the genus *Echinocactus* for the plants. Dr Carlos Spegazzini in Argentina and Dr. José Arechavaleta in Uruguay

Author	Publication	Date	Number of Accepted Species
DeCandolle, A.P.	Prodromus Systematis Naturalis	1828	2
Förster, C.F.	Handbuch der Cacteenkunde	1846	6
Labouret, J.	Monographie des Cactées	1853	9
Rümpler, T.	Handbuch der Cacteenkunde	1886	9
Schumann, K.	Gesamtbeschreibung der Kakteen	1903	12
Britton, N.L. & Rose, J.	The Cactaceae, Vol.3	1922	23
Berger, A.	Kakteen	1929	24
Backeberg, C.	Das Kakteenlexikon	1966	83
Pilbeam, J.	Gymnocalycium. A Collector's Guide	1995	80
Anderson, E.	The Cactus Family	2001	71
Anderson (Eggli, U.)	Das Grosse Kakteen Lexikon	2005	81
Hunt et al.	The New Cactus Lexicon	2006	49
Till, H.	Gymnocalycium Sonderausgabe	2008	103
Charles, G.	Gymnocalycium in Habitat and Culture	2009	56

Fig.4 Table of the numbers of accepted species attributable to *Gymnocalycium*

Fig.5 Examples of catalogues from the firm H. Winter that offered seeds from Ritter's explorations (1950s and 1960s)

published works in 1905. At around the same time the collecting activities of Fric resulted in the description of *Echinocactus mihanovichii* by Prof. Max Gürke.

Britton and Rose (1922) recognised 23 species of *Gymnocalycium*, publishing a number of new combinations of taxa originally published as species of *Echinocactus*. Their account includes many illustrations reproduced from other publications or obtained from specialists of the period. Even so, some of these were probably mis-identified, a problem which continues to confuse the understanding of the genus to this day.

Another significant botanist working in Córdoba, Argentina was Dr. Carlos Hosseus who collected large quantities of local cacti and sent them to the nursery of Adolf Haage jn. in Erfurt, Germany. There was great interest in the cactus hobby during the late 1920s and 1930s in Germany and this was encouraged by a stream of novelties from South America (Fig.9). There were many new *Gymnocalycium* names in the catalogues of A. V. Fric who was a very active explorer (Fig.3), although the descriptions of his finds were left to others.

This period also saw the commencement of the activities of Curt Backeberg who was to become the most prolific author of the century on the subject of cacti. Backeberg made a trip to Peru, Bolivia and northern Argentina in 1930-31 commissioned by the Haage nursery. He collected a number of new taxa himself, as well as buying habitat plants from the commercial collector Stümer who lived in Buenos Aires. He went on to publish many of these as new species, as well as describing plants he found in German nurseries which lacked accurate habitat data. The discovery of a number of the Paraguayan species was achieved by Adolfo Maria Friedrich, who was employed as a photographer during the Chaco war between Bolivia and Paraguay (1933-1935) and was able to utilise new roads built through the Chaco at that time. He sent plants to Germany including to the Blossfeld nursery at Potsdam.

The next period of activity came after World War II when the Bolivian botanist Martin Cárdenas described many cacti from his country. Exploration continued into the 1960s and 1970s with new Brazilian species being found by Buining and Horst, and further novelties resulting from the collections of Fechser, Rausch, Knoll and the most accomplished explorer of all, Friedrich Ritter. Around this time, Alfred Lau made extensive collections in South America and made many little-known plants available to enthusiasts through various European nurseries. We also have Jörg and Brigitte Piltz to thank for making seeds available of a wide range of *Gymnocalycium* species harvested from plants collected in habitat with disclosed localities. Since then, the strict laws controlling the collection of plants in the wild have resulted in habitat-collected specimens rarely being available in Europe. At the same time, there has been a great increase in the number of enthusiasts visiting the habitats, resulting in the publication of many new names.

This appetite for exploration and growing plants with location data created a demand for a specialist journal, first satisfied by 'GYMNOS', a German A5 publication which appeared in 1984 and continued for 26 issues until 1997. Another German language publication, *Gymnocalycium*, the journal of an Austrian specialist group, was first issued in 1988 and still continues today. It is a well-illustrated A4 loose-leaf journal with articles that evaluate old names and often publish new ones.

Many recent enthusiasts have made useful contributions to our understanding of the genus. Notable among these are Hans Till and his son Walter from Austria who have published many articles putting forward their opinions of the old, often confused, names. Amongst many authors who have written on the genus in recent years, I should also acknowledge Detlev Metzing (Germany), Massimo Meregalli (Italy), Wolfgang Papsch (Austria), Gert Neuhuber (Austria), Walter Rausch (Austria), Volker Scädlich (Germany) and Bohumil Schütz (Czech) for their significant contributions.

Gymnocalycium has long been one of the most popular genera for cactus enthusiasts to grow, so it is perhaps surprising that more books have not been written on the subject. The only comprehensive account written in English has been John Pilbeam's *Gymnocalycium A Collector's Guide* published in 1995. Since going out of print, it has become difficult to find and now commands a premium price.

Characteristics and Classification

Fig.6 The horizontally splitting fruit of *G. pflanzii pflanzii* GC858.05 in habitat south of Camiri, Bolivia

The genus *Gymnocalycium* is one of the most clearly defined in the family Cactaceae. There has been little disagreement about the bounds of the genus, one notable exception being the attempt by Paul Hutchison (1957, 1959) to include species of *Weingartia* and *Neowerdermannia*, an approach which has since received little support.

The fruits of *Gymnocalycium* vary considerably in shape and size, the largest probably being those of *G. horstii* which can attain the size of a small plum. They all lack spines, bristles or hair on the pericarp, having just a few naked scales. When completely ripe, the fruits usually split to reveal a sticky mass of seeds and pulp inside that in nature is soon visited by ants that clean out the contents remarkably quickly. However, the seeds are often ripe before the fruits actually split so it is possible to collect them before the ants get a chance. Almost all fruits split vertically but the interesting exception is *G. pflanzii* subsp. *pflanzii* where the split is horizontal, revealing unusually bright red pulp.

It is the seed which has become the most important tool in defining species relationships. Since most seeds are more than one millimetre in diameter, a hand lens is usually adequate to see to which subgenus the seed belongs. Using the images and descriptions given here for each subgenus should enable you to determine to which subgenus any *Gymnocalycium* seed belongs. The seed coat characteristics are always inherited from the female parent so you do not need to have true seed to examine. Seed from fruits on any plant will be characteristic of that species, irrespective of what pollinated it.

Classification

There have been various attempts to segregate the genus into groups of species based on seed characters. The first, founded on the work of Fric and Kreuzinger (1935), and still the one preferred by most, is that of Dr. Bohumil Schütz (1962). There was an alternative system proposed later by Buxbaum (1968) and this is still favoured by some. As I was completing work on this book, H. Till, H. Amerhauser and W. Till (2008) published the full version of their new organisation of the genus. It is radically different from historic treatments in classification and nomenclature, and is here rejected for a number of reasons. It is, in part, based on the application of the name *G. quehlianum* to a species of the subgenus *Gymnocalycium*, together with the unsubstantiated application of old unattributable names to plants which already have well-typified and well-known names. This attempted destabilisation of the taxonomy of the genus is to be regretted and serves only to cause confusion.

As discussed previously, there are many opinions on what constitutes a species. I have taken a broad view and accepted just 56 species and another 17 heterotypic subspecies. In contrast, the latest treatment of Till has accepted 103 species, with 68 further subspecies, together with many more varieties and forms. This is the splitter's view, where almost all the names ever published are accepted at some rank, even if the chosen application is little more than a guess. It is a fact that the more interest there is in a genus, the more names are created. Enthusiasts tend to look for differences rather than similarities, so they describe every form as something new. This is not a reasonable scientific approach and to be consistent across the whole of the

Fig.7 The naked bud, characteristic of *Gymnocalycium*, here seen on *G. spegazzinii*, *Vereb*17 from Cafayate. Photographed in the collection of Elisabeth Charles.

Fig.8 A group of *Gymnocalycium monvillei* GC390.02 grown from seeds collected in habitat from a single population in the Sierra Chica, Prov. Cordoba, Argentina

Cactaceae, species of *Gymnocalycium* should include a range of variation as they do in other, less popular genera.

I have here broadly followed the system of Schütz, modified in the light of recent discoveries and in minor ways to maintain simplicity. His approach was first published in the first issue of *Friciana* in 1962. The translated title of the article is 'Towards the Systematics of the Genus *Gymnocalycium*'. A more accessible version for English speakers appeared in the *National Cactus and Succulent Journal* of December 1969 called 'The Genus *Gymnocalycium*'. His five subgenera are convenient groupings of species that help to make a large genus more understandable. Subgenus *Microsemineum* has tended to become a repository for everything that doesn't fit in the other subgenera, so within it there are identifiable groupings of related species. The identification of *G. gibbosum* as the type species of the genus by Eggli & Metzing (1992)) happened after Schütz created his subgenera at which time the type was thought to be *G. denudatum*. Since the subgenus containing the type has to repeat the generic name, we now have to change the name of Schütz's subgenus *Ovatisemineum* to *Gymnocalycium*.

The list of accepted species and subspecies for each subgenus, together with their synonyms, can be found at the beginning of each section where the taxa are treated in detail.

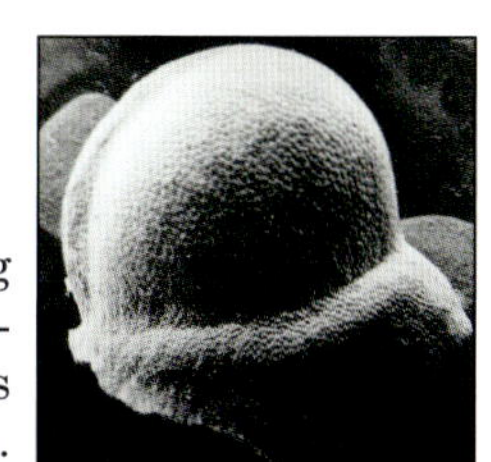

Macrosemineum Page 31
Description according to Schütz**:** Fruits still green when ripe, not always splitting open but sometimes softening and disintegrating. Seeds 1 to 3mm diameter, subglobose, a little compressed, noticeably broader toward the aril. Testa black, more or less dull. Hilum elongated and slightly depressed. Aril dark or pale around the hilum. Habitat Uruguay, E. Paraguay, S. Brazil and N.E. Argentina.

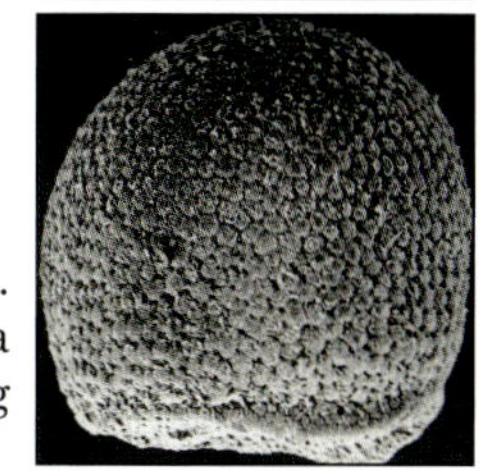

Gymnocalycum Page 57
Description according to Schütz: Fruit opening at maturity by a vertical split. Seeds c. 1mm diam., spherical or nearly spherical, apparently truncate at the aril end. Testa black, dull, fully or partly covered with a brownish film. Hilum round; aril protruding only slightly. Habitat Argentina (Patagonia, Córdoba).

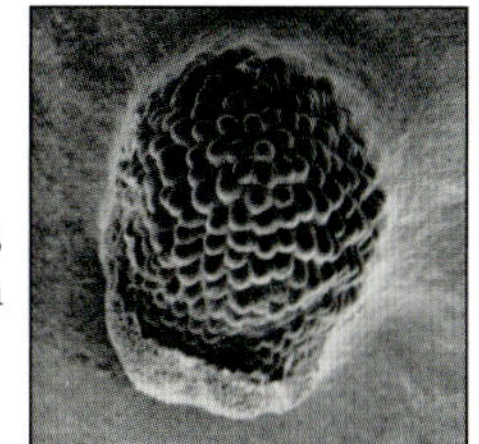

Microsemineum Page 129
Description according to Schütz: Plants of various sizes. Fruits splitting vertically or, in one species, horizontally so that the top of the fruit lifts up. Seeds small, less than 1mm diam. Habitat Argentina, S. Bolivia, W. Paraguay.

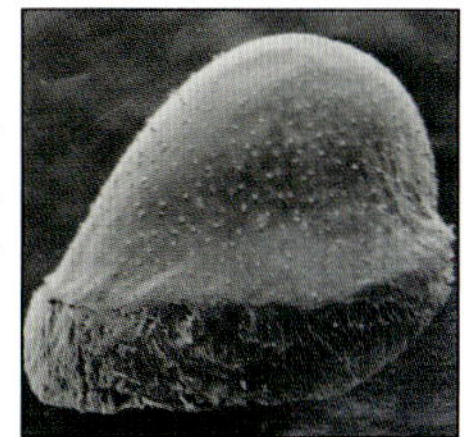

Trichomosemineum Page 208
Description according to Schütz: Small to medium-sized plants, hardly exceeding 15cm diam., very flat, mostly solitary. Fruits club-shaped, splitting vertically. seeds up to 1mm diam., hemispherical., a little depressed, shell-shaped, getting broader towards the aril. Testa light to dark brown, very shiny as if lacquered. Hilum basal, elliptical. Aril very prominent, mostly light. Habitat Argentina (Cordoba, La Rioja, Catamarca, Tucuman, San Juan).

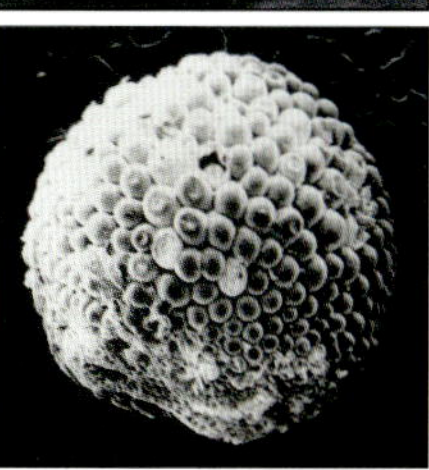

Muscosemineum Page 226
Description according to Schütz: Habit various: some species small, others up to 30cm diam. or more; some flat, others columnar with age, but usually un-branched. Fruit club-shaped, splitting vertically. Seeds up to 1mm diam., spherical. Testa light brown, dull, pulverulent. Hilum small. Aril narrow, not very prominent. Habitat Paraguay (Chaco); E. Bolivia; Argentina (Chaco, Cordillera). [Now known to occur in S.W. Brazil]

Recognizable Species and Subspecies

For the amateur enthusiast, the classification and nomenclature of plants can be a source of frustration due to the frequent name changes where familiar names are replaced, resulting in many 'alternative' names for the same plant. There can be a number of reasons for these changes, but whether you accept the changes can be a matter of personal belief or choice. I hope it will be helpful here to outline the process for naming plants and why some name changes are necessary.

The description of a new taxon is controlled by the 'International Code of Botanical Nomenclature', a complicated set of rules which must be followed for the publication of a name to be valid. The rules have been modified several times since they were first published in the 19th century. A convenient way to refer to these rules is to use the on-line version of the code at http//:ibot.sav.sk/icbn/main.htm where the text of the articles can be consulted to explain why a name is illegitimate or invalid.

Fig.9 Commercial catalogues from the 1920s and 1930s

Today, an essential part of the description is the designation of a 'type', a preserved specimen permanently stored in a recognized institution and intended to fix the use of a name forever. Types were not a requirement before 1st January 1958 and could be an illustration before 2006. Descriptions after 1st January 1935 must have a Latin diagnosis. Names not in accordance certain rules are designated as illegitimate (nom. illeg.), such as a name published including an older name at the same rank or its type in synonymy (that should have been used as the accepted name of the taxon in question).

When plants were named a long time ago, such as in the 18th century, the descriptions were often brief, maybe just sufficient to differentiate the new plant from the few other similar ones known at the time. The habitat location was often very vague or totally wrong so it can therefore be difficult or impossible to unequivocally equate the description, and hence the name, to a particular taxon known today. Lack of effective communication between botanists and personal competition in those early days meant that the same plant could be described by more than one individual using different names.

Later, plants would be imported into Europe by commercial nurseries like Haage and Uhlig from collectors working in the field. Authors, such as Backeberg, would see these plants and erect new species based on selections from the importations. There would often be little indication of where the plants had been collected and no understanding of the extent of natural variation in the taxon being described. Again, these taxa are often difficult to equate with certainty to particular plants we see in habitat today. Where such uncertainty exists, there are two possible courses of action. We can use what information we have to associate the name with a particular taxon known in the wild today. There are many examples of this being attempted over the years, but unfortunately different authors have sometimes chosen different taxa thus leading to confusion. The alternative is to dismiss the name as confused and unattributable (nom. confus.), thus allowing the use of a later but better-defined name.

Some botanists favour the former approach even if the association is little more than a guess. Others prefer to dismiss these confused old names and use a later, preferably typified, name. There are examples of both approaches in this book, the reason being that I have largely followed the choices made for the *New Cactus Lexicon* (2006) in an attempt to achieve a stabilised nomenclature for *Gymnocalycium*. This may appear somewhat inconsistent, but supports most of the reasonable long-used applications of names and minimises the introduction of further confusion. A list of names of uncertain application and an explanation of their history can be found starting on page 260.

When there was no holotype designated in the original description and no original material exists, which is often the case with old descriptions, a neotype can be designated by a later author. This should be a specimen or illustration that supports the interpretation of the original name which the author making the designation believes most likely to be the original author's intention. This practice is intended to fix the use of the name, but a neotype can be challenged if there is evidence that it is in conflict with details in the original publication. Names where a neotype has been designated have been used here in the sense commensurate with that neotype, for instance, *G. quehlianum*, a name abandoned in the *New Cactus Lexicon* but for which there is a neotype.

List of Species and Subspecies recognized in this book.

Individual lists will be found at the beginning of the section dealing in detail with the plants of each subgenus. Names treated as synonyms are included with those lists.

Cultivation

The following notes are intended to give a general understanding of the cultural requirements of Gymnocalyciums, especially when grown under glass in temperate climates.

Temperature and Light

To grow the best quality plants, location is the first and probably the most important consideration. One key aspect of location is where you live. Generally speaking, the warmer your climate is, the easier it is to provide the conditions the plants will enjoy. However, *Gymnocalycium* have evolved in climates which have seasons which may be characterised by periods of drought or cold or both. So, they are difficult to grow well in places with regular high humidity and warm nights.

For most of us the challenge is to keep the plants sufficiently warm and receiving enough sunshine. In Europe, the ideal climate is around the Mediterranean such as in southern Spain, southern France and Italy where many species thrive outdoors. A similar climate can be found in parts of Australia and notably the south west of the USA where California has become a centre of commercial cactus growing because of its near-ideal environment. Here, the reliably sunny days and cool nights enable Gymnocalyciums to be grown so long as they are given an appropriate level of protection from rainfall and sunshine. Seeing the results obtained by skilled growers with these idyllic conditions are quite astonishing.

Succulent growing is very popular in the countries of northern Europe where a glasshouse is essential for the serious hobbyist. A smaller collection can be housed in a conservatory or even on a sunny window-ledge, but eventually the plants outgrow their surroundings and a glasshouse becomes the only answer. There is a long history of growing Gymnocalyciums in eastern Europe where the winters are extremely cold and heating a glasshouse is too expensive, even for those which are partly underground, a design which was once popular in cold climates. In such circumstances, once the plants are dry and have stopped growing for the winter, they can be brought into the house and plants survive indoors away from bright light until the spring, when they can be returned to the glasshouse.

Fortunately, for those Europeans who live further west, the winters are generally less severe and they can leave their plants in a glasshouse all the year round so long as it has adequate heating. The siting of the glasshouse is very important although, with the smaller gardens provided with today's new houses, there may not be much choice. It should be erected in the sunniest place possible, unshaded by trees or buildings. Winter sun is particularly valuable, so bear in mind that when the sun is lower in the sky obstructions will be more of a consideration. This advice is based on the principal that you can always shade plants in a sunny glasshouse but you cannot make a shaded glasshouse sunny. Sun is not only important for the strong growth of spines but it encourages flowering and generally heats the glasshouse, so allowing you to ventilate more whilst maintaining the temperature. A common problem in cultivation is lack of light and this is often seen with plants grown in a house, the new growth gets drawn up and unhealthily thin, a condition known as etiolation. If you have the option of orientating your glasshouse in any direction, then it is generally considered best to have the ridge running in a north-south direction to maximise the sun's energy passing through the glass and lean-tos are best on a south facing wall in the northern hemisphere. Whichever type of glasshouse you choose, make sure that you specify the maximum amount of ventilators possible, both in number and size.

As your collection grows due to new acquisitions and successful cultivation, then the temptation to erect shelves can be irresistible. If you place sun-loving plants on these shelves, being near to the glass, they will probably thrive. The problem is that plants beneath the shelves will receive less light and this can be detrimental to their development. Shelves in a glasshouse can be a benefit but too many will result in some plants suffering from lack of light, so it is a matter of judgement as to how many shelves to construct and where to put them to avoid creating too much shade. It is often Gymnocalyciums that are placed in the shadier parts of a mixed collection, because the owner wants to give the sunniest locations to Mexican cacti which would certainly suffer in the shade. Whilst Gymnocalyciums will tolerate this situation, the best results are achieved in a sunny, well-ventilated location where plants will grow strong spines and be less prone to growing unnaturally tall.

Heating

The minimum temperature you have to maintain depends on which species you want to grow. Most collectors are satisfied to keep the temperature above 5°C which is sufficiently warm for most Gymnocalyciums and the majority of cacti and other succulents. Those requiring warmer conditions can be accommodated in a propagator within the glasshouse that can easily be kept at a higher temperature. It

Fig.10 Gymnocalyciums (in the foreground) as part of a mixed cactus collection.

actually doesn't cost any more since the escaping heat contributes to the overall heating, so a cabinet with glass or plastic panels and a thermostatically controlled tubular heater with a fan to circulate the air can easily be constructed. In summer, the panels can be removed to improve ventilation and the light intensity reaching the plants. Generally speaking, every 5 deg C increase in the minimum temperature of the overall glasshouse doubles the cost, but none of this is really an issue for Gymnocalyciums since only a few species from the Bolivian and Paraguayan Chaco benefit from winter temperatures above 5°C. Please refer to the cultivation notes on individual species to see which these are.

Ventilation

Just as important as heating, ventilation is often neglected by succulent enthusiasts. In habitat, plants are used to moving air, and stagnant air, particularly if it is hot, can be damaging even to the most arid-adapted species. If you use a fan heater in winter then leave the fan on at all times, even when the heating elements are off. It can be beneficial to leave the fan on in summer, although better still is to have a large fan which turns on when the temperature rises above 20 or 25°C. In propagators and heated cabinets, the small fans recycled from discarded computers are ideal to keep the air moving, particularly when used in conjunction with electric tubular heaters.

The use of the glasshouse ventilators is also a vital consideration. Most growers fit automatic controllers to their ventilators which open them proportionally to the internal temperature. These are a great benefit and remove the risk of forgetting to open the vents on a warm sunny day which could result in scorching of the plants. The only problem is that they close the windows during cool summer nights to maintain the temperature when, in fact, the plants would grow better if the temperature was allowed to drop. Diurnal variation in temperature is beneficial to Gymnocalyciums since it is what they experience in habitat and what they have evolved to exploit using their their Crassulacean Acid Metabolism (CAM). So, if the night temperature is forecast to remain above 10°C, then the ventilators should be left open because, if the internal temperature stays above 20°C, growth of these CAM plants will be adversely affected.

Scorching of plants, which in effect is the burning of the epidermis, will scar plants for years or even result in death. It is caused by sunburn and is most likely to happen in early spring on a sunny day when the plants are still weakened from the long winter rest. Small glasshouses are more prone to this problem than larger ones where the bigger volume appears to facilitate air movement. Plants which are dry and near to the glass in stagnant air are most likely to be affected, so the solution is to have a fan running and the windows open. Alternatively, the risk can be reduced by painting a light temporary shade to the outside of the glass using one of the products sold for this purpose. Rain will wash it off slowly during the summer, but once plants are in active growth, it can be wiped off to maximise light intensity during the growing season.

Fig.11 *Mammillaria* is the one genus claimed to be more popular than *Gymnocalycium.*

Soils and Potting

Successful cultivation can only be achieved by an effective combination of all the factors involved and there is no simple foolproof formula. If you get the basics right, you will be able to grow the plants and gain the experience which will enable you to improve your technique. When visiting collections it is clear that there are many different approaches which can work. There are many and varied soil mixtures but so long as other factors such as locality and watering are appropriate then plants will thrive. Cactus growers are always willing to share their expertise including the soil mixes they use. Although at first they seem quite different, there are many common features. Considering the diversity of habitats where *Gymnocalycium* grow naturally, it is remarkable that almost all will grow in the same basic mixture, the most important considerations are texture, pH and nutrients.

In nature, most Gymnocalyciums grow in largely mineral soils formed from weathered rock and containing little humus, but they can often be high in nutrients. Some growers use a similar substrate in cultivation consisting of a mixture of gravel, coarse sand, crushed lava or other mineral materials, whilst avoiding fine particles which could clog. When using this sort of medium, the watering and feeding becomes very influential with little margin for error with regard to nutrients and pH; in effect it is like hydroponics. The benefit with this approach is that the texture of the medium remains consistent over time and so the need to re-pot plants is reduced, only becoming necessary when the plant outgrows its container. However, small pots dry out quickly, not ideal for Gymnocalyciums which thrive when kept just moist through the growing season.

So-called soil-less composts have been very popular in recent times. They are manufactured from peat or peat substitutes such as ground bark or coconut fibre (coir) mixed with lime and a fertiliser. These composts can be used for the cultivation of cacti, and in fact are extensively used by commercial growers who usually have special formulations made up which also contain grit, less or no lime and perhaps a slow-release fertiliser. For the amateur, these composts have a number of disadvantages, the main one being their short life. As the organic matter ages, it disintegrates and the open texture breaks down so that the small particles pack together resulting in root loss and, once dry, it can be very difficult to re-wet. Another difficulty is that plants in this type of compost need to be constantly fed since fertiliser is quickly washed out. So, although growth can be very good when plants are first put in these soils, their long-term maintenance is problematic. They can, however, be useful for raising seedlings when the ericaceous (lime free) formula is probably the best.

The usual ingredients used in soil preparations are loam with peat or substitutes, to which is added grit, coarse sand or soil conditioners like vermiculite or perlite. These ingredients can be purchased in bags from a garden centre or DIY shop, the quality of which is variable and any manufacturer is often unable to supply a consistent product from year to year. Using these ingredients, the following soil mixes are a good basis which will enable you to start growing plants, after which your own experience can make appropriate modifications for your circumstances.

The ingredients:

Loam nowadays, is sterilised topsoil which has been sieved to remove stones. It should be rich in organic matter and free draining with good structure. John Innes 3 compost can be substituted but only if it contains the correct proportion of loam. Many commercially produced JI composts have too high a proportion of peat so bagged sterilised loam is recommended.

Peat or substitutes are useful for opening up the soil and providing air pockets. They do not contain any significant quantity of nutrients, but peat in particular is useful for lowering the pH of the soil. The best peats for succulents are those that are fibrous, since these keep their structure for the longest time. Because of the environmental considerations of peat extraction, various alternatives are being offered to gardeners. Coir, a by-product of coconut production, has been tried by some growers but results have been mixed, probably due to the inconsistency of the product. A better alternative seems to be peat free composts made from ground bark which have the advantage of retaining their structure for longer than peat but they are less acidic.

Grit is a key part of any succulent compost since it helps to maintain sharp drainage. It should be sharp (not round) with an ideal size of up to 5mm but with particles less than 1mm sieved out to prevent clogging. It is also important that it be alkaline-free, so materials like acid washed quartzite or flint are commonly used. To reduce weight, some of the grit can be replaced with perlite or horticultural grade vermiculite, both of which absorb water and release it slowly for the plant to use.

The following is my recommended compost which will suit Gymnocalyciums. Minor modifications can then be made to this mix to accommodate the few species with specific requirements such as extra grit for those which are particularly sensitive to organic material. Ingredients should be just moist when mixed.

1 part by volume of sterilised loam (or JI3)

1 part by volume of peat or a substitute

1 part by volume of grit

A complete fertiliser including trace elements should be added as recommended by the manufacturer but, a few weeks after potting, regular liquid feeding will be required to achieve the best results.

The Importance of pH

pH is a measure of the acidity or alkalinity of a compost and maintaining the right value is critical to success. Most Gymnocalyciums are extemely sensitive to pH value and if it is too high (alkaline) or too low (acid) then growth will be reduced or completely stopped. You can test the pH of your soil by using a soil test kit which uses a liquid indicator with a different colour for each step in pH. There are also small electronic devices for the same purpose. If your soil in neutral, you will get a reading of pH = 7, if acid then the reading will be less than 7, and alkaline greater than 7, each single number increase representing ten times more acid or alkaline. The importance of pH was first brought to the attention of most cactus growers by Franz Buxbaum in his landmark book *Cactus Culture based on Biology* published in 1958. Before that time, many species, which are easily grown today, were grafted because cultivation on their own roots had proved impossible in the alkaline soils growers then used. He illustrated the increase in weight in *Cereus* seedlings at various values of pH, proving that the ideal value was about 6, that is slightly acid, and that a seedling at neutral hardly grew at all. This result has since been found to be applicable to a wide range of cactus species, particularly those from South America. The first thing is to get the pH of the soil right when it is mixed, then it needs to be maintained by using water and fertiliser which does not increase the alkalinity.

Containers

Plastic pots with adequate drainage holes are the most popular choice to use, being available in a wide range of sizes, shapes and colours. The overall appearance of a collection can be greatly enhanced by using plastic pots of the same colour (Fig.12), some growers choose black or green rather than the usual terracotta. Square pots are available in the smaller sizes and these pack together neatly on the greenhouse benches but no pot smaller than 6cm in diameter should be used, because anything smaller will dry out too quickly for plants to thrive. The main problem for cactus growers is that most pots are made for the volume market which is for short-term use, disposable pots, made as cheaply as possible so tend to be flimsy with thin walls. Pots are used in a cactus collection for years and so need to be strong and resistant to the degrading effect of sunshine as well having adequate drainage holes. For larger plants, shallow pots are the best choice since the root system of Gymnocalyciums tends to be shallow and if there is a lot of unused soil in the bottom of the pot, it will stay wet too long and lead to problems. For this reason, growers use half-pots or pans for sizes larger than 10cm diameter

Since cacti can often live for a very long time and never get too big, individual plants can remain in a collection for years, even decades. As the collection grows it becomes increasingly important to label the plants, not only to record the name but to keep track of where the plant came from. With restrictions on plant material collected in the wild getting ever tighter, what we have in cultivation today in collections like yours becomes even more important. Serious collectors keep records of all their plants, usually giving them acquisition numbers which they write on the labels. They keep a note of where each plant came from and if it was grown from seed collected in habitat or if it has any other authentification. This data is invaluable if the plant is ever used for scientific study, so if you assemble a collection of Gymnocalyciums, try to get plants with a collector's number and keep these data with the plant. Most people use the usual white plastic labels pushed down the side of the pot. The main difficulty with such a label is that the marker pens used to write on them are never completely fade-proof and the information slowly disappears under the influence of the sun. To avoid the complete loss of the information, you can write your acquisition number on the end of the label which is pushed under the soil where it will remain readable.

Potting

It is often advised that plants should be re-potted every year, but in practice this is too much of a chore for most collectors and is not actually necessary so long as the soil structure remains good and the plant is fed regularly. Ideally, a plant should only be potted when it has outgrown its pot, so to do that you will need suitable compost and enough pots in a range of sizes to allow you to move a plant into a larger size as it grows. The increase in pot size should not be too great, since putting a plant into a pot which is too large can lead to problems since it will not fill the new space with roots quickly enough to prevent it remaining wet and risking root rot.

Fig.12 Matching round and square pots creates a pleasing effect in part of Elisabeth Charles' collection.

For easy maintenance, it is desirable that all your plants are growing in your preferred soil mix so that they all behave similarly with respect to such things as the time taken to dry out after watering. So it follows that all new acquisitions should be re-potted prior to being placed in the collection. These new plants could be growing in anything from pure grit to just peat, so if you introduced them without re-potting, they would almost certainly lose their roots in a growing regime designed for plants in the soil mix like the one recommended here. Re-potting is also a good idea because it gives you the chance to check for pests and treat any that are found.

Watering and Feeding

Once you have potted your plants in a suitable compost and placed them in your glasshouse, they will need regular watering and feeding. This is a pleasurable activity and gives you the chance to look at your plants and enjoy them. It is often while you are watering that you first notice flower buds appearing or a new offset, but it also is a good opportunity to watch for pests or signs of disease. The earlier you spot a problem and deal with it, the better. The organisation of plants in the glasshouse is worth some thought. Some growers put related plants together since this makes the provision of any special treatment they need to be provided more easily, but other people prefer to mix the plants up to create an attractive display. For ease of maintenance, small pots (less than 9cm) should be kept together where they can be watered more often, probably at least three times more frequently than large pots.

During the growing season, the objective is to provide water soon after the pots have dried out from the previous watering. It is best to thoroughly wet the whole pot of soil each time you water, rather than giving them a partial wetting. It takes a lot of practice to perfect this technique because the time taken for a pot to dry out depends on the weather, the vigour of the plant and the size of the pot. If you use the same soil for all plants, experience helps you tell how dry a pot is by picking it up and feeling the weight. It doesn't really matter if you leave plants dry for a while, it is one of the benefits of growing cacti that if you leave them dry, they don't die, they just stop growing and wait.

Generally speaking, most growers don't feed their plants enough. Because of the porous nature of the soil we use, nutrients are rapidly leached out with watering and the plants will then stop growing. Once the plant has a good root system spread throughout the pot then regular feeding will be necessary to optimise growth. When plants are growing quickly, they are particularly susceptible to lack of light so you don't want to feed too much late in the year when light levels are falling and the plants need to be hardening off ready for the winter. The recommendation is to feed with half strength liquid fertiliser every time you water established plants from early spring through to late summer. A low nitrate liquid feed (12.5N-25P-25K) which is acid in solution and contains trace elements is the best one to use. Whichever fertiliser you choose, make sure it has trace elements because these are vital for healthy growth and cactus soils can suffer from deficiencies leading to symptoms like browning of the growing point and chlorosis.

The importance of pH has already been explained, so the quality of the water you use is very important. The best water to use is rainwater saved from the roof of the glasshouse or your house so long as it does not get polluted by anything harmful like chemical sprays from nearby farming activities. The roof, pipes and storage tank also need to be kept clean to avoid the build-up of fungal infections from rotting leaves, dead animals or birds. If collecting rainwater is impractical, then tap water can be used, but if you live in a hard

Fig.13 Gymnocalyciums have a pleasing assortment of spine colours and they flower easily.

water area it will need acidifying to prevent the hardness slowly making the soil in the pots alkaline. This build up of alkalinity is a major reason for Gymnocalyciums losing their roots and is probably the single most common cause of failure with this genus. Rainwater can be stored in a water butt inside the glasshouse and acidified by the addition of a small amount of acidifying agent such as potassium dihydrogen phosphate that is used in aquaria. The first time you treat the water, you should use a pH indicator or test strip to ascertain the amount to use to adjust the water to pH = 5.5 or 6.

Propagation

One of the most rewarding things about collecting and growing plants is that you can always make more individuals by propagation. It is also an important practical thing you can do for conservation, and selling spare plants not only helps other collectors but the money can go towards your expenses such as heating, pots, soil etc. Specialist *Gymnocalycium* enthusiasts like to grow true species rather than hybrids, particularly if they have some provenance, a pedigree if you like. They are always looking for plants of newly-discovered species, so to reduce the temptation to steal plants illegally from the wild, skilled propagators endeavour to satisfy the demand by using a number of techniques.

Seed Raising

Probably the most satisfying aspect of the hobby is seed raising. If the seed is from the wild, or a number of cultivated plants, then the diversity in the seedlings is sometimes remarkably large and only by seed raising can you appreciate the variation in a species (Fig.8). Seed raising has the reputation for being difficult but so long as a few simple principles are followed, results can be very satisfactory. The first thing is to start with good seed, which you can either collect from your plants or buy from a cactus seed dealer. Your own seed will often germinate best because you can sow it fresh, but of course, you only get more plants of those you already have. In most cases you will also need two plants of the same species to set seed, but they do not necessarily need to flower at the same time, since the pollen can be collected and stored for later use. Pollination of cactus flowers is usually easy to do by transferring the pollen from the stamens of one flower to the stigma of another plant. It is worth repeating the exercise at different times of the day since flowers are not always receptive, even when they are open. Once the pods are ripe the seed should be harvested, washed to remove the sticky pulp, and dried.

Seed can be stored in cool, dry conditions for some time, even many years in some cases, but as a general rule it is best sown within two years. The best time to sow *Gymnocalycium* seeds is in spring when a temperature of 20°C to 25°C can be provided in the day for good germination. This temperature can be provided by artificial heating like soil cables in sand under the trays or simply by waiting until early May when the sun will do the job for you.

The containers can either be small pots or seed trays divided up into segments for each seed batch. The latter are easier to keep evenly moist since the larger volume of soil dries out more slowly. Having filled the container with a commercial seed compost, the most suitable being those made from fine peat or peat substitute to which can be added some fine vermiculite or perlite, a thin layer of grit should be placed over the top before sowing. This gives the seeds a place to settle where they still receive light, which is necessary for germination, and some protection from drying of the compost surface. As the seeds grow the gravel will also reduce the tendency for moss to develop. The tray should be soaked in tap water and drained, then the seeds sown thinly so that the resulting seedlings can be left undisturbed for up to a year without them becoming overcrowded.

Fig.14
The best results are achieved from propagation if you dedicate a glasshouse or special area to the process

Once sown, the seeds need to be lightly sprayed with a fungicide such as chinosol to wash them into the gravel where they can germinate. To maximise germination, the seeds need to be kept in a humid atmosphere so the tray should be placed under a tight-fitting sheet of white (not clear) polythene which needs to be only a few centimetres above the soil and then put in a bright place in the greenhouse protected from direct sun. Daytime temperatures above 20°C are necessary for success but night temperatures should be lower, the variation encourages germination which will occur from a few days through to a few weeks depending on the species and the freshness of the seed. The white polythene can be left in place until all the seeds that are likely to germinate have done so, then it can be raised like a tent to give more air, eventually in summer being replaced by fleece which allows more air movement but still maintains some higher humidity. Once this stage is reached the seedlings should look like miniature versions of what they will eventually grow into and they will need watering to keep the compost just moist. During the first winter the seedlings should be lightly watered in fine weather to stop them drying out, and when the new growing season starts around April the young plants will start to grow rapidly, at which time they can be pricked out into another tray of normal cactus compost for growing on. Light shade at this time is still desirable until their third year when they can be treated as adult plants.

Cuttings

For some *Gymnocalycium* species, propagation by cuttings is a good way of increasing your stock. Since cuttings are genetically identical to the plant they were removed from, they have the same appearance, flower colour etc., so if you have a single plant of habitat origin you want to multiply, then cuttings are the only way to do it. For those that produce offsets naturally, these can be cut off and rooted as cuttings. If, for instance, a species is naturally solitary then the growing point can be cut out, resulting in the production of offsets suitable for use as cuttings. Such a plant can then become a stock plant which will continue to produce cuttings for years. Cut surfaces should be dusted with sulphur and left to dry for a few days to form a callus before being placed on a moist rooting medium. When taking a cutting, always cut in the place that leaves the smallest cut surface possible, like the junction of the offset with the main stem. Allow the cutting to dry for a few days in a warm shady place, then sit it on a mixture of grit and perlite which is just kept moist and warm.

Although grafting is a good way to grow difficult cactus species, it is rarely necessary for Gymnocalyciums, except for completely variegated clones. See page 280 for some examples.

Natural Habitats and Distribution

The species of *Gymnocalycium* can be found in a diverse range of habitats in the southern part of South America. Most are native to Argentina but some grow in Bolivia, Paraguay, Uruguay and just into Brazil. Their distribution is bounded by the Andes on the west and expansion further to the south and north has probably been stopped by unsuitable climate.

They have evolved to live in various biogeographic zones, each with their own combination of characteristics like climate and elevation which results in the typical plant species of each zone. Each *Gymnocalycium* has a preferred zone but its actual locality will also be influenced by factors like the slope of the land, the aspect and the nature of the substrate. Competition from other plants is a major influence, some species appearing to cope with a cover of trees or grass more effectively than others. Like many other cacti, Gymnocalyciums often favour sloping, rocky ground, especially bare patches where the layer of gravel and stones is too shallow to support trees or a dense growth of grass. Mountain chains are often separated by large expanses of flat land which is unsuitable for the slope-dwelling species and prevents their spread. Some species, however, such as *G. schickendantzii* thrive in the flat lands, perhaps the reason why this species is so widespread.

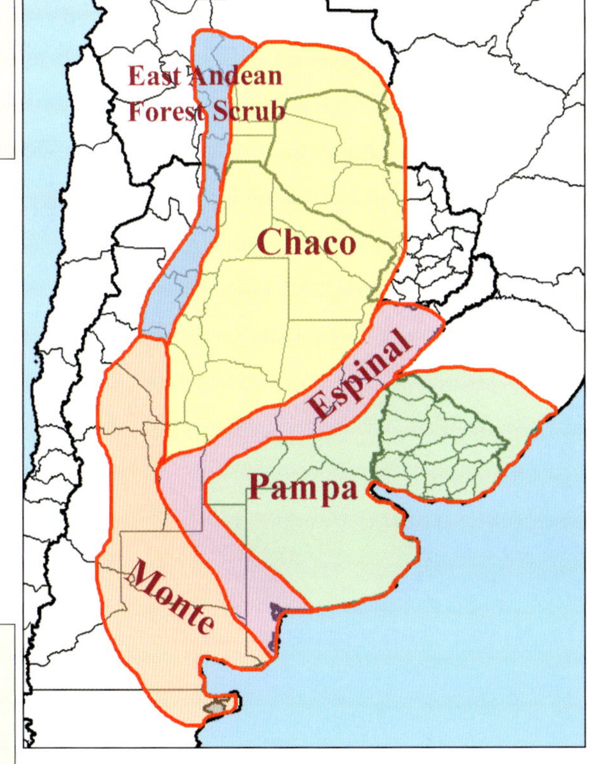

Map 1
Simplified map of the biogeographic regions where Gymnocalyciums can be found.

Fig.15
A view of the hilly Pampa near Masoller, Rivera, Uruaguay

Photo: Massimo Meregalli

All species are affected to a greater or lesser degree by destructive influences which threaten their future. Particularly vulnerable are those that have a very limited distribution, especially if they have a specific habitat preference such as open rocky or sandy patches in forest. This is the case with *G. paraguayense* which is now very rare, if not already exterminated in the wild. This example is due to agriculture, a process that mainly affects plants in the biogeographic regions which are suitable for farming. Other threats include excessive grazing, particularly goats, and the expansion of human activity especially near to towns. It is sad to report that I have seen a gradual decline in the condition of many cactus habitats over the nearly 20 years I have been visiting them. Several other enthusiasts report a similar experience, only occasionally a consequence of cactus collecting, more often due to over-grazing or intentional habitat destruction. Of course, there are still many remote places where the environment is largely unaffected by human activity and the creation of protected reserves can certainly help conservation. In my opinion, the only long-term solution is if the local people can learn to respect their natural heritage. The assumption that the plants growing near where you live can be destroyed because they grow elsewhere can have disastrous consequences for a very localised species.

Here is a simplified categorisation of the biogeographic zones in which *Gymnocalycium* can be found in the wild. They are not clearly demarcated, merging into each other and sometimes resulting in areas with characteristics of both neighbouring zones.

a) Pampa

The Pampa region lies to the east of the Espinal region, mainly in Province Buenos Aires and extends into Uruguay where it is more hilly. It is predominantly an extensive plain with some isolated low hills.

It comprises a vast area of grassland, extensively used for farming and is the most densely populated region of Argentina. It enjoys a mild to warm climate with rainfall throughout the year diminishing in quantity from north to south. Gymnocalyciums can only be found in certain parts of this region, either on rocky hills or raised gravelly areas.

b) Monte

The Monte region runs along the east side of the Andes, getting wider towards the south and curving eastwards to the Atlantic Ocean. It is characterised by shrub steppe with occasional patches of open woodland where water is more regularly available. It is interesting to note that there are some non-cactaceous plant genera in common with the Sonoran desert of Mexico and the USA. The climate is hot and dry with rainfall commonly between 80 and 200mm per year mainly during the summer. The pronounced dry season can extend up to nine months. Many Gymnocalyciums occur in this region, preferring to grow under the bushes on dry slopes or among rocks. This is a rich environment for cacti with a great diversity of species.

Fig.16 Monte vegetation near the Sierra Famatina, Prov. La Rioja, Argentina

c) Chaco

The Chaco region occupies a large part of the north of Argentina, extending into western Paraguay and eastern Bolivia. The landscape is predominently plains with scattered hills and occasional low-lying areas which may form salt lakes. The vegetation includes large trees, palms, grassland and halophilous steppe. The climate is warm and dry, the north becoming hot in summer. Rainfall is restricted to the summer, heavy in the east but becoming more infrequent to the west. Another favoured region for cacti, Gymnocalyciums are widespread, often growing on rocky slopes or under the protection of bushes.

Photo: Volker Schädlich

Fig.17 Dense Chaco vegetation in Alto Paraguay, Paraguay

d) Espinal

The Espinal region lies in a band to the east of the Monte and Chaco regions. It is mainly plains with a few scattered hills. The vegetation consists of large trees, shrubby steppe, palms and grassland. It has been modified by the activities of man so reducing the natural areas left intact. The climate varies from mild and dry to warm and humid, with most of the rain in the summer. The number of *Gymnocalycium* species increases to the north of this region where they often grow on steep rocky slopes.

e) Eastern Andean forest and scrub

A region in the north west of Argentina that extends into the eastern mountains of Bolivia. It comprises steep-sided mountains with deep valleys. There are extensive forests thriving on the ample rainfall which falls mainly in the summer. The climate varies from warm and dry to hot and humid at different times of the year. Here the cacti favour rocky places where there is less competition from other plants and where the shade from trees is less.

Fig.18 Espinal vegetation in the Sierra San Luis. Prov. San Luis, Argentina

Fig.19
East Andean forest scrub near the Rio Grande, Prov. Santa Cruz, Bolivia with *Vatricania.*

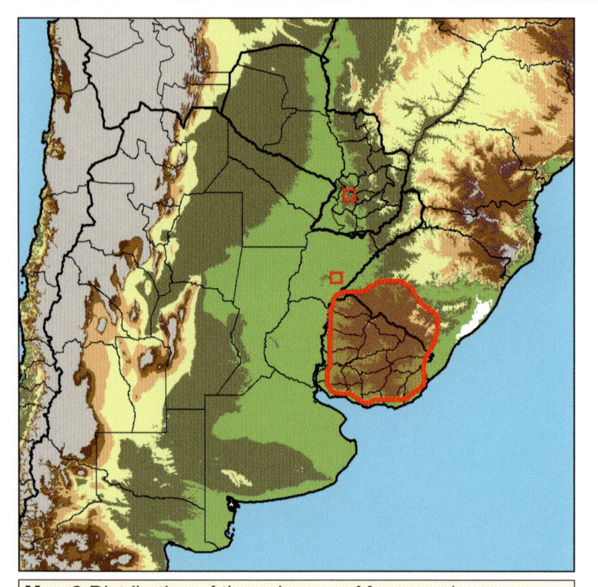

Map 2 Distribution of the subgenus *Macrosemineum*

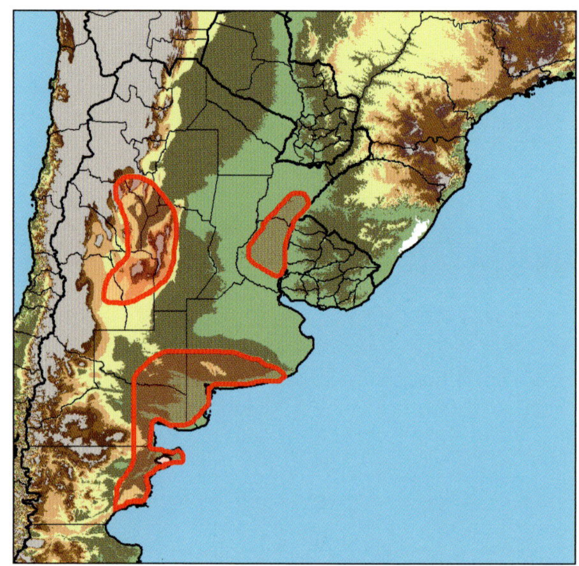

Map 3 Distribution of the subgenus *Gymnocalycium*

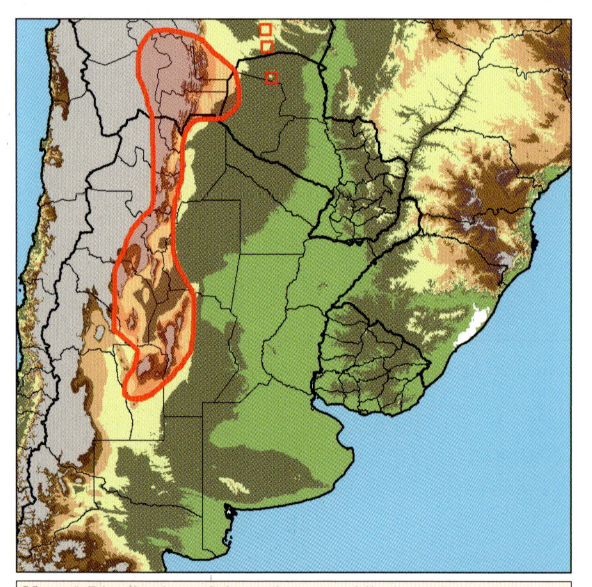

Map 4 Distribution of the subgenus *Microsemineum*

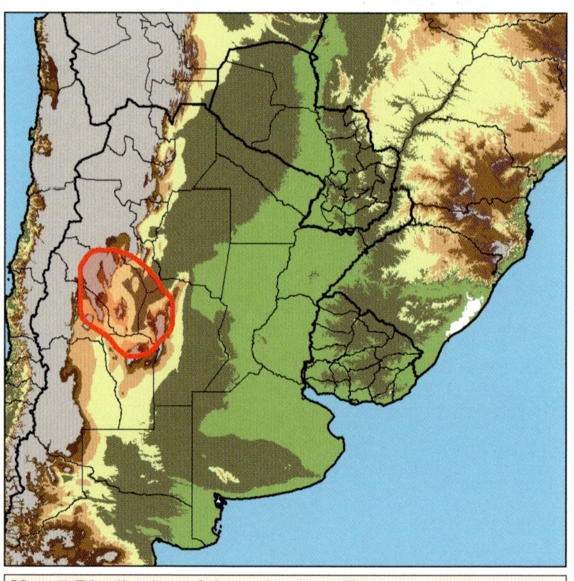

Map 5 Distribution of the subgenus *Trichomosemineum*

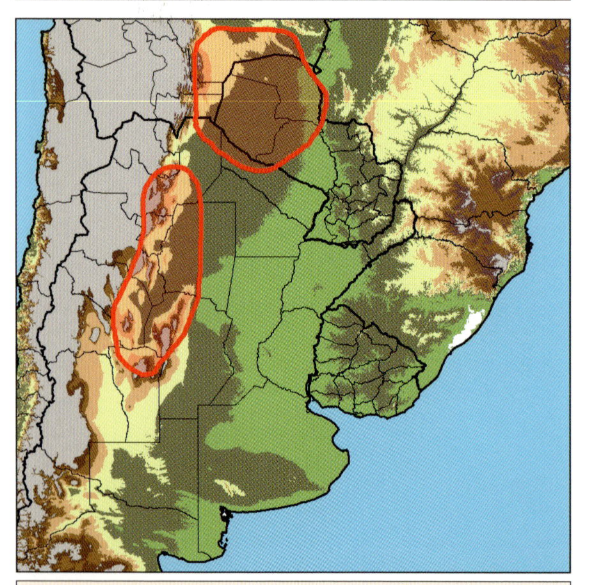

Map 6 Distribution of the subgenus *Muscosemineum*

Distribution of the subgenera

The maps on this page show the approximate distribution of the five seed groups used in this book to divide *Gymnocalycium*. The sum of these does not make a single uninterrupted area, there are gaps where none of the speces grows.

There are many examples of plants from different subgenera growing close together, sometimes even side by side. The subgenus *Microsemineum* has become a repository for species that do not fit in the other subgenera, so it includes a number of groups of related species. Sometimes, more than one species of *Microsemineum* can grow sympatrically.

Descriptions of these subgenera and images of the seeds are on page 15.

Treatment of Recognizable Species and Subspecies

Contents

In this chapter the accepted species and subspecies are organised by their subgenera which are based on their seed structure as defined on page 15:

Subgenus *Macrosemineum*
Subgenus *Gymnocalycium*
Subgenus *Microsemineum*
Subgenus *Trichomosemineum*
Subgenus *Muscosemineum*

For each accepted species, the text is organised under the following headings:

First Description

This is a transcript of the original first description of the taxon, translated into English where necessary. When this was particularly brief, the next published description has also been reproduced. The original text is printed on a yellow-tinted panel. There is always a risk that subsequent authors who amplify descriptions may misinterpret the original author's concept of the taxon, especially where there is no extant type. However, the few later descriptions reproduced here are supportive of the present-day concept of the species and reflect the best available interpretation of the first author's intention.

Etymology

The origin of the name.

Other Names

These are other published names (combinations) of the species based on the same type, for instance, when the species has been transferred to another genus or its taxonomic rank was changed.

Synonyms

Taxa originally published as species or subspecies but here treated as synonyms of the accepted species name are listed here. Synonyms published at species rank have their first descriptions reproduced here, in translation where necessary. This synonymy is the opinion of the author based on personal experience and the suggestions of other open-minded specialists who generously shared their knowledge.

Distribution

The known distribution of the species and synonyms is defined. This information relies on the observations of many individuals including their correct identification of the plant they saw in habitat when reporting the location. The distribution maps are based on these observations and show the area within which the species is believed to occur. This does not mean that it grows everywhere within the indicated area, since populations will be found only in locations where the environment suits the plant. For instance, some taxa will grow only on sloping ground with a particular aspect, among rocks, at a particular altitude, or on a particular substrate. Very localised species are indicated with squares on the map. These localities should be regarded as approximate since they are based on generalised information, exact co-ordinates usually being withheld by the observer to protect the plants.

Conservation Status

This is the author's assessment of the conservation status of the taxon based on personal observations or previously published information. It follows the categories and criteria of the Species Survival Commission of the International Union for the Conservation of Nature (IUCN).

History

This is a synopsis of the discovery of the species and treatment by subsequent authors following its initial description.

Commentary

Information and opinion which does not fit into any of the other headings is given here, including support for the taxonomic standing of the species.

Cultivation

If the species benefits from any cultural treatment different from the norm, such advice will be found here. The general cultural requirements of Gymnocalyciums are described in the chapter on cultivation (See page 18. The frequency or rarity of the species in northern European collections is also discussed.

Illustrations

References to published pictures which compliment those in this book, including those that cannot be reproduced here because of copyright considerations.

Subgenus *Macrosemineum*

Type Species
G. denudatum (Link & Otto) Pfeiffer ex Mittler

Gymnocalycium angelae Meregalli

First Description (Meregalli)

KuaS 49(12) pp.283-290 (1998)

Gymnocalycium angelae spec. nov., a new species from Argentina

Plant: Semi-globular, up to 10cm diameter, 8cm high, offsetting from the base areoles. Epidermis light to mid-green, slightly shiny, smooth. Root stringy, with 2-3 thick main roots. Ribs 7 to, in age, up to 10, blunt, wide, with straight vertical furrows, these not strongly notched; with deep crosscuts above the areoles, the rib sides hardly distinguishable. Humps above the areoles slightly projecting, more or less round, often with a bronze coloured tint.

Areoles oval, 5mm long, about 1cm apart, with yellowish to whitish wool-felt. Spines, 7, all radial, above consistent yellowish, then becoming light grey, yellowish at the base, darker in the crown, thin, in cross section flattened, straight or irregularly curved, lying on the body, 1.5-2cm long; the top pair of spines thinner, pointing to the side, also sometimes missing, the second pair longer and stronger, pointing sideways, sometimes curved downwards, the third pair pointing downwards, the seventh spine pointing straight down; central spines missing.

Flowers from near the apex of the plant, Buds round, green to dark green with a tint of purple, at maximum anthesis 40mm long, 50mm diameter. Pericarp funnel-shaped, 20mm long, 4mm at the base, 8.5mm above, with shiny green pink tipped scales, these are 2.5mm x 3mm. Petals c. 40, almost straight, pointed, up to 25mm long, inner petals 5mm wide, white with a pinkish-red base, outer petals 6.5mm wide, white with a narrow pinkish-brown middle stripe on the inside, greenish on the outside, pinkish-brown at the tip. Many filaments inserted in a few rows, upper third pale yellowish, below pale pink.

Anthers with yellow pollen. Style white, thin, overhang the anthers with 8 stigma lobes. Pollen trizonocolpate (NPC-Formel 343), pollen diameter 43 μ (polar) relative to 49 μ (equatorial). Exine with few thin spinulae and ringed with dots (less than 10 in 5 μ^{2}). Fruit green, elongated, when ripe 3cm long, splitting vertically. Flower parts remaining.

Seeds globular, 1.9-2.0mm long, HMR 2.0-2.1mm wide, in the middle 1.2mm thick. Testa semi-matt, reddish-brown, with a keel, the border around of the HMR enlarged, curved round, by the micropyle sloped downwards. Cells flat, anticlinal borders almost straight, barely sunken, the transitional area of the cells finely dotted. The cuticle in fine folds, creases at the central part of the periclinal wall above the anticlinal border running parallel; HMR basal, irregular oval, flattened, maximum width of hilum 1.5-1.6mm, clearly narrow for the micropyle. Lastly, profile clearly strong in cross section, yellowish, the whole surface is uniform and not extending over the edge.

Chromosome count; 2n = 22.

Type: Argentina, Prov. Corrientes, Tres Cerros, R. Kiesling, ex cult. coll. Meregalli (vegetatively propagated from Kiesling original collected plant) (TO-HG No. 2705).

Photo: Alfredo Lorenzini

Fig.20 *G. angelae*, ex-habitat plant

Etymology

Named for Angeles Gracilea Lopez, the wife of Roberto Kiesling who accompanied him on his trip to Tres Cerros.

Other Names

G. denudatum subsp. *angelae* (Meregalli) Prestlé in *Gymnocalycium* 17(2): p.577 (2004)

Distribution (Map 7)

It is known only from the type locality, which is a series of low hills in an agricultural area, and there are no more stony, uncultivated places nearby. Annual rainfall is high at about 1400mm and is spread around the year. The few plants are difficult to see among the grass and rocks.

An interesting article by Hans Till (1999) recounts how Rausch brought back some plants from Uruguay (HK404) which appear to be the same as *G. angelae*. However, Meregalli (pers. com.) does not believe this collection to be *G. angelae* but rather *G. denudatum* subsp. *angulatum*.

Map 7
Type locality of *G. angelae* at Tres Cerros

Photo: Norbert Gerlof

Fig.21 *G. angelae* in habitat at the type locality:Tres Cerros, Prov. Corrientes, Argentina

Photo: Alfredo Lorenzini

Fig.22 *G. angelae*, ex-habitat plant in culture

Conservation Status

Critically Endangered. Total area of occupancy only a few hundred square metres. It is certainly known only from the type locality where it is rare and the population may consist of only a few individuals.

History

First discovered by Roberto Kiesling in the 1970s on the Cerro de Susini, one of the 'Tres Cerros' in the Prov. Corrientes, Argentina, it was first thought to be *Gymnocalycium denudatum* growing outside its known distribution in nearby Brazil. This was not an entirely unexpected finding, since it had been suggested by Spegazzini (1905) that *G. denudatum* grew near Santa Ana, Prov. Misiones, Argentina, 300km away.

Commentary

This plant looks a lot like *Gymnocalycium denudatum* with which it was first confused. However, the epidermis is a paler, duller green, the flower, whilst of a similar shape, has a red throat. The seeds, which also belong to the subgenus *Macrosemineum*, are smaller with a differently shaped hylum-micropylar region. The flowers and seeds are, therefore, more like those of *G. mesopotamicum,* which also has a chromosome count of 2n=22 (*G. denudatum* is 2n=44). From this and other evidence, Meregalli (1998) concludes that this species is most closely related to *G. mesopotamicum*.

Meregalli (pers. com.) explains that his view is supported by phylogenetic analysis where around 40 different characters were examined. As a result of this, *G. angelae* was treated a synonym of *G. mesopotamicum* in the *New Cactus Lexicon* (2006). It should be said, however, that Detlev Metzing (pers. com.) does not share this view, believing it to be more closely related to *G. denudatum*. This opinion was shared by Karl Heinz Prestlé (2004) who made it a subspecies of *G. denudatum*. He pointed out the similarities of its flowers and seeds to those of *G. denudatum* populations in the west of Rio Grande do Sul.

When a taxon has such a limited distribution and is not a strong grower in culture, one wonders if it is of hybrid origin. Meregalli (pers. com.) refutes this suggestion based on the analysis of its electrophoretic bands. Of course, it is possible that it was more widely distributed once but, like many other taxa from the area, it may have been adversely effected by human activity, especially farming. Norbert Gerloff has recently told me that he has visited the location three times and found only a few plants each time. If this is the only locality then it must be critically endangered.

Cultivation

Rare in cultivation and said to be difficult to grow, resulting in plants often dying. The temperature in habitat does not fall below 10°C and there are no prolonged periods of drought so the plant probably should not be kept completely dry at any time of the year.

It readily offsets so providing a convenient method of propagation. It has proved difficult to set true seed on the few plants in cultivation. This is maybe because there are so few clones in cultivation, many believed to be descended from the three clones of the original collection maintained by Omar Ferrari in Buenos Aires.

The first plant to flower for Meregalli was used to illustrate the first description. Seeds were produced by crossing it with *G. mesopotamicum*. Seeds produced this way have the characteristic appearance of the female parent's seeds, so allowing them to be used as representative of *G. angelae*. He repeated the procedure two years later and produced identical seeds. Attempts to cross the plant with *G. hyptiacanthum* and *G. denudatum* have been unsuccessful.

Illustration

Prestlé (2002), *Gymnocalycium* 17(2): pp. 576-7, Abb. 17-20.

Gymnocalycium denudatum (Link & Otto) Pfeiffer ex Mittler

Fig.23 *Echinocactus denudatus*, Plate 9 from *Icones plantarum rariorum* Link & Otto (1828). This fine illustration accompanied the first description and shows a rather weak root system suggesting that the plant had been imported without roots and the illustrated roots had grown since.

Fig.24 *Gymnocalycium denudatum* GC758.03 near Minas do Camaqua, Rio Grande do Sul, Brazil

First Description (Link & Otto)

Icones plantarum rariorum (*Abbildung neuer und seltener Gewächse*) 2:17, + plate 9 (1828) [see Fig.23]

Echinocactus denudatus sp. nov.

An Echinocactus with a green, almost spherical body, ribs 6-8 blunt, all are widely spreading, flower tube with few scales.

Habitat in southern Brazil.

[from the Latin]
Body 3"-4" (76-101mm) high, 3"-4" (76-101mm) wide. Ribs with distinct projections, separated by shallow grooves on either side and having narrow cross furrows. Spines 5-8, all 1/3"-2/3" (8.5-17mm) long, very wide spreading almost adpressed, twisted, surrounded with short wool. Depressed in the crown, lacking wool. Flowers 2"-3" (51-76mm) high, rising from the ribs among the spines in the crown. Receptacle cup-shaped with few scales, often arranged in a circle, small, blunt, oval, with a point, scales more numerous near the top, longer, more pointed finally resolving into the corolla. Corolla with white acute linear petals. Style having many stigma lobes, scarcely longer than the stamens. Fruit not known.

It is clear that Sellow sent this plant because of the unusual nature of the flower tube.

[From the German]
Plant 3"-4" (76-101mm) high, 3"-4" (76-101mm) wide. Ribs with distinct projections, flattened at the sides, blunt with narrow cross-grooves. Spines are in groups of 4-6-8, and vary from 1/3"-2/3" (8.5-17mm) long, spreading lying close to the body, curved to and from, having a little wool at the base. Centre of crown is sunken. Flowers appear on the ribs near the centre of the crown from among the spines. The pericarp region is cylindrical, carrying small, blunt, oval scales, mostly arranged in a circle, these scales increase in number towards the top, becoming longer and more sharply pointed, resolving in to the flower petals, which are lanceolate, pointed, white. Stamens occur in large numbers, shorter than the petals. The style is about the same length as the stamens, with many stigma lobes. We have not seen the fruit.

Etymology

'Denuded' referring to the few appressed spines.

Other Names

Cereus denudatus (Link & Otto) Pfeiffer (1837)
Echinocactus denudatus Link & Otto

Synonyms

G. megalothelos (Sencke ex K. Schumann) Britton and Rose in *The Cactaceae* 3: p.162 (1922). Basionym: *Echinocactus megalothelos* Senke ex Schumann (1898). The spelling was later changed to

Fig.25 *Gymnocalycium denudatum,* GC766.03 near Caçapava do Sul, Rio Grande do Sul, Brazil

Photo: Norbert Gerlof

Fig.26 *Gymnocalycium denudatum, Gerloff* 612 Bagé, Rio Grande do Sul, Brazil

G. megalothelon by H. Till (2005) to agree with the neuter *Gymnocalycium*.

First description of *Echinocactus megalothelos* Senke ex K.Schumann in *Gesamtbeschreibung der Kakteen* p.415 (1898):

Body simple, later offsetting from the base, depressed, becoming taller, short column forming, round on top, sunken in the crown, sometimes covered with small tufts of woolly felt, no spines on the top, new growth light green, eventually darker, up to 16cm diameter, nearly as high. Ribs 10-12, separated by sharp grooves, up to 1.5cm high, blunt with cross grooves over plump chins, below there are often deep cross-cuts lying in the flattened warts. Areoles 1-1.5cm apart, sitting in deep sunken grooves, round, 2.5cm diameter, covered with a little dirty white felt, soon becoming bald. Radial spines 7-8, radiating horizontally, awl-shaped, straight or slightly curved, sometimes a few smaller spines at the top, the bottom pair the longest, up to 1.5cm long, in the new growth clear yellow, soon becoming horn-coloured, central spines few, a little stronger, standing straight out. The flowers are unknown to me; they are reddish-white and look similar to those of *E. denudatus*. Flower tube covered with scales, no hairs.

Geographical location: Paraguay.

Commentary on *G. megalothelon*

First mentioned in the catalogue of Sencke, Schumann later made his description from plants in the collection of Gruson which may have originated from the firm of Haage jr. There is no original picture and the illustrations in Britton & Rose of a plant collected by Prof. Robert Chodat in Paraguay are not a particularly good match for the description, probably representing a form of *G. paraguayense*. Because the plants being named by Schumann at that time originated from Paraguay, Meregalli (pers.com.) believes that this name could well be the first description of *G. paraguayense*.

Frank in Krainz *Die Kakteen* CVIf (1968) expanded the description and then H. Till (2005) concluded that strong-spined forms of *G. denudatum* with prominent humps collected in Rio Grande do Sul such as HU7 and HU28 are re-collections of *G. megalothelos*. He recounts reports of pink-flowered forms in the literature and illustrates

Fig.27 *G. denudatum, Hofacker* 204 from Torrinhas, Rio Grande do Sul, Brazil

Fig.28 *G. denudatum, Hofacker* 36 from Minas do Camaqua, Rio Grande do Sul, BR

Fig.29 *G. denudatum, Hofacker* 381 from Torrinhas, Rio Grande do Sul, Brazil

Fig.30 *G. denudatum, Hofacker* 73 from Santana da Boa Vista, Rio Grande do Sul, Brazil

Fig.31 *G. denudatum, Hofacker* 381 from Bagé, Rio Grande do Sul, Brazil

Fig.32 *G. denudatum (*angulatum), *Meregalli* 418 from the Serra do Cavera, Rio Grande do Sul, Brazil

an example of plants said to be from Catalán, Artigas, Uruguay which were given to Rausch and passed onto Till. Claiming these to be a stable colour form, he gives them the name *G. megalothelon* var. *susannae*.

Map 8 Distribution of *G. denudatum* in the south east part of Rio Grande do Sul, Brazil and neighbouring Uruguay in the Sierra de los Rios. There is also a population north of Porto Alegre to the east of this map.

In the same article H. Till (2005) describes a freely-offsetting subspecies based on collections of K. H. Prestlé from Capivarita, Rio Grande do Sul, Brazil: *Gymnocalycium megalothelon* subsp. *prestleanum* H.Till that is not recognised here.

Although treated here as a synonym of *G. denudatum*, this name would probably be better regarded as a name of uncertain attribution and abandoned.

Gymnocalycium denudatum subsp. *angulatum* Prestlé, *Gymnocalycium* 17(2): p.573 (2004)

Commentary on *G. denudatum* subsp. *angulatum* (Fig.32)

The recognition of this form from the Serra do Cavera (left shaded area on the above map) as a separate subspecies is supported by Meregalli (pers. com.) although he tells me that intermediate forms can be found north of Dom Pedrito.

Gymnocalycium denudatum subsp. *neocopinum* Prestlé, *Gymnocalycium* 17(2): p.575 (2004)

Gymnocalycium megalothelon subsp. *prestleanum* H.Till, *Gymnocalycium* 18(2): p. 625 (2005)

Distribution (Map 8)

Classically described as growing in Rio Grande do Sul (Brazil), northern Uruguay, Paraguay and also, by Spegazzini (1905), in Prov. Misiones, Argentina.

Modern reports are limited to Rio Grande do Sul, Brazil, and the Sierra de los Rios in Uruguay, suggesting that the other observations were not this taxon.

Conservation Status

Least Concern. Widespread and often abundant in its many reported habitats.

Fig.33 *G. denudatum, Bercht* 814 from Pedras Altas, Rio Grande do Sul, Brazil

Fig.34 *G. denudatum, Hofacker* 204 from Torrinhas, Rio Grande do Sul, Brazil

Fig.35 *G. denudatum, Hofacker* 43 from Vila Sao José, Rio Grande do Sul, Brazil

History

Originally it was collected in Rio Grande do Sul, Brazil by Sellow and sent to the Berlin Botanical Gardens about 1825. Ritter (1979) tells us that Sellow actually collected the plants between Caçapava do Sul and Bagé in 1824. The species was subsequently described as *Echinocactus denudatus* by Link and Otto in 1828. Pfeiffer transferred it to *Cereus* in 1837. It was one of the first species to be included in the genus *Gymnocalycium* when Mittler (1844) published his description of Pfeiffer's new generic name. It was assumed by Britton and Rose to be the type species of the genus, but recently Pfeiffer's original choice of *G. gibbosum* has been reinstated as the type. In CSI21: p.8 (2006), Hunt and Taylor designated the Link & Otto plate as the lectotype of *Echinocactus denudatus* (Fig.23)

By the end of the century, a great many varieties had been described, many by the nurseryman F. Haage jun. Perhaps this is an early example of commercial interests driving the creation of new names. Backeberg (1959) considered these to be hybrids rather than wild variants, stating that he had never seen collected plants which corresponded to them. It was stated that plants of *G. denudatum* were coming from Paraguay and some of these varieties later proved to be a different species, *Gymnocalycium paraguayense*.

Commentary

The great variability of this species has resulted in many names for minor variants, historically as varieties, but recently Prestlé (2004) described two local forms as subspecies. The spination, number of ribs, size of bodies and flower petal shape can all vary, with regional forms sometimes exhibiting a reasonably consistent appearance.

The spidery spines often associated with this species are not always a diagnostic feature. However, the spines are always adpressed against the body and this feature is one of the easiest ways to differentiate this species from the similar-looking *G. horstii* which has stronger spines standing out from the body. The two species sometimes grow in close proximity near the town of Minas da Camaquã.

Cultivation

This is one of the most frequently seen species in cultivation and is very variable. The large showy flowers have clear white petals, not the dirty white of so many other species. It is easy to propagate from seed or by removing offsets which are usually freely produced and root easily.

There is a long-known hybrid with *G. baldianum* called 'Jan Suba' which has lovely pink flowers which, according to Elsner (1970), was created by Pazout. H.Till (2005b) questions this origin, suggesting confusion with imported plants of a pink-flowered form by Fric around 1929.

Illustrations

Schumann in Martius *Flora Brasiliensis* Vol. IV. Part II (1890) A reproduction of the original plate of Link & Otto

Schumann, *Gesamtbescheibung der Kakteen* (1899) Fig.72, p.414: A line drawing by T. Gürke also based on the original plate

Britton & Rose, *The Cactaceae* Vol.III, Plate XVIII(1) as *G. megalothelos*: and Fig.173 p.162 (1922): perhaps misidentified.

Kupper, W. *Das Kakteenbuch*: p.116 (1929)

Hofacker, KuaS 50(7);: pp.166-168 (1999)

Gymnocalycium horstii Buining subsp. *horstii*

First Description (Buining)

KuaS 21(9): pp.162-5 (1970)

Plant body solitary up to 11cm diameter and to 7cm high. Very old examples sometimes somewhat taller. Fresh green, shining green when the plants are in full growth, roots fibrous. Ribs 5, sometimes 6, up to 7cm broad at the base, blunt but not completely flat, sometimes raised in the middle region, tubercles, none or only weak.

Areoles per rib 3, sometimes a few more, somewhat woolly, oval, up to 5mm long and 4mm wide, about 3cm apart. Spines hard, rigid, standing out obliquely, not adpressed, as a rule 5 in number, a pair on either side and a single one below, lacking central spines, pale yellow or whitish yellow, up to 3cm long.

Fig.36 *G. horstii horstii, Horst-Uebelmann* 79 from Guaritas, Caçapava do Sul, Rio Grande do Sul, Brazil

Fig.37 *G. horstii horstii, Hofacker* 67, the form from Cordilheira, Rio Grande do Sul, Brazil

Flowers up to 11cm long and just as wide, opening fully in direct sunlight, from morning until early evening. Pericarpel measures up to 25mm long, 12mm diameter (outside). Ovary cavity is up to 18mm long and 8mm diameter, covered with small pink scales.

Receptacle up to 25mm long, funnel shaped, green, bearing 12 pink coloured scales. Petals up to 6cm long, and up to 14mm wide, pointed, lilac-pink to creamy-white with a deep pink mid stripe, outer petals a deep pink. Nectar chamber with only one nectary, very small, almost non existent. Primary stamens inserted close to the base of the style, with the secondary stamens inserted further away, distributed over the whole of the receptacle, pale yellow in colour as are the anthers. Style up to 3cm long and 4mm thick, light yellow, with 9 stigma lobes.

Fruit oval, 5-6cm long and 3-4cm diameter, with 10-12 scales, 8-12mm wide, green with a bluish tinge. Flesh of the fruit watery and pure white, the bloom on the fruit lasts noticeably longer than is the case with previously known *Gymnocalycium* species from Brazil, but like them, the fruits split longitudinally.

Seed cap-shaped, 1.2mm wide, 1.3mm long, the texture being faintly lustrous; with small flattened, round, chestnut-coloured projections with black in between. Hilum is filled with spongy tissue of an ochre colour. The point of attachment to the funicle and the region of the micropyle is somewhat depressed.

Place of origin: From a single location near Caçapava, Rio Grande do Sul, Brazil.

The type plant under the collection number HU79 is placed in the Herbarium of the University of Utrecht, Holland. I name it after the discoverer, Leopoldo Horst in Brazil.

Amerhauser (2008) stated that the holotype cannot be found at Utrecht, so he designated a lectotype, another specimen of HU79 from the first collection, deposited at the herbarium of the Instituto de Botânica, Universidad de São Paulo (SP)

Synonym

Gymnocalycium horstii subsp. *megalanthum* Amerhauser in *Gymnocalycium* 21(3): p.800 (2008)

Commentary on *G. horstii* subsp. *megalanthum*

A casual reading of this description leads one to believe that it is the valid name for the form of *G. horstii* subsp. *horstii* which grows near Cordilheira (Fig.37), a place about 60km east of Caçapava and

Fig.38 *G. horstii*, *Gerloff* 175 from Minas de Camaquã, Rio Grande do Sul, Brazil

Fig.39 *G. horstii*, *Gerloff* 105 from Cordilheira, Rio Grande do Sul, Brazil

Fig.40 *G. horstii*, *Horst-Uebelmann* 79 from Guaritas, Caçapava do Sul, RS, Brazil

Map 9 Habitats of *G. horstii* in Rio Grande do Sul, Brazil
black square = *G. horstii* subsp. *horstii*; red square = *G. horstii* subsp. *buenekeri*

Fig.41 Red bud of *G. horstii*, *Gerloff* 105 from Cordilheira, RS, Brazil

Fig.42 *G. horstii*, *Horst-Uebelmann* 1290, Cordilheira, Rio Grande do Sul, Brazil

shown as the top right black square on Map 9. Clearly, this was Amerhauser's intention, however, there is a problem with his designated type. Instead of preserving a recently collected plant with good provenance, he designated an old plant from Till's collection HT922 which has no locality data. Since the Cordilheira plant was not discovered until 1990, it is likely that the type plant is not this form.

The Cordilheira plant (the top right black square on the map) certainly differs from the type of *G. horstii* in a number of respects. Its most obvious distinction is the form of the flower buds (Fig.41) which are red when young, maturing to a flower that is larger than the type. The fruit (Fig.42) is also a different shape, rather more round. It has been known under various provisional names including *G. horstii* var. *rubrogemmatus*.

Amerhauser (2008) suggests that the wild population of this form may have been destroyed by a combination of rock extraction and illegal collecting. It is usually encountered in collections with the collection numbers AH67 or Gf105.

Etymology

Named for Leopoldo Horst (1918-1987), a German-born cactus enthusiast who lived in Rio Grande do Sul in Brazil and discovered many new taxa with the Swiss nurseryman Werner Uebelmann.

Distribution (Map 9)

This taxon is known only from a few small widespread localities, usually isolated patches of rock in a region of flat agricultural land. The plants do not actually grow on the rock but rather in the shallow soil between them. Meregalli (pers. com.) tells me that further habitats have been discovered by local

Photo: Marlon Machado

Fig.43 *G. horstii, Machado* 746 Minas de Camaquã Rio Grande do Sul Brazil

Photo: Massimo Meregalli

Fig.44 The large fruits of *G. horstii horstii, Gerloff* 175 from Minas de Camaquã, Rio Grande do Sul, Brazil

botanists on the extensive ranches in the area, and every effort is being made to protect them from damage.

Conservation Status

Vulnerable. Not only are the localities under threat from farming activities, but cactus enthusiasts still visit the known places and illegally collect plants. Fortunately there are still little known places where the plant grows and probably more yet to be discovered

History

This is one of the many new cactus species found in Brazil by the team of Horst and Uebelmann who did so much exploration in the sixties and seventies. Buining also travelled with them sometimes and described this species in 1970.

Amerhauser (2008) explains how the first imported plants, with the collection number HU79, arrived at the nursery of Werner Uebelmann in 1965. Three of the plants, with 6, 7 and 8 ribs, were bought by Fritz Peham, an enthusiast from Saltzburg who subsequently passed the smallest 6-ribbed plant to Hans Till.

Commentary

Similar in appearance to *G. denudatum*, it can be distinguished by its stronger, outstanding spines, less flattened body, larger flowers and larger fruits (Fig.44). Although this species looks superficially like *G. denudatum*, its characters are actually consistently different and Gerloff (1999) says that the two grow within 10m of each other near to Minas de Camaquã.

Fig.45 *G. horstii horstii, Hofacker* 89, in habitat at Minas de Camaquã, Rio Grande do Sul, Brazil

Fig.46 *Gymnocalycium horstii horstii,Gerloff* 1634 Minas de Camaquã, Rio Grande do Sul, Brazil

Fig.47 *Gymnocalycium horstii horstii,Gerloff* 1634 Minas de Camaquã, RGdS, Brazil. An extreme form with eight ribs.

Fig.48 *Gymnocalycium horstii horstii,Gerloff* 1634 Minas de Camaquã, Rio Grande do Sul, Brazil

Cultivation

Eventually making large clusters, this plant has a shiny bright green body and just a few strong spines giving it a very green and often bloated appearance. It is easy to grow and needs just 5°C in winter when an occasional watering is beneficial to stop excessive shrivelling.

The flowers are large, pure white and, if pollinated, produce very large fruits (Fig.44). Gerloff (1999) tells us that these fruits take 6 months to ripen in contrast to those of *G. denudatum* which take only 6 to 8 weeks. He also reports that a female sterile plant in his collection became fertile and produced viable seeds.

Surprisingly rare in cultivation, this taxon is not seen in collections as often as the subsp. *buenekeri*. Propagation is usually by offsets, since plants are usually either male or female sterile and getting one of each to flower at the same time to produce seed is difficult to achieve. In fact, Gerloff (1999) stated that he had never managed to obtain seeds of this taxon from any dealer,

Many plants in cultivation carry the type collection number HU79 but HU numbers are not necessarily a single location. Horst and Uebelmann would give the same number to plants at various localities if they thought they were the same species. However, it is stated that HU79 was indeed from a single locality near Caçapava do Sul.

Illustrations

New Cactus Lexicon Atlas (2006) Illn. 268.3

Gymnocalycium horstii subsp. *buenekeri* (Swales) Braun & Hofacker

First description (Swales)

Gymnocalycium buenekeri sp. nov.

CSJGB 40(4) pp.97-100 (1978)

Plant body hemispherical at first, later becoming more cylindrical up to 10cm or more in diameter, in old specimens, dark green with characteristic matt surface; ribs usually 5, broadly rounded, bearing areoles situated in shallow notches, the notches very shallow or entirely absent in young specimens; areolar wool almost white to pale yellow, darkening with age and becoming less conspicuous on older areoles. Central spines 0; radial spines 3-5, usually 3, sometimes exceeding 2.5cm long, on old plants usually somewhat shorter, c. 1mm diameter at the base, moderately well developed, slightly curved and standing out to some extent and slightly irregular, pale yellow at first, darkening in age to pale brown.

Flower buds arising at young areoles near the apex, pinkish green at first with numerous scales, later greenish, slowly developing, 4-5cm long 6.5cm in diameter, when fully open, very pale peach with pink mid-stripe to uniform rose-pink, somewhat deeper in the throat; stamens with very pale pink filaments, pale pinkish-fawn anthers and pale fawnish-yellow pollen (N.B. some of the specimens appear to be male sterile); style pale green; stigma lobes c. 9 white. Fruit sub-cylindrical at first, later ovoid, when ripe mid-green, with few, pale green scales and scarcely any waxy bloom, crowned by the persistent withered flower remains, splitting when ripe to expose the seeds embedded in juicy white pulp composed of the succulent funicles; seeds c.1 x 1.5mm, testa very dark brown-black with conspicuous interstitial lacunae.

Holotype: Cult. Swales G.261 (K)

Etymology

Named after Rudolf Heinz Büneker, the discoverer of the plant and brother-in-law of Leopoldo Horst.

Other Names

Gymnocalycium horstii var. *buenekeri* Buining nom. inval. (Art 37.1) in KuaS 21(9): pp.162-165 (1970)

Distribution (Map 9 on page 39)

Known only from a few localities on the steep sides of flat-topped hills near Sao Francisco de Assis, Rio Grande do Sul, Brazil, about 200km from the type locality of ssp. *horstii* near Caçapava do Sul.

Conservation Status

Critically Endangered. One of the most threatened *Gymnocalycium* taxa because of its very restricted natural habitat range and its desirability in cultivation. It has been heavily collected since its discovery and now plants are hard to find in the wild.

History

In his article about *G. horstii*, Amerhauser (2008) published Till's recollections of a meeting when Buining visited Till in July 1969. It is suggested that this was the first time Buining had

Fig.49 *Gymnocalycium horstii buenekeri*, *Gerloff* 260, N. São Francisco de Assis, Rio Grande do Sul, Brazil

Fig.50 *Gymnocalycium horstii buenekeri* ex Waras

Fig.51 *Gymnocalycium horstii buenekeri* ex Waras

Photo: Norbert Gerlof

Fig.52 The habitat of *Gymnocalycium horstii buenekeri*, *Gerloff* 260, north of São Francisco de Assis, RGdS, Brazil

seen plants which Till had received from Uhlig in 1968 called *G. denudatum* var. *pentacanthum n.n.* that were later to be described by him as *G. horstii* var. *buenekeri*.

This taxon was subsequently described invalidly by Buining (1970) as a variety of *G. horstii*. It was invalid because he did not designate a type. Till felt that the description was inaccurate, suggesting that Buining had not properly researched it. There was no illustration accompanying the brief description of the variety. Buining commented on various differences from var. *horstii*, not all supported by observations of large numbers of plants by subsequent workers. Interestingly, its flowers were said by Buining to last only 2 days in Brazil, compared to 5-6 days for the var. *horstii*.

Eight years after the name appeared, Swales (1978) expanded the description and validly published the name as a separate species, *G. buenekeri*. His justification for this decision was based on the differences from *G. horstii*, i.e. the body colour and texture, flower colour, fruits, and seeds.

Buining had allocated the seeds to Buxbaum's section *Mostiana* but Swales showed this to be an error, and it is now considered to belong to the subgenus *Macrosemineum*.

Commentary

Some *Gymnocalycium* specialists agree with Swales and still consider this to be a separate species, different from *G. horstii*. It is an example of a *Gymnocalycium* where individuals are usually either just male or female. Gerloff (1998) reports

Photo: Norbert Gerlof

Fig.53 *Gymnocalycium horstii buenekeri, Gerloff* 260, north of São Francisco de Assis, Rio Grande do Sul, Brazil

Photo: Norbert Gerlof

Fig.54 *Gymnocalycium horstii buenekeri, Gerloff* 260, north of São Francisco de Assis, Rio Grande do Sul, Brazil

Photo: Norbert Gerlof

Fig.55 *G.horstii buenekeri*, *Gerloff* 260, north of São Francisco de Assis, Rio Grande do Sul, Brazil

that he has successfully cross-pollinated it with the type subspecies.

Cultivation

It is very easy to grow, quickly making impressive clumps of stems and often seen as big specimens in shows. It enjoys liberal watering in summer and frequent repotting will be rewarded with fast growth. The pretty pink flowers have many petals which vary between pale and dark pink, depending on the clone.

The very large fruits are difficult to produce since pollination of flowers from both a fertile male and a fertile female plant is needed. Even if pollen is transferred, a fruit does not always result. A possible way to be successful is to store the pollen from the flower of a male fertile plant until a female fertile plant opens a flower to receive it.

Most plants in culture are propagations from habitat plants, many having been sent to Europe by Waras in the 1970s. The offsets are often produced underground, turning green only as they grow above the soil. In habitat, they can travel underground for some time before emerging into the light, often giving the plants an untidy appearance.

Photo: Massimo Meregalli

Fig.56 *G. horstii buenekeri*, *Horst-Uebelmann* 363 from the type locality near São Francisco de Assis, Rio Grande do Sul, Brazil

Gymnocalycium hyptiacanthum subsp. *hyptiacanthum* (Lemaire) Britton & Rose

First Description (Lemaire)

Echinocactus? *hyptiacanthus* Lemaire in *Cactearum genera nova. speciesque novae* (1839): pp.21-22

Diagnosis: Body slightly elongated, crown sunken, rich green, 11 tubercled ribs; ribs straight, tubercles six sided; areoles oval; spines, seven, short, slender, rigid, clearly curved towards the body, brownish-yellow to yellow, four laterals in two rows.

Tubercles arranged in 11 rows, 6 sided at the base, 8-10 mm wide, separated by short shallow cross-cuts; these separated by long curved downwards running furrows, giving the effect of a green line, flattening towards the base, areoles elliptical, covered with short hair, not very woolly, white, persistent, later becoming grey; carrying 6-7 spines, the top 1-2 erect, hardly bristle-like, 2-4 mm long, 2 pairs radiating each side, the lower pair a little stronger, 6-8 mm long, the one pointing straight down being the longest, 8-10mm long, all thin, very rigid, curved towards the body, brownish-yellow, deep purple at the base and tips.

Clearly a distinct species related to *Ech. gibbosus*, yet totally different. The described plants were 2.5" (63.5mm) high, 2" (51mm) broad, all mature plants.

Habitat, flowers and fruit, unknown.

Neotype: Uruguay, Dep. Florida, Ruta 5, Cerro Pelado, 17 Jun 1990, *Kiesling & Kroenlein* 8505 designated by Kiesling in *Catálogo de las Plantas Vasculares de la República Argentina 2*. Monogr. Syt. Bot. Missouri Bot. Gard. 74 (1999)

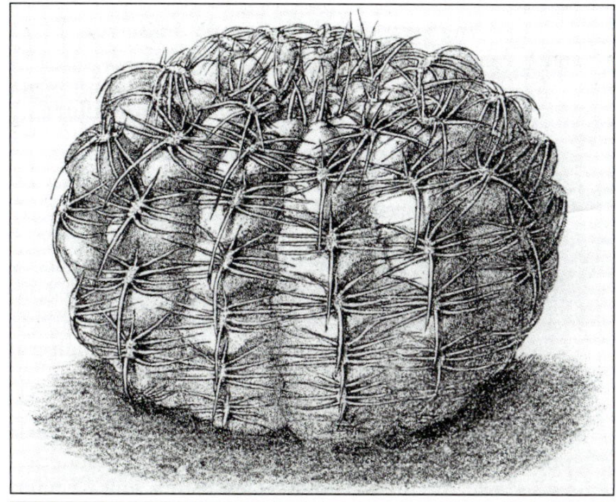

Fig.58 The illustration of *Echinocactus hyptiacanthus* Lem. from Schumann 'Gesamtbeschreibung der Kakteen', Fig.70 (1899)

Photo: Massimo Meregalli

Fig.59 *G. hyptiacanthum hyptiacanthum, Meregalli* 366 from Cerro Mosquitos, Canelones, Uruguay

Fig.57 A habitat of *G. hyptiacanthum hyptiacanthum* at Cerro Mosquitos, Canelones, Uruguay

Photo: Massimo Meregalli

Photo: Massimo Meregalli

Fig.60 *G.hyptiacanthum hyptiacanthum, Meregalli* 450 from Cerro Mosquitos, Canelones, Uruguay

Etymology

From the Greek 'hyptios' meaning recurved and 'akantha' for spine.

Distribution (Map 10)

The neotype is from a population from hills south of the town of Florida but it can be found at a number of places in the very south of Uruguay, in the Provinces of Maldonado and Canelones. The centre of distribution is in the Cerros Mosquitos according to Meregalli (pers. com.).

Conservation Status

Near Threatened. Although several fragmented populations (sometimes with few plants) still exist, they are being exterminated by habitat destruction. The reasons include increased cultivation and invasion by alien plants such as Eucalyptus.

History

Lemaire's original description was based on plants of unknown source and without information about the flower or fruit. Later authors added to the description, but with observations based on their interpretation of the name that was not necessarily the same as that of Lemaire. Labouret (1853) misquoted Lemaire's description of the spines, saying they were yellow with a red apex rather than yellow with a red base and apex. The idea that the plant had a white flower was perpetuated by some authorities, although it was generally accepted that this taxon came from southern Uruguay.

Papsch (2001a) attempted to neotypify the name *Echinocactus hyptiacanthus* with a plant from the Sierra Bayas, more recently described as *G. schroederianum* subsp. *bayensis*. However, as pointed out by Kiesling et al. (2002b), the name had already been validly neotypified by Kiesling (1999) with this Uruguayan taxon, even though he cited the specimen in error as a lectotype rather than a neotype. The neotype was derived from a plant of a population living on a Cerro south of the town of Florida; a population that is now practically extinct (Meregalli pers. com.).

Kiesling's view was first proposed by Schlosser & Schütz (1982) who also believed that *Echinocactus hyptiacanthus* was a yellow-flowered plant from southern Uruguay.

This means that the name *Gymnocalycium hyptiacanthum* has to be applied in accordance with Kiesling's neotype to this widespread yellow-flowered taxon from Uruguay that is better known as *G. uruguayense*.

Commentary

The old name *Echinocactus hyptiacanthus* Lemaire has recently been regarded as indeterminate and it was treated as such in the *New Cactus Lexicon* (2006). Following the rules of the ICBN, I have used it for this taxon because of the neotypification by Kiesling (1999). I personally think it is more likely that the application of the name by Papsch (2001a) is correct. He put up a convincing argument that the plant described as *G. schroederianum* subsp. *bayensis* is the original *E. hyptiacanthus*.

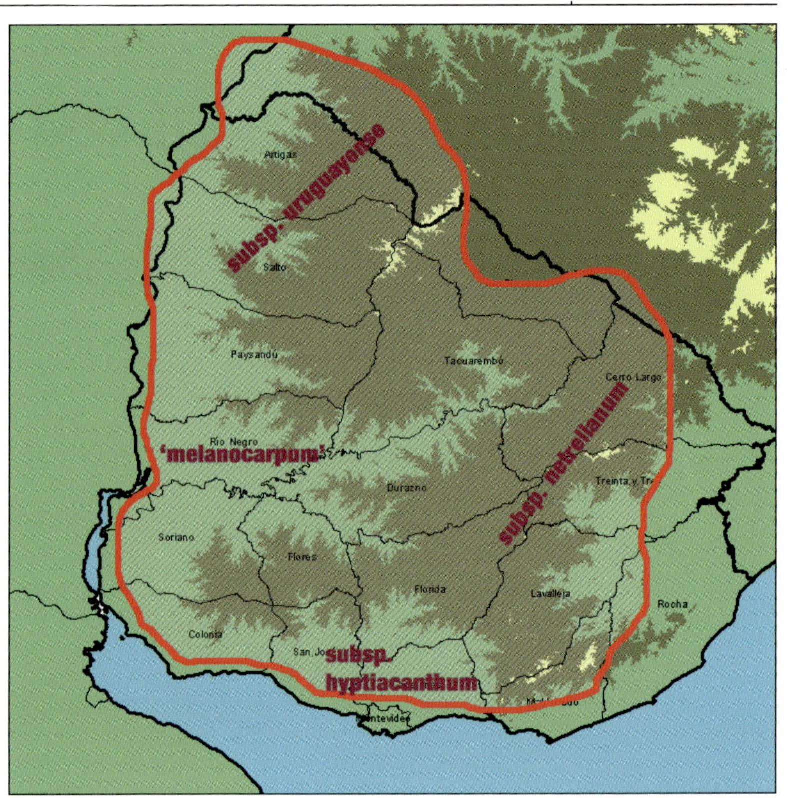

Map 10 The distribution of *G. hyptiacanthum* extends over most of Uruguay and small parts of neighbouring countries

It is a pity that if one accepts that there is just one species of yellow-flowered *Gymnocalycium* in Uruguay, then one must now use the name *G. hyptiacanthum* for a species which has been known for so long as *G. uruguayense*. At least, the familiar name can be retained at subspecific rank. Meregalli (pers. com.) claims that there are plants in southern Uruguay that match the original description, or rather do not contradict it, so supporting Kiesling's neotypification.

Echinocactus leeanus was included in this taxon by Schumann but that approach was rejected by Britton and Rose. See *Gymnocalycium reductum* subsp. *leeanum* on page 110.

Cultivation

This yellow-flowered species from Uruguay is a good-looking plant to grow. It makes neat clusters of stems and it flowers early in the year, in fact among the earliest of all Gymnocalyciums. The flowers are usually self-sterile or of a single sex so propagation from cuttings is the easiest method.

Gymnocalycium hyptiacanthum subsp. *netrelianum* (Monville ex Labouret) Meregalli

First Description (Monville ex Labouret)

Labouret, *Monographie de la Famille Des Cactées* p.248 (1853)

Echinocactus Netrelianus (Monv.).

Synonym. *Echinoc. Netrelianus* Monv.

Habitat ?

Diagnosis: Body globular, slightly depressed in the crown, somewhat greyish-green; ribs 14, rounded, slightly projecting; intercostal grooves in the upper part of the plant, missing in the lower part; ribs tuberculate; areoles inserted on the crown of the tubercles, round, furnished with yellowish-white down, later bare; spines inserted in the lower part of the areoles, 5-7, twisted, unequal, adpressed, the two upper spines (sometimes missing) are very short; all tawny coloured at the base, dull grey at the tip.

Flowering in May and June. Flowers large, with a sheen, tubes without spines, made up of scales that in elongating little by little change into shiny citron-yellow petals, with a greenish median streak, very prolific.

Neotype: UY, Prov. Maldonado, hills north of Piriápolis, 95m, 9 Feb 2004. Meregalli MM 0452 (TO-HG). Designated by Meregalli in CSI24: p.22 (2008) (Fig.61)

Etymology

Named for Mr. Netrel.

Distribution (Map 10 on page 46)

This subspecies occurs in numerous populations in eastern Uruguay. It intergrades with subsp. *hyptiacanthum* in the south and subsp. *uruguayense* in the north.

Photo: Massimo Meregalli

Fig.61 The plant from which a part was preserved as the neotype of *Echinocactus netrelianus, Meregalli* 452

Conservation Status

Least Concern. The plant is widespread and is known from a large number of habitats. Some are threatened by excessive grazing or destruction of the habitat but not to a significant degree.

Fig.62 *G. hyptiacanthum netrelianum, Van Vliet* s.n. Punta Ballena, Uruguay

Fig.63 *G. hyptiacanthum netrelianum, Hofacker* 182 from Piriápolis, Maldonado, Uruguay

Fig.64 *G. hyptiacanthum netrelianum, Hofacker* 167 from south of Zapican, Zavalleja, Uruguay

Fig.65 A habitat of *G. hyptiacanthum netrelianum, Meregalli* 452, north of Piriápolis, Maldonado, Uruguay

History

The name *Echinocactus Netrelianus* first appeared, without any description, in the 1846 sale catalogue of the Monville Collection. A description followed in Labouret's *Monographie de la Famille Des Cactées* (1853). Britton & Rose (1922) made the combination in *Gymnocalycium*.

Although the first publication gives no origin for the plant, the description allows it to be identified as of eastern Uruguayan origin where plants with yellow flowers, and yellow spines without a red base can be found.

Meregalli neotypified *Echinocactus netrelianus* and made it a subspecies of *G. hyptiacanthum* in CSI 24: p.22 (2008).

Commentary

The name *Echinocactus netrelianus* was rejected as indeterminate in the *New Cactus Lexicon* (2006) but following the work of Meregalli, it is here recognised as a subspecies of *G. hyptiacanthum*. It differs from the type subspecies by its completely yellow spines.

Cultivation

This is one of the prettiest of the yellow-flowered Gymnocalyciums from Uruguay . It is easy to cultivate, soon growing into a cluster of stems and produces its flowers early in the year. It is easily reproduced from seeds or cuttings and is readily available, often as propagations of habitat clones. The flowers are usually dioecious so you need male and female plants to flower at the same time to set pure seeds of the subspecies.

Fig.66 *G. hyptiacanthum netrelianum, Meregalli* 483, east of Sarandy del Yi, Florida, Uruguay

Fig.67 *G. hyptiacanthum netrelianum, Knize* 153 from Las Minas, Prov. Lavelleja, Uruguay

Gymnocalycium hyptiacanthum subsp. *uruguayense* (Arechavaleta) Meregalli

Fig.68 *Echinocactus uruguayensis* (left) and *E. melanocarpus* (right), pictures published with the first descriptions (1905)

First Description (Arechavaleta)

Arechavaleta, J. *Flora Uruguaya - Anales del Museo Nacional de Montivideo V*: pp.218-220 (1905)

Echinocactus uruguayensis Arech. nov. sp.

Globose, tapering below, upper part has a flat appearance, depressed in the crown. Ribs vertical, 12-14, separated by straight grooves, with tubercles, six sided at the base, with humps below the areoles, in habitat. Areoles round, covered with short grey hair. Spines usually 3, seldom more, 1.5-2cm long, lying flat, yellowish when young, whitish in age, very rough or scaly. Flowers near the crown, globular before opening; bell-shaped when open, ± 4cm high, 5.5-6cm diameter, devoid of hair, but has short tufts of short white hair at the base. The lower scales the same as those on the ovary, small, oval, few, greenish-brown, the upper scales green with a dark or medium brown mid-stripe on the back.

Perianth tube yellowish-green, outside pale green, inside whitish. Petals long lanceolate, white, fleshy, numerous. Stamens in two series, set in the wall of the perianth tube, filaments pale, anthers small yellow. Style lower than the stamens: stigma with 8-10 lobes. Ovary green, 4mm high. Fruit oblong, compressed sideways, narrower at the bottom, carrying small flower remains.

Habitat: Paso de los Toros. It has been found to be very difficult to set fruit on these plants in cultivation. Blooms in November, fruiting the following month.

Etymology

Occurs in Uruguay.

Fig.69 *G. hyptiacanthum uruguayense ,Schlosser* 102 from Tambores, Tacuarembo, Uruguay

Synonyms

G. artigas Herter in *Revista Sudamericana de Botanica* 10(1): pp.1-2 (1951)

A new Uruguayan cactus and observations on the flowering of a Uruguayan cactus by G. Herter

***Gymnocalycium artigas* Herter sp. nov.**

Body globular when young, then becoming conical, almost cylindrical, strongly depressed, 6-8cm diameter, 2-3cm high, without the underground near napiform parts, colour green (neither glaucous nor greenish-grey). Ribs at first 6-8, later 10, large, glabrous, with distinct tubercles, semi-globular, humped or nearly hexagonal, sub-confluent, irregularly arranged in vertical series. Spines 3-5, rarely 6, radiating or almost pectinate, bristly, quite hard, but flexible. Plants monoecious. Flowers large, solitary or in groups, usually of 3 or 4, 5cm long, 5cm diameter. The outer petals are scale-like, green, embellished with a red mid-stripe, the smallest 5mm by 5mm, giving way to the interior petals. Petals citron yellow, bright, 6-8mm wide, 3-4cm long. Stigma lobes white, gathered together in male plants, separated in female plants. Anthers yellow.

Habitat:- Uruguay, Dept. Duranzo. Blanquillo, in the rocks at 150m. altitude. IX, 1947 Herter (H.H. 99. 773.), type specimen grown in Montevideo, in flower x 1947. Plantas Uruguayenses exsiccatae 1722a.

Fruits are ovoid in form, 8-15mm diameter, 10-18mm long, colour olive green. They ripen until the end of the year, opening lengthwise, splitting in 2, rarely 3 places, which run from the base reaching nearly to the tip of the fruit, releasing numerous (c. 1000) polyhedric seeds, black, 1mm diameter. The seeds are tangled in a spongy, pink coloured mass of insipid tasting pulp.

Commentary on G. artigas

Named in the memory of General Artigas, founder of the Uruguayan nation.[Not the place with this name]

Herter (1951) explains that he described this plant because it did not agree with previously described species from Uruguay. Even though he accepts that *G. uruguayense* came from the same valley as his plant, he says that is a different taxon because of its white (or pinkish) flowers and 3 spines. It is interesting that *G. uruguayense* was described with white flowers, a colour which in fact is found in only a few of its widespread populations.

Fig.70 *G. hyptiacanthum uruguayense* ,GC779.02, west of Santana do Livramento, Rio Grande do Sul, Brazil

The description of *G. artigas* includes characters which are intermediate between subsp. *uruguayense* and subsp. *netrelianum*.

G. guerkeanum (Heese) Britton and Rose in *The Cactaceae* 3: pp.154-155 (1922). Basionym: *Echinocactus guerkeanus* Heese (1911)

First description of *Echinocactus Gürkeanus* from MfK 21(9): p.132 (19111)

Echinocactus guerkeanus belongs to the dwarf forms from which one only obtains larger specimens by grafting. The examples imported from Bolivia during 1904 now in my collection are c. 5cm diameter, the height only 3.5cm, offsetting from the base. Ribs 9, with slight chin-shaped protruding humps. Crown sunken, sparsely covered with small flecks of wool. Areoles c. 8cm apart, elliptical, 2-3mm long, covered with yellowish felt, becoming bald with age. Spines 5, the two shortest spines at the top, pointing sideways, c. 5mm long. Two bigger, spreading apart to 12mm long. Central spine pointing down, c. 10mm long, in youth pointing straight out, rough, yellowish, brownish-red at the base. Body colour matt blue-green.

Flowers coming from the crown, up to 5cm long, nearly 4cm in diameter. Ovary greenish, with longish rounded scales of the same colour. Receptacle funnel-shaped, the short greenish scales, which cover the bottom of the tube become long spatulate towards the top, then into very shiny light yellow petals on which a dark central stripe runs down the outside. Many stamens, half as long as the scales of the receptacle. Carpel deep inside with nine segments. Anthers neapolitan yellow. Filaments a delicate light green.

Commentary on *G. guerkeanum*

Heese had received the plant he described from Fiebrig in 1904. It was said to be in a consignment from Bolivia, some of which was sent to the Royal

Photo: Massimo Meregalli

Fig.71 *G. hyptiacanthum uruguayense* 'melanocarpum', *Meregalli* 515, Arroyo Sánchez, Río Negro, Uruguay

Fig.72 The habitat of *G. hyptiacanthum uruguayense,* GC779, west of Santana do Livramento, Rio Grande do Sul, Brazil

Botanical Gardens, Berlin. Even though Bolivia was larger then than it is now, plants like this are not found in that country.

G. melanocarpum (Arechavaleta) Britton and Rose in *The Cactaceae* 3: pp.161-162 (1922). Basionym: *Echinocactus melanocarpus* Arechavaleta (1905)

First description of *Echinocactus melanocarpus* from *Flora Uruguaya - Anales del Museo Nacional de Montivideo* V: p.220 (See Fig.68 on page 49)

Echinocactus melanocarpus Arech. nov. sp.

Plant olive colour, globose, almost spherical, 7-8cm high, 8-9cm diameter, tubercled in the sunken crown. Ribs 15, with tubercles, which are many sided at the base, wavy furrows form a hexagonal shape at the base of the tubercles. The long areoles sit over a hump on the tubercles, covered with short whitish woolly hair, balding in age. Spines radiating, arranged in two rows of 5-6, like the legs of a spider, all lying flat, the longest being 2-2.5cm long, pale yellow, reddish at the base, ash grey in age. Flowers from the shoulder of the plant, ovary dark green, with scales, scales are broad with pale edges. Berries egg-shape, ± 2cm high, 1.5-2cm diameter, dark olive green with few broad white edged scales. Habitat: Paysandú, Uruguay.

Photo: Massimo Meregalli

Fig.73 The habitat of *'G. melanocarpum'* ,*Meregalli* 515, Arroyo Sánchez, Río Negro, Uruguay

Fig.74 *G. hyptiacanthum uruguayense, Schlosser* 140, from south of Artigas, Uruguay. A form with white flowers.

Commentary on *G. melanocarpum*

Arechavaleta named this taxon because of its dark olive-green fruit. There is no description of the flower here but later Herter (1954) published a line drawing of a flower section without telling us the origin of the material illustrated. Plants resembling the original photograph had never been found again in habitat until Prestlé (2001) claimed to have found it. He expanded the original description based on his collection but gave no information on where he found it..

Meregalli (pers. com.) also reports finding plants matching the original description in the Provinces of Paysandu and Río Negro where he says that the plants are often found near rivers in white clay soil as mentioned by Arechavaleta (Figs.71 & 73).

Distribution (Map 10 on page 46)

This subspecies is widely distributed in north-western Uruguay and a distinctive form occurs just across the border in Rio Grande do Sul, Brazil but interestingly, never overlaps with *G. denudatum*. Meregalli (pers. com.) reports one population south of Concordia in Argentina. In the south of its range it was named *G. melanocarpum* by Arechavaleta.

Populations with white or pink flowers can be found in the north of the distribution area, often mixed in one location, even with yellow-flowered plants as well.

Conservation Status

Least Concern. It is widespread and can colonise nearly all possible habitats, being absent only from fields with deep soil. Most often it is found in gravelly soil, with grasses, very frequently on the slopes of small streams where the soil is shallow and there are rocks. It can be found in fields with gravelly soil, where the plants can be very difficult or nearly impossible to find in summer, since the grass is high and the plants are partly buried in the soil.

Nowadays, it can often be found only below the fences, in the moss, where the sheep cannot reach it, since in the open, overgrazed fields it has been destroyed.

Fig.75 *G. hyptiacanthum uruguayense, Herm* 25, from Masoller, border of Salto/Rivera, Uruguay. Pink flower form.

History

It was Arechavaleta who described *Echinocactus uruguayensis* based on a plant from Paso de los Toros in central Uruguay, which is curious because Meregalli (pers. com.) says that plants from there are usually yellow-flowered, not white as in the description.

Commentary

Plants in culture with the name '*G. uruguayense*' on the label may in fact belong to one of the other subspecies of *G. hyptiacanthum*. Here, *G. uruguayense* is applied to just the northern subspecies rather than in the wider sense used in the *New Cactus Lexicon* and other recent works.

Cultivation

There are not many Gymnocalyciums that have yellow flowers so, although the colour is less intense than that of *G. andreae*, this plant makes a more attractive specimen, often forming a clump of stems. The pink-flowered form is sought out by collectors although the colour is not strong. White-flowered clones are even less commonly seen in cultivation.

The plant is of very easy culture but should be kept moist during the growing season and, even in the winter, a drop of water occasionally will prevent excessive shrivelling. It is easily reproduced from seeds or cuttings but the flowers are usually dioecious so you need male and female plants to flower at the same time to set true seeds.

Illustrations

MfK 21(9): p.133 (1911) as *Echinocactus Gürkeanus*
Vaupel, F. *Blühende Kakteen* Tafel 144 (1912) as *Echinocactus Gürkeanus*

Gymnocalycium mesopotamicum Kiesling

Fig.76 *G. mesopotamicum, Ferrari* s.n., from Mercedes, Corrientes, Argentina

First Description (Kiesling)

CSJGB 42(2) pp.39-42 (1980)

Stem solitary, globose, 1.2-2.3cm high and c. 4cm diameter, epidermis dark green, apex umbilicate, the depression 0.5cm deep and 1cm diameter, with pronounced tubercles. *Ribs* c. 7-9, broad, low, obtuse, 1.5cm broad at the base; *tubercles* not very prominent, separated by shallow transverse grooves. *Areoles* c. 5 to each rib, oval, large, 3mm long and 1mm broad, with abundant yellowish-white wool (grey when adult), somewhat sunken, sited on the upper part of the tubercles. *Spines* 9-12, the smallest only 2-3mm long, the longer 9mm long, setaceous, flexible, adpressed, arching over the body, subpectinate, arising at the edges of the areoles, reddish brown when young, later greyish white or somewhat pinkish.

Flowers arising from the adaxial side of the areoles bordering the umbilicus, 6-7cm long, 2cm max. diameter when closed, 5.5cm long, 6.5cm diameter when open; *receptacle* obconic, 3cm long, 0.5cm diameter at base and 1cm diameter above, light opaque green, fleshy, the wall c. 1mm thick; *scales* c. 10, prominent, broad, obtuse, the lower smaller, 2-5mm long, 4-5mm broad, mucronate, pinkish green; *outer perianth-segments* spatulate, 1-3cm long, 6-7mm broad, mucronate, somewhat fleshy, pinkish green to pinkish white. *Inner perianth-segments* lanceolate, 3.1-3.3cm long, 6mm broad, white, tinged pink near the tip, reddish at the base with a pinkish green mid-stripe, *innermost perianth-segments* narrowly lanceolate, very acute, 3cm long, 4-5mm broad, pure white; *tube* 1.8cm long, pale red within *nectar chamber* at the base of tube, cylindric, 1mm long, 2-3mm diameter; *filaments* erect, 8-13mm long, white with a pinkish base; *anthers* small, 1mm long, 0.5mm broad, yellow, curving over the style; *pollen* abundant, yellow, globose, tricolpate with low fertility, tectum perforate-spinulate; *ovary* obconic-fusiform, truncate, 13mm long, 3.5-4mm diameter, with numerous small white ovules; *style* cylindric, 1.8cm long, 1mm diameter, creamy-white; *stigma* white; papillose, with some 8 radiating sub-cylindric lobes, c. 5-6mm long 1mm diameter;

Fruit clavate, 2-3cm long, 7-8mm diameter, dull green, with the dry perianth persistent, brown, *scales* few, about 6, similar in shape to those of the flower, greenish at the base, pale pink above. Seeds globose-truncate, 1.7-1.8mm diameter, with a large convex hilum, c. 1.8mm long, 1.2mm broad, oval, curved; *testa* black with cells superficially smooth and slightly convex but each one with an eccentric or more often capitate projection, 0.025mm high and 0.015mm diameter at the base. *Chromosome no.* 2n=22 (observed at mitosis in ovary-wall cells, M.A.T. Johnson, unpublished data).

Photo: Ludwig Bercht

Fig.77 Habitat of *G. mesopotamicum, Bercht* 2298, Rte 123, near Mercedes, Corrientes, Argentina

Fig.78 *G. mesopotamicum, Piltz* 241, from Mercedes, Corrientes, Argentina

Map 11 Localities of *G. mesopotamicum* in Prov. Corrientes, Argentina

Type: *Cutler & Lonsdale* 126-552, Nov. 1978 (K, holo; SI, iso) Argentina, prov. Corrientes, Mercedes.

Etymology

Named after the region Mesopotamia, between the Río Paraná and Río Uruguay in Argentina where it was discovered.

Distribution

The type locality is a few kilometres south of the town of Mercedes, Prov. Corrientes. There are other known populations in the area and further populations about 20 km to the east. It grows somewhat concealed in grassy places with stones where the soil is shallow and near to cultivation and grazing animals.

Conservation Status

Endangered. Although the presently known populations are not endangered, the plant has a limited range in an agricultural region. The total number of plants in the largest population is probably less than 1000.

History

The plant was found during a trip to northern Argentina in the late nineteen seventies organised by the Royal Botanic Gardens, Kew and the Darwinion Institute, Buenos Aires. Since its description in 1980, it has had an uneventful existence, neither having been lumped nor split.

Commentary

It is stated in the original publication that one individual was male, another female, where others were normal hermaphroditic plants. This is an example of subdioecious behaviour, where separation of the sexes is only partial. Kiesling says the seed has characteristics of *Trichomosemineum* as well as *Macrosemineum* where it is placed here. However, Meregalli (pers. com.) suggests that the papillae on the seeds have evolved independently and that the plant has no relationship to *Trichomosemineum*. Its closest relative is probably *G. angelae* which grows nearby.

Fig.79 *G. mesopotamicum, Ferrari* s.n. from Mercedes, Corrientes, Argentina

Cultivation

This is a beautiful small species to grow and soon makes a pleasing clump of stems which have a shiny epidermis. The pretty white flowers are easily produced on small plants. It should be protected from strong sun and never left completely dry for very long even in winter when the temperature should be kept above 5°C. Many plants in cultivation are descendants of the *Piltz* 241 collection.

Gymnocalycium paraguayense (Schumann) Hosseus

First Description (Schumann)

Bulletin de l'Herbier Boiss. 2, 3: p.252 (1903)

The Latin diagnosis by Prof. Dr. Karl Schumann of *Echinocactus paraguayensis* was included in the text of Dr. Hassler's travel report:

White petals, purple on the inner face at the base. Found between the rocks, in the prov. Chodat, in the river valleys of Chololó.

Lectotype: Paraguay, "inter rupes propre Chololó, in valle fluminis Y-aca, ex cult. Vallertes", 4.IV.1902, E.Hassler 6693 (G!). Designated by Meregalli et al. (2002).

Etymology

Occurs in Paraguay.

Other Names

Echinocactus paraguayensis Schumann in *Bulletin de l'Herbier Boiss.* 2, 3: p.252 (1903)

Map 12 Localities of *G. paraguayense* in Paraguay

Synonym

G. fleischerianum (Meregalli, Metzing and Kiesling) Vala in *Gymnocalycium* 16(3): p.520 (2003). Basionym: *G. paraguayense* forma *fleischerianum* Meregalli, Metzing & Kiesling (2002)

The original invalid naming of *Gymnocalycium fleischerianum* from Backeberg *Die Cactaceae* pp.1703-1706 (1959):

Globular, shiny light green, sometimes offsetting, sunken crown. Ribs c. 8, rounded, 2.5cm wide at the base, slightly humped. Areoles round, c. 5mm wide with thick brown-white felt, c. 20 spines, not very stiff, bristly, elastic-like, spreading like rays of the sun, whitish-yellow and brown, later grey, up to 2.5cm long, centrals difficult to define. Flowers funnel shaped, white with a fluorescent pink throat, up to 4cm long, 3.5cm wide. Paraguay.

(Named after the Czech cactus grower Zdenek Fleischer)

Fig.80 *G. paraguayense, Moser* s.n.

Fig.81 *G. paraguayense, Metzing* 35, from Cerro Pero, east of Piribebuy, Cordillera, Paraguay

Fig.82 *G. paraguayense* (fleischerianum), *Piltz* 453, from Chololo-i, Piribebuy, Cordillera, Paraguay

Photo: Ludwig Bercht

Fig.83 *G. paraguayense, Bercht* 2266, in habitat south of Piribebuy, Cordillera, Paraguay

Distribution (Map 12)

Occurs on isolated and shady patches in woodland, steep river valley sides or among cultivated land in the south of Department Cordillera and the north of Department Paraguari, Paraguay. Bercht (pers. com.), who believes there are two species, tells me that the typical *G paraguayense* prefers sandy patches while *G. fleischerianum* is a rock-plate dweller. It has been suggested [Meregalli et al. 2002] that recent changes in the climate of this region, from dry and cool to warmer and more humid, has resulted in more woodland and the restriction of the cacti to a few rocky locations where trees cannot establish. Moser (1970) tells us that the plants in habitat can sit in water for prolonged periods.

Conservation Status

Critically Endangered. Known from only a few isolated habitats, often with few specimens and threatened by agricultural expansion, grazing and the increased use of the land by the local people. It

Photo: Ludwig Bercht

Fig.85 *G. paraguayense, Bercht* 2266, in habitat south of Piribebuy, Cordillera, Paraguay

may be that the typical form of *G. paraguayense* is eradicated in the wild.

History

In 1897, the Nomenclature Commission of the DKG considered a plant collected in Paraguay and submitted by W. Mundt to judge if it was a new species. On their advice, Mundt (1897) described it as *Echinocactus denudatus* var. *Golziana*. This publication was subsequently referred to as var. *paraguayense* by Schumann and others. It was six years later that Karl Schumann (1903), a member of the Commission, described the plant as a separate

Photo: Ludwig Bercht

Fig.86 *G. paraguayense* (fleischerianum), *Bercht* 3118 in habitat between Chololó and Piribebuy, Paragarí, Paraguay

Photo: Ludwig Bercht

Fig.87 *G. paraguayense* (fleischerianum), *Bercht* 3118 in habitat between Chololó and Piribebuy, Paragarí, Paraguay

Photo: Ludwig Bercht

Fig.84 The habitat of *G. paraguayense* (fleischerianum), *Bercht* 3118 between Chololó and Piribebuy, Paragarí, Paraguay

Fig.88 *G. paraguayense* (fleischerianum) *Piltz* 416, from Pirareta, Piribebuy, Cordillera, Paraguay

Fig.89 *G. paraguayense* (fleischerianum), *Piltz* 453, from Chololo-i, Piribebuy, Cordillera, Paraguay

species calling it *Echinocactus paraguayensis*. It was later imported into Europe again by Fric who was given plants by a local woman in Paraguay. Subsequently, collections were made by A. M. Friedrich and Willy Knoll.

Following the finding of *E. paraguayensis* by Prof. Hassler, plants were sent to the Berlin-Dahlem Garden around 1895. They were also collected by Herman Grosse towards the end of the 19th Century and sent to Haage's nursery. Because of its similarity to *Echinocactus denudatus*, the first published names were varieties of that species (Haage 1898).

The first description of *G. fleischerianum* was published by Backeberg (1936) in *Kaktus-ABC* when he was under the false impression that Jajó had already published it in Kaktusar, a journal for which he was the editor. The Latin diagnosis did not appear until 1959 when Backeberg again published the name invalidly in *Die Cactaceae* (1959), this time without designating a type. He had seen a specimen which had been given the provisional name *Gymnocalycium denudatum* var. *anisitsii* by Fric in his 1929 catalogue *Kakteenjager*.

Fig.90 *G. paraguayense* (fleischerianum), *Arzberger* 25, from Ita Moroti, Cordillera, Paraguay

The first valid description of the name was as a form of *G. paraguayense* (Meregalli et al. 2002). In 2003, *G. fleischerianum* was raised to specific rank by Ladislav Vala (2003) as a result of dismissing the findings of Meregalli et al.

Commentary

Although the plants usually seen in cultivation as *G. fleischerianum* look different from those of *G. paraguayense*, it is suggested by micromorphological and DNA analyses that these are extremes of a variable species (Meregalli et al. 2002). The former usually have more rounded ribs, often lacking transverse furrows with more and longer spines. An impression of the variability of this taxon is effectively conveyed by the many pictures in Günther Moser's article in the NCSS journal (1970). Moser had obtained a large number of collected plants from Paraguay while co-operating with A. M. Friedrich, which he then cultivated in Austria.

Cultivation

This is a distinctive species often with a shiny body and attractive spination. The best way to grow this plant is to protect it from strong sunshine and provide constant moisture during the summer months. Most clones offset, although typical *G. fleischerianum* is more prolific and is often propagated that way.

Illustrations

Backeberg, C. *Die Cactaceae* III: pp.1703-1705 (1959) as *G. fleischerianum*

Moser, G. NCSSJ 25(1): pp.34-37 (1970) as *G. 21*

Subgenus *Gymnocalycium*

Type Species

G. gibbosum (Haworth) Pfeiffer ex Mittler

Species and Subspecies **Page**

(synonyms indented)

Echinocactus gibbosus P. D. C.
Tafel 55.

Gymnocalycium amerhauseri H.Till

Fig.91 Habitat of *G. amerhauseri*, GC1002.01 in the Sierra Chica west of Ascochinga, Prov Córdoba, Argentina

First Description (H. Till)

Gymnocalycium 7(3) pp.131-134 (1994)

The above ground part of the plant is semi-globular, 25mm high, 50-60mm diameter, chubby; epidermis dark green to blue-grey-green, smooth, somewhat shiny. The part below ground is tapered when standing in rich humus soil, up to 8 times as long as the above ground part, with conspicuous areole and spine remains, often with a longer stake-like root.

Ribs mostly 8, quite flat at the base of the plant, hardly raised, 19-20mm wide, lightly raised near the middle, with flat rounded humps, formed by shallow cross-grooves. Areoles sunken, oval, 2.5-3 x 2mm, 11-12mm apart, young areoles with yellowish-white, short felt, soon disappearing. Spines (5-) 7, radiating, curved slightly downwards, sometimes straight, the top and bottom pair 6-8mm long, the middle pair 11-12mm long, all needle-like, dark brown at the base, otherwise ivory to white, not piercing. In age (seldom on young plants) a central spine appears, 12-14mm long.

Fig.92 *G. amerhauseri*, GC1002.01 in the Sierra Chica west of Ascochinga, Prov Córdoba, Argentina

Flowers funnel shaped, at full anthesis the petals are recurved, cream-white to delicate pink, with a deeper pink throat. Pericarpel thin, slightly conical, up to 19mm long, 6mm diameter at the base, 8mm diameter at the top, dark green, with half-round, green, white-edged scales, the top scales 7mm wide.

Outer petals spatulate, 24mm long, 10mm wide at the widest part, cream-white to delicate pink, with a greenish middle stripe on the outside, darker at the base, only 2-3mm wide at the base. Inner petals spatulate, often with a small point or slightly fringed, 22mm long, 12mm wide at their widest part, quite narrow at the base, the same colour as the outer petals. Sometimes a ring of innermost petals appear, lanceolate, 14-16mm long, at the widest part 8-9mm wide.

Filaments long, thin, greenish-white to greenish-yellow, inclined to overhang the style, (covering the style). Anthers small, round, 1-1.2mm long, yellow. Pollen light yellow. Style 15-16mm long, c. 2mm thick, white at the base, becoming greenish towards the top. Stigma with 8 short branches. Nectar chamber slightly "V" shaped c. 4mm long, and as wide at the top, delicate pink, enclosed by the lowest stamens. Fruit cavity longish, c. 15mm long, 3-4mm wide, white walled.

Fig.93
G. amerhauseri, GC1002.01 in the Sierra Chica west of Ascochinga, Prov Córdoba, Argentina

Fruit longish-oval to spindle shape, 32-33mm long, 19mm thick, dark green, strongly bloomed, splitting lengthwise.

Seeds globular, 1.1-1.2mm diameter, with almost round, sometimes shiny black, slightly sunken, hilum-micropyle area. Testa ± arillate.

Distribution: In the north-west part of the Prov. Córdoba, in the higher regions of the mountains, mostly between 1000-1600m, seldom lower.

Type: H.Till 88-229 (cult H.Till) Argentina, prov Córdoba, Sierra Chica, between Ascochinga and La Cumbre, 1450-1600m (Holotype WU)

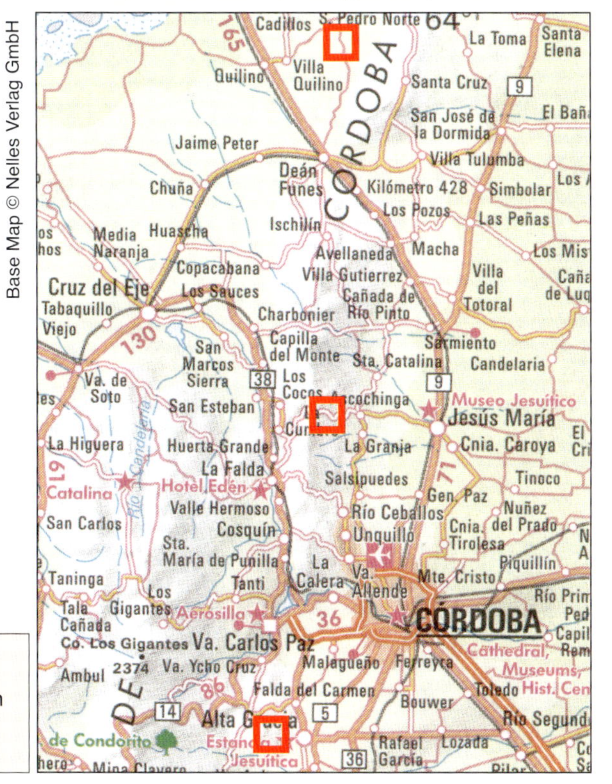

Map 13
Localities of *G. amerhauseri* in Prov. Córdoba, Argentina

Etymology

Named after the Austrian *Gymnocalycium* specialist Helmut Amerhauser, co-founder of the specialist group 'Arbeitsgruppe Gymnocalycium'

Synonym

G. amerhauseri subsp. *altagraciense* H. Till & Amerhauser in *Gymnocalycium* 20(3): p.735 (2007)

Commentary on *G. amerhauseri* subsp. *altagraciense*

This plant was found growing in long grass by the Austrian Franz Bozsing early in 1979 and given the provisional name *G. altagraciense* referring to the nearby town of Alta Gracia. It remained without a valid name until described as a subspecies of *G. amerhauseri* by Till and Amerhauser in 2007. There is some doubt about whether this taxon belongs with *G. amerhauseri*.

G. walteri (H. Till) in *Gymnocalycium* 16(4): pp.541-544 (2003)

First Description of *G. walteri* H.Till spec. nov.

Body semi globular, with 2-3 strong branched roots, which owing to the conditions and the stony ground, are partly vertical and partly sloping downwards. In the natural habitat 20-30mm diameter and only slightly raised above the ground. In cultivation the flowering size of plants is 35-50(-65)mm diameter and 20-25mm high, in habitat no offsets were observed, on the seedlings in cultivation occasionally, seldom strongly offsetting.

Crown slightly, to hardly sunken, and lightly armed, Epidermis matt, fine grained, dark green. Ribs (5-) mostly 7(-9), at the base 20-23mm broad, flat, near the crown 3-5mm high and wide, divided into mammi-form humps. Areoles large, round to oval-round, 4-4.5mm diameter, with white wool that is lost on the lower areoles. Spines mostly radials, 2-3 pairs pointing sideways of which the top pair are sloping towards the crown, variable in length, 6-8mm long and a shorter spine pointing straight down. All spines needle-like, straight to slightly curved, recurved, these however are not adpressed, elastic, yellowish to light brown, at the base sometimes darker.

Flowers coming from the young crown areoles, funnel shaped, 45-50mm long when closed, 40-45mm diameter at anthesis, white with a pale pink throat. Pericarp conical, c. 15mm long, 8-9mm

Fig.94 *G. amerhauseri* STO88-229, Sierra Chica, between Ascochinga and La Cumbre, Prov Córdoba, Argentina

Fig.95 *G. amerhauseri* STO88-229, Sierra Chica, between Ascochinga and La Cumbre, Prov Córdoba, Argentina

diameter at the top, dark green, with small, half round, dark green, white edged scales, these resolving into spatulate outer petals, these are white on the inside with a dark green middle stripe, the outside is dark green with a white edge, up to 25mm long and 4-5mm wide. Inner petals lanceolate, 19-21mm long, 3-3.5mm wide at the top, at the base narrower, 2mm wide, white, the base pale pink. One or two rows of stamens leaning onto the style, which is not free-standing, (with no space) in the perianth, filaments thin, 4-6mm long, the lowest pale pink at the base, the rest greenish-white, all with small 1mm diameter, round anthers and golden-yellow pollen. Style thin, greenish-white c. 18mm long, 21mm long with stigma, c. 1.3mm diameter, as a rule the anthers overhanging the stigma. Stigma light green with mostly 6 light claw-like lobes.

Fruit spindle-shaped, in cross section 30mm high, 15mm diameter, dark grass green, it becomes yellow-green before it splits (lengthwise?). Seeds almost globular, (1.3 x 1.3mm) mostly entirely covered in loose skin and with a cut off, oval, somewhat sunken hilum-micropyle-region, without a definite swollen edge.

After the reclassification of the genus *Gymnocalycium*, this new species belongs to Section A. *Gymnocalycium*, Series C. Quehliana. It stands near to *G. amerhauseri*.

Type: Argentina, Prov. Córdoba, near to San Pedro del Norte, 770m, 22 Oct. 1996, H.Till 96-1139 (Holotype: CORD)

Commentary on *G. walteri*

G. walteri, described by Till (2003b), is here regarded as a form of *G. kieslingii* growing near to San Pedro del Norte in the north east of Prov. Córdoba. This form is said to grow with *G. erinaceum* which supports the view that they are different species.

Distribution (Map 13 on page 59)

It was originally known only from a small area in the Sierra Chica, between La Cumbre and Ascochinga. Till (1994) also mentions populations in the north east side of the Sierra Grande and further north in the Sierra Ischilín but the latter may belong to *G. erinaceum*, a similar taxon.

At its type locality, *G. amerhauseri* grows in humus-rich soil with grass on south facing slopes, often associated with rocks. When out of flower, it can be difficult to find, which probably accounts for why it took so long to be discovered on this well-used road over the mountains.

The inclusion of the subsp. *altagraciense* extends the distribution further to the south, to the west of the town of Alta Gracia.

Conservation Status

Least Concern. Although quite limited in distribution, it grows in a place where there is little threat to its survival and it is probably more widespread than currently known.

History

This species was found by Helmut Amerhauser when travelling with Till in 1988. The formal description appeared in 1994 and it has generally been accepted as a good species.

Commentary

It is the most southerly member of a group of species including *G. baldianum* and *G. uebelmannianum*. In fact, *G. amerhauseri* is very much like a white-flowered form of *G. baldianum* and grows in a similar type of habitat. It is said to flower at the end of October in habitat, one of the earliest Gymnocalyciums. Some specialists consider it has a close relationship with *G. erinaceum*.

Cultivation

A good plant to cultivate, this species grows easily and flowers when young, looking much the same as it does in habitat. It is still quite rare in collections but deserves to be better known.

The subsp. *altagraciense* is usually seen in collections as *G. altagraciense* Plants originating from seed of *Piltz* 119 identified with this name are from much further north in Córdoba and probably do not belong here.

Illustrations

Schütz, in *Kaktusy* 17(3): p.59 (1981), a colour picture of subsp. *altagraciense*.

Gymnocalycium andreae (Bödeker) Backeberg

Fig.96 *G. andreae, Piltz* 213 from the Cerro los Gigantes, Prov. Córdoba, Argentina

Fig.97 The illustration with the first description of *Echinocactus andreae* from Monatsschrift der DKG (1930)

First Description (Bödeker)

Monatsschrift der DKG 2(10):p.210-212 (1930)

Echinocactus Andreae Böd., sp. n.

Body globular, according to the plants in front of me, mostly flattened, up to 4.5cm diameter, readily offsetting from the base, body colour dark blue-green to black-green, semi-matt, crown a little sunken with humps, nearly bald, a few small spines in the crown. Ribs 8, flat round, up to 1.5cm broad at the base, separated into humps by small cross-grooves. Areoles in the centre of the humps, up to 2cm diameter, with white wool, the wool thicker in the crown. Radial spines 7, with 3 laterals each side and 1 pointing downward, all thin, needle-like, somewhat curved, up to 8mm long, rough, matt white, somewhat brownish at the bottom, central spines 1-3, same as the radials, sometimes a little shorter, pointing straight out or upwards, also with brownish-black colouring, rough.

Flowers mostly in the crown, coming from younger areoles, often short funnel shaped, 3cm long and up to 4.5cm diameter. Ovary cylindrical, 6-12mm diameter. Scales 3 sided, 4mm wide, grey-green, white edged, set spirally on the tube, 8mm apart, lengthening gradually becoming the outer petals, 5-25mm long, to 6mm wide, long oblong with a very short rounded point, shiny, light greenish-yellow, then darker to a clear greenish tip, with an

Fig.98 The habitat of *G. andreae,* GC942.01, in the Sierra Grande near El Cóndor, Prov. Córdoba, Argentina

Fig.99 *G. andreae,* GC394.01, pink-flowered form from Los Gigantes. Prov. Cordoba, Argentina

olive to olive-green mid stripe on the back, inner petals of the same shape to a little wider with short sharp points, 5-25mm long, clear sulphur yellow. Stamens and style light yellow; the anthers and the 6 overhanging stigma lobes, whitish-yellow. Fruit and seeds are yet unknown to me.

Origin: Argentina, Sierra de Córdoba and in the Pampa de la Esquina, near the Cerro los Gigantes, 1500-2000 m, where *Gymnocalycium bruchii* Hoss. (= *E. lafaldense* Vaup?) grows. I first saw the plant in the collection of Herr F. Laflor, Duisburg 1927.

Lectotype: Illustration in Bödeker (1930) accompanying the above description and clearly referring to the original collection. Designated by Metzing et al. (1995) *Allionia* 37: p.192

Synonym

G. andreae subsp. *maznetteri* Rausch in *Gymnocalycium* 13(4): p.379 (2000)

Etymology

Named after Wilhelm Andreae (1895-1970), German cactus specialist and nurseryman.

Other Names

Echinocactus Andreae Böd. in *Monatsschrift der DKG* 2(10):p.210-212 (1930)

Distribution (Map 14)

This small plant grows in pockets of soil between rocks high in the Sierra Grande, Prov. Córdoba, Argentina. The environment looks like moorland with rocky outcrops.

Conservation Status

Least Concern. Although with a limited distribution in an area grazed by animals, it appears to be surviving without obvious damage to its environment.

Fig.100 *G. andreae* (doppianum), *Piltz* 378 from the Pampa de San Luis, Prov. Córdoba Argentina

History

First collected by Hosseus and sent to Europe around 1927. It was described by Bödeker as a new species of *Echinocactus* in 1930. Backeberg then made the combination in *Gymnocalycium* invalidly in *Neue Kakteen* (1931). The valid publication was eventually made by him in *Kaktus-ABC*: p.285 (1936).

A long spined form with larger flowers was sent by Hosseus in 1932 to the nurseryman Wilhelm Andreae in Bensheim and was distributed as var. *grandiflorum*, a name eventually published by Krainz & Andreae in *Die Kakteen* (1957).

A white flowered form was invalidly described in 1960 as var. *svecianum* by Pazout, named after the Prague cactus grower Frantisek Svec who propagated the plant from originals sent by Hosseus. The name was validated as a form when Till (1992) designated *Rausch* 108 as the type.

Map 14 Localities of *G. andreae* in Prov. Córdoba, Argentina

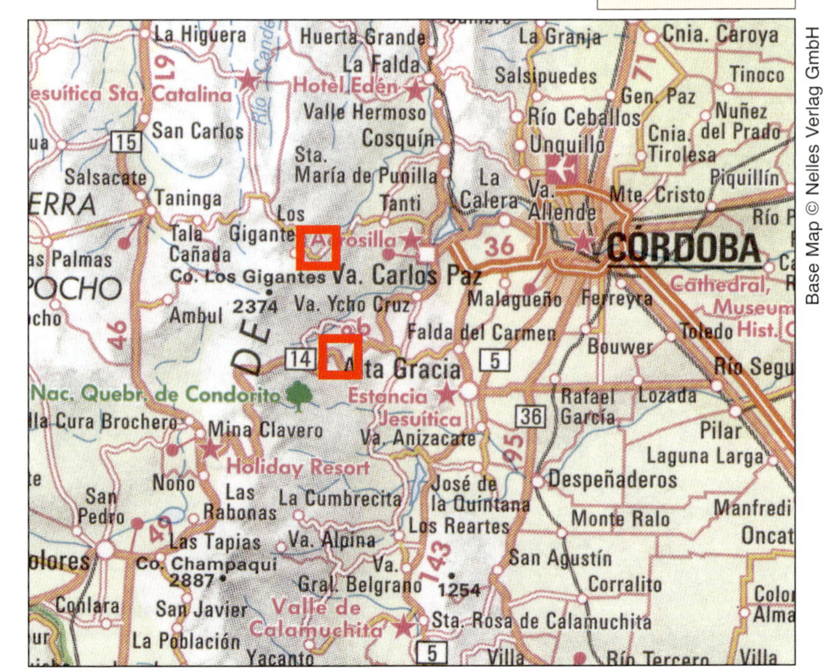

Fig.101 *G. andreae,* GC942.01, in habitat in the Sierra Grande near El Cóndor, Prov. Córdoba, Argentina

Commentary

This taxon is very variable in its habitat high in the Sierra Grande in Prov. Córdoba. The spines can be long or short and the flowers vary in colour from sulphur yellow to pale yellow and occasionally white. A form with more cylindrical bodies sometimes seen as *G. doppianum* was described as var. *fechseri* by Till in *Gymnocalycium* 13(4): p.378

Cultivation

Here is an easy to grow species with the best yellow flowers in the genus, much stronger coloured than the *G. hyptiacanthum* forms. It easily grows into a cluster of heads and the best clones have nice long spines and large bodies. Plants under the names var. *svecianum* and *subsp. maznetteri* usually have white flowers.

Being a high altitude plant, it will take cold in winter and prefers not to be exposed to strong sun and excessive heat. A soil with organic material and a plentiful supply of water in summer will produce a fine specimen.

Fig.102 *G. andreae, Piltz* 213 from Cerro Los Gigantes, Prov Córdoba, Argentina

Fig.103 *G. andreae,* GC942.01, in habitat in the Sierra Grande near El Cóndor, Prov. Córdoba, Argentina

Fig.104 *G. andreae, Kiesling* s.n. from between Mina Clavero and El Cóndor, Prov. Córdoba, Argentina

Gymnocalycium baldianum (Spegazzini) Spegazzini

First Description (Spegazzini)

Cactacearum Platensium Tentamen - Anales del Museo Nacional de Buenos Aires Serie 3, 11(4) pp.505-506 (1905)

Echinocactus baldianus Speg. (n.sp.)

Description: Hybocactus, small, flattened-globular, dark grey-greenish. Ribs 9-11, wide and blunt, almost lost in humps, which are formed by sharp ± deep furrows. Areoles small, with mostly 5 delicate spines, all radiating and pressed onto the body, a dirty pale grey. Flowers coming from the crown, erect, average size, outside dark blue-green, completely without hairs, loosely covered with scales, scales gradually resolving into intense purple coloured petals, with 6 whitish ochre-yellow stigma lobes.

Habitat: Rare in the mountains near Ancasti, Prov. Catamarca.

Observations: Body simple, 4-7cm diameter 2.5-4cm high, crown slightly sunken, without wool. Ribs straight, 5-8mm high, 10-15mm wide at the base. Tubercles 4-6, the topmost blending in to each other, lower down separated, uneven. Areoles sunken, somewhat elliptical, 3mm long, 1.5mm wide, 5-7mm apart. Spines 3-7, straight to recurved, 7-12mm long, clearly no centrals. Flowers solitary or 2-3 growing together from near the centre of the crown, 35-40mm long, free from hairs, dark green; petals a beautiful purple to pinkish-purple. Stamens, filaments and style pinkish-purple, anthers and fine stigma lobes, pale ochre.

Lectotype: Neuhuber & Till (1999) designated in *Gymnocalycium* 12(1): p.282 as the photograph from LPS by Spegazzini reproduced as Fig.178 in Britton & Rose *The Cactaceae* Vol.III page 165.

Etymology

Named after J. Baldi, a sponsor of Carlos Spegazzini.

Other Names

Echinocactus baldianus Speg. in *Cactacearum Platensium Tentamen - Anales del Museo Nacional de Buenos Aires* Serie 3 (1905)

Synonyms

G. raineri H. Till in *Gymnocalycium* 20(2): p.722 (2007)

First Description of *Gymnocalycium raineri*

Body semi globular with a sunken, weakly spined crown, 60-70mm diam., 25-30mm high, in age richly offsetting, young plants from 20-25mm diam. and already capable of nice flowering, with 6-8 ribs and 5-7 radial spines per areole. Root evenly branched, no taproots. Epidermis fine-granular, matt, in habitat bluish grey-green, in cultivation dark green. Ribs 9-11, straight, 13-15mm wide at the base, 6-7mm high, often angular, divided into small thick humps with blunt chins separated by horizontal cross grooves. Areoles on wild plants ± deep to sitting up, roundish to oval, c. 3-4mm diam. with a little short felt, on cultivated plants sitting up, oval, 3x4 mm diam. with greyish-white wool-felt. Spines mostly 7, radiating sideways and straight down, unequal in length 7-12mm, the lowest pair of spines are the longest, straight to slightly curved, occasionally twisted, grey, darker at the base, lying onto the body.

Fig.105 *G. baldianum,* GC35.02 in habitat near Singuil, Prov. Catamarca, Argentina

Fig.106 *G. baldianum,* GC35.02 in habitat near Singuil, Prov. Catamarca, Argentina

Flowers coming from the areoles in the crown, funnel shaped, when warmer, large wheel shaped, a delicate lilac-pink with a green throat, only single sex (masculine) examples observed. Ovary short funnel shaped 8-10mm high 6-8mm thick. Outer petals up to 30mm long, 7-8mm broad, a delicate lilac-pink with a darker middle stripe, towards the centre gradually becoming shorter, the innermost petals 17-18mm long and 4-5 mm wide a delicate lilac-pink.

Fruit dry when ripe, club shaped, 20mm long, 9-10mm thick, splitting lengthwise. Seeds large globular, 1.5mm diam., with a layer of fine grained black cuticle. Hilum-micropyle-region somewhat sunken, having the form of a keyhole, with rounded bulging edge, without elaiosom.

Habitat: Argentina, Prov. Catamarca, on the west side of the Sierra de Graciana.

Fig.107 *Echinocactus sanguiniflorus,* Tafel 33 from 'Blühende Kakteen' by Werdermann (1932)

Type: Argentina, Prov. Catamarca, on the west side of the Sierra de Graciana, 20.12.1993, leg H. Amerhauser HA 93-780 = coll. H.Till 3548, Holotype: BA. Isotype: WU.

Commentary on *G. raineri*

This is a form of *G. baldianum* from the west side of the Sierra de Graciana. It is said by the author to differ by its numerous offsets and larger, pink flowers.

G. sanguiniflorum (Werdermann) Werdermann in *Kakteenkunde* 4(10): p.183 (1936). Basionym: *Echinocactus sanguiniflorus* Werdermann (1932)

First description of *Echinocactus sanguiniflorus* Werd. in *Feddes Repert*. 30: pp.56-57 (1932)

Body simple or offsetting from the base, rounded or ± upside down egg shape, (according to the example at hand), 7-8cm high and up to 8cm diameter, body colour matt dark green, later a little lighter, becoming blue-green. Crown a little sunken, with some wool in the young areoles, no spines coming from the younger areoles in the centre of the crown. Ribs 10-12, in the crown rather high and narrow, later spreading, broad to almost flat, ribs already showing humps in the crown, later the ribs are lost in humps but are still recognisable as ribs. Humps ± 4-6 sided with forward protruding chins below the areoles. Areoles sunken, longish, 3.5-5cm long, 2.5cm wide, with short white wool, primarily never entirely bald.

Spines 7-9, all radials, radiating, straight or a slightly curved, strong needle-like to thin awl-shape, sharp, up to 1.5cm long, in pairs with the odd one pointing down, whitish to pale reddish-brown, rough, scaly, mostly with a darker point, brown and a little thicker at the base.

Flowers from near the crown, c. 4.5cm long, ovary moderately thin, c. 1.5cm long, dark olive green, with relatively thick, upright, wide heart shaped scales, from c. 3.5cm wide; scales with hardly noticeable dark tips and white edges. tube short, wide bell-shaped. Outer petals wide spatulate to ± wide rhomboid shape, the tips being obtuse triangular, somewhat pointed, dark olive green with pink edges, inner petals oblong to spatulate, c. 2cm long, 0.8cm at the widest part, the top running into a triangular point, deep blood red, shiny, paler at the tip. Stamens light yellow; style red, stigma lobes whitish, c. 11, 2cm long, spreading, shorter than the tallest stamens.

(Later in his Blühende Kakteen) *E. sanguiniflorus* Werd. belongs to the subgenus *Gymnocalycium* and it apparently stands next to *E. gibbosus* (Haw.) D.C. It differs from all other known species of this subgenus by way of the deep blood red metallic shiny petals.

Habitat: Argentina. Location unknown.

Commentary on *G. sanguiniflorum*

Werdermann (1932) tells us that in 1926 Prof. Hosseus of Córdoba, Argentina, sent the original plant, number 179, to the Botanical Garden Berlin Dahlem, where it flowered for the first time in June 1930.

Till (1972) gives us more information about the plant sent by Hosseus. He says it was a two-headed specimen found in Córdoba [?]. Frau Muhr sent similar plants collected by Genser to Herr. Schatzl at Linz Botanical Garden in 1969 with the field number B80.

G. Venturianum Fric ex Backeberg in *Kaktusar* 5: pp.61-63 (1934)

First description from *Kaktusar* 5: pp.61-63 (1934)

***Gymnocalycium Venturianum* (Fric) Bckbg**

Echinocactus sanguiniflorus Werd. DKG. 1932. without habitat.

Flat-globular, simple, colour fresh blue-green, somewhat sunken in the crown. Ribs 9 with shin-like humps below the areoles. Radial spines 5, with two pointing to each side and one pointing downwards, 6-7mm long, all yellowish, brown at the foot; central spines missing. Flowers with a thin tube, up to 4cm long,

Fig.108 *G. baldianum* (sanguiniflorum) *Rausch* 765 from Alumbrera, Prov. Catamarca, Argentina

Fig.109 *G. baldianum* ,GC35.02 from near Singuil, Prov. Catamarca, Argentina

Fig.110 *G. baldianum, Ferguson* 406 from east of Andalgalá, Prov. Catamarca, Argentina

sepals greenish-red, petals blood red to carmine-red, scales pink. Fruit relatively small, longish. Style and stamens yellow.

Habitat: Uruguay, to the extreme north of Montevideo.

Said to have been collected by Müller-Melchers in Uruguay but this habitat was shown to be wrong by Pazout (1943).

The name '*G. Venturii*, Fric' first appeared in the Fric catalogue of 1929, named for Don Santiago Venturi from Tucumán. Fric found it on his last trip to South America but, although he did not give a locality, we can deduce that it was probably from the border region of Catamarca and Tucumán.

Backeberg (1959) himself agrees with Dölz (1938) that this is a synonym of *Echinocactus sanguiniflorus*, which was described by Werdermann two years before Backeberg validated Fric's description of *G. venturianum*.

Distribution (Map 15)

This high altitude plant is often found among grass in the Province of Catamarca. It is known from the Sierra Ancasti, Sierra de Graciana, Sierra

Fig.111 *G. baldianum*, GC29.07 from El Portezuela, Prov. Catamarca, Argentina

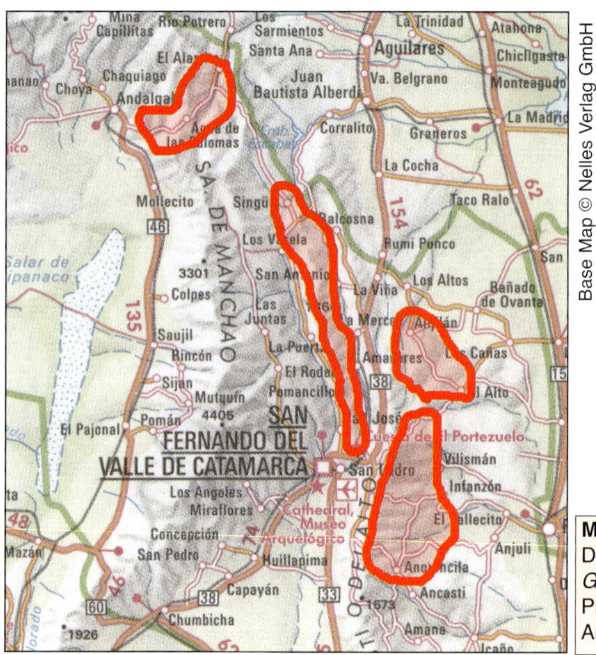

Map 15 Distribution of *G. baldianum* in Prov. Catamarca, Argentina

de Manchao and on the mountains east of Andalgalá.

Bercht (pers. com.) has observed that the plants below about 800m in the south of the Sierra Ancasti have pinkish flowers, whilst those from the north at higher altitudes have red flowers with a shorter tube.

Conservation Status

Least Concern. The taxon is widespread and not subject to any immediate threats.

History

First described as an *Echinocactus* by Spegazzini in 1905, it was placed by him in *Gymnocalycium* twenty years later when he commented that Britton and Rose were wrong in making it a synonym of his *G. platense*.

In 1962, Fechser sent a consignment of plants to Frank that agreed with Spegazzini's description but without a locality. It included white-flowered examples.

Commentary

Although it has been redescribed twice since its first publication, it is now accepted that these names only represent minor forms of this variable plant. It is related to *G. amerhauseri* and *G. uebelmannianum* which grow in similar environments.

During his trip in 1992, Bercht found a white flowered plant north of El Alto, Catamarca, which he named *G. baldianum* var. *albiflorum* Bercht (1994). Till & Neuhuber (1998) considered that this taxon does not belong to *G. baldianum*. Later, Neuhuber & W. Till (1999) formally made it a synonym of *G. rosae* which is here considered a form of *G. kieslingii*.

Neuhuber & W. Till (1999) created the name *Gymnocalycium* x *hiediae* for a hybrid between *G.*

Fig.112 *G. baldianum, Neuhuber* 740-2154 from the Sierra Ancasti, Prov. Catamarca, Argentina

Fig.113 *G. baldianum* STO135/2 from Agua de Paloma, Andalgala, Prov. Catamarca, Argentina

baldianum and a white-flowered plant of the same seed group. They maintain that many plants previously considered to be *G. baldianum* that have larger bell-shaped flowers, a longer pericarp and a larger, fleshy body are in fact this hybrid.

Cultivation

This is one of the most popular *Gymnocalycium* species because of its lovely red flowers. The colour can be a deep red or sometimes more orange and it changes as the flower ages. In habitat it often grows amongst grass and in organic-rich soil, much like its southern relative *G. amerhauseri*. This gives a clue to its preferences in cultivation where it enjoys a rich soil, protection from strong sun and steady supply of water in the growing season. Without regular watering it will go into a shrunken state from which it can take some time to recover. It is tolerant of cold conditions in winter.

Illustrations

Britton & Rose (1922) *The Cactaceae* Vol. III: Fig.178 on page 165 contributed by Spegazzini and mis-captioned as *G. platense*.

Werdermann (1932): Tafel 33 *Blühende Kakteen* (as *E. sanguiniflorus*)

Pazout, F. (1943) *Kakteenkunde* 11(2): p.40 (as *E. sanguiniflorus*)

Till, H. (1972) KuaS 23(9): pp.238-240 pictures illustrating various forms.

Fig.114 *G. baldianum,* GC35.02 from near Singuil, Prov. Catamarca, Argentina

Fig.115 *G. baldianum,* GC35.02 from near Singuil, Prov. Catamarca, Argentina

Gymnocalycium berchtii Neuhuber

First Description (Neuhuber)

Gymnocalycium 10(3): pp.217-220 (1997)

Body simple, flat to discoid, up to 20mm high, 40-60mm diameter, stake-like root, 30-50mm long, sometimes branched, small roots mostly in the lower parts. Crown slightly sunken, epidermis dull blackish-grey or blackish-brown, with a bloom, under strong UV. light, grey-violet. Ribs flat, (7-)9, almost straight, quickly widening towards the base, there 17-20mm wide. Humps hardly discernible, mostly only in the upper part of the body, then round and separated by slight cross-cuts. Areoles oval, up to 3mm long, 2mm wide, 9(-12)mm apart, with white to yellowish wool, soon becoming bald. Spines straight, seldom slightly curved 3(-5), one pair pointing sideways and distinctly slightly downwards, a further spine pointing straight down, when 5 spines the top pair pointing directly sideways, thin, 7-10mm long, mostly lying onto the body, dark brown to black, towards the tip red-brown or lighter, soon becoming grey, no central spine.

Flowers coming from the area of the crown, funnel shaped, 53-79mm long, 43-60mm diameter, pale pink to dirty white with a pinkish-brown throat, outer petals lanceolate, white with a delicate pink mid stripe, the outside with a pale green to pale pink centre and occasionally with a grey-green tip, pale pink at the base, up to 38mm long, 9mm wide at the widest part, 4mm wide at the base, inner petals lanceolate to trapeziform, sometimes slightly fringed, mother of pearl colour, the lower parts pale green, with a fine pink middle stripe, base pink, up to 25mm long, 7.5mm at the widest part, 2mm wide at the base. Receptacle pale pink, darker at the bottom, up to 11.9mm high, 11.7mm diameter.

Fig.116 *G. berchtii,* GC939.01 near to Los Chañares, Prov. San Luis, Argentina

Nectar chamber orange-pink, 1.8-2.4mm high, 3.5-3.8mm diameter. Filaments pale yellow, 8-11mm long, 1-2 rows at the base, short, leaning onto the style, a further 3-4 series inserted on the receptacle wall, at first the topmost curved towards the centre. Anthers yellow, 1.5mm long, 0.6mm wide, leaning inwards at c. 90°, for the first few days covering the stigma then opening out to form a ring. Style pale green at the base, becoming white towards the stigma, 12.9mm long, 2mm diameter. Stigma yellow, with 7-9 lobes, 5mm long, "middle standing". Ovary white to pale pink, 16-19mm high, 5mm diameter, with a suggestion of an undeveloped ovary extension, 3-8mm long. Pericarpel dark green, with a green to blue bloom, 18-29mm long, 9mm diameter, 4-5mm diameter at the base. Scales half round, light brown to pale pink, with a grey-green to grey-blue centre, some with a pink point, 2.5-3.5mm high, 4-6mm wide, which resolve into the outer

Fig.117 The habitat of *G. berchtii,* GC939.01 near to Los Chañares, Prov. San Luis, Argentina.
The rock strata are near the surface and almost vertical, providing an ideal habitat where little other vegetation can grow

Fig.118 *G. berchtii,* GC939.01 near to Los Chañares, Prov. San Luis, Argentina

petals. Elongated scales are green to blue-green, with a pale pink or light brown edge, stiff or sometimes curling outwards. Fruit short club-shaped, seldom grey-green, mainly with a bluish bloom, 21(-42)mm high, 9.5(-20)mm diameter, splitting vertically, the size of the fruit and the stigma is determined by the gender of the flower. The flower remains on the fruit 20-24mm long.

Seeds 1.1-1.3mm diameter, with a brownish aril, black underneath. Hilum micropyle region mostly wide, tear-drop shaped, slightly sunken, the centre raised, black.

Typus: Argentina, Provincia San Luis, Los Chanares, 690m, 16 December 1989 leg. G. Neuhuber 89-158 (Holotype: WU).

Synonym

G. nataliae Neuhuber in *Gymnocalycium* 18(3): pp.635-638 (2005)

First Description of *Gymnocalycium nataliae* Neuhuber spec. nov.

Body solitary, not offsetting, semi-globular, 25mm high, 35mm diameter, tap-root up to 60mm long, not ramified. Crown hardly sunken, armed. Epidermis dark green to blackish-brown, often with a strong grey bloom. Ribs flat, 9-12, straight. Humps round flat, separated by shallow horizontal grooves. Areoles round, 2mm diam., c. 6.5mm apart, mostly sitting on or slightly sunken, with white wool, seldom becoming bald. Spines mostly 7(-9) , straight, needle-like, thin, lying on the body, light brown to red-brown, becoming very dark towards the base, the topmost pair horizontal or slightly pointing upwards, the second pair are the longest spines (4.5-8.5mm), pointing sideways to downwards, of the three lowest spines the middle one 2.5-5.5mm long the shortest, all three are set very close and are pointing to the base, never becoming grey, no central spine.

Flowers coming from the region of the crown, 58-75mm long, 44-54mm diameter, white to pink, mostly bell-shaped, little but unpleasant scent. Outer petals white with a narrow, light pink middle stripe and a greenish tip on the outside of the petal, lanceolate, 30-35mm long, 9mm at the widest part, inner petals white with a light pink middle stripe or a light pink tip, lanceolate to narrow lanceolate, 20-24mm long and up to 8mm wide. Receptacle wall thick fleshy, pink, light pink at the top, up to 13mm high, 9-10mm diameter, Nectar chamber pink, base white, 2.5-3mm high, 3-4mm diameter. Filaments, white to yellow, the primary rows that close the nectar chamber 4mm long, pink at the base, most of the 5 secondary rows 5-8mm long lying on the flower wall. Anthers white to pale yellow, 1.1mm long, 0.7mm wide. Pollen yellow. Style white

Photo: Massimo Meregalli

Fig.119 *G.. nataliae, Meregalli* 611 from north of Membrillos, Prov. San Luis, Argentina

to light green, 12-15mm long 1.5-2mm diameter, occasionally channelled, never free standing, style base always straight, white and thick walled. stigma yellow 2.5-3mm high with 9-12 stigma lobes, stigma base mostly level with the 3 secondary rows of anthers. Ovary white, elongated funnel shaped or narrow long-oval, 20-25mm high, 3.5-5.5mm diameter. Pericarp wedge-shaped, mostly light green with a silky sheen, occasionally with a light grey bloom, 20-35mm high, 8-10mm diam. Scales light pink, rather numerous, half round, somewhat upstanding and hardly recurved, occasionally with a hint of a small point.

Fruit light green or olive, club-shaped, 25-38mm high, 12-14mm diameter, splitting vertically. Seeds 1.2-1.3mm long, 1.2mm diameter, narrower towards the hilum-micropyle region (HMR), testa black with well rounded structures, covered with a dried white skin, HMR sunken, small, round to wide tear-drop shape., (Subgenus *Gymnocalycium*).

Habitat: Argentina, Province San Luis, north of Villa de Praga, 890m ASL.

Type: Argentina. Province San Luis, north of Villa de Praga, 890m ASL., leg. G. Neuhuber GN88-94, 18th December 1988. Holotype: BA ex WU 1894 = GN88-94/235.

This new species is named after Natalia Schelkunova, Moscow Russia. Secretary of the publication Kaktus-Klub.

Photo: Massimo Meregalli

Fig.120 *G.. nataliae, Meregalli* 606 from north of Cerros del Rosario, Prov. San Luis, Argentina

Commentary on *G. nataliae*

Neuhuber (2005b) compares this plant with *G. berchtii* which was found not far away to the north east. The differences are slight and so this name is here treated as a local form of *G. berchtii*.

Etymology

Named after Dr. Ludwig Bercht, Dutch *Gymnocalycium* specialist and seed merchant.

Distribution (Map 16)

Originally described only from the type locality but there are few roads in that part of Prov. San Luis so it may grow elsewhere. The inclusion of *G. nataliae* extends the distribution some way to the south-west. Bercht (pers. com.) considers that a plant from Sierra del Morro could also be this species which would extend the distribution even further south. Meregalli has found this plant in a number of places which begin to fill the gap.

The type grows among or near rocks, particularly in bedrock with vertical strata and where the soil is shallow. The distribution area is covered with dense shrubs and small trees, the plants growing both under these and occasionally out in the open.

Conservation Status

Least Concern. At its type locality, the fact that it grows in shallow soil near rocks on grazing land makes this taxon vulnerable to further degradation of its habitat. However, reports of a much greater distribution significantly reduces the risk to its future.

History

It was found by Bercht when travelling with Neuhuber on 16th December 1989 and eventually described as a new species in 1997.

Commentary

The type locality of this species is near a ford on the dirt road from Santa Rosa del Conclara over the

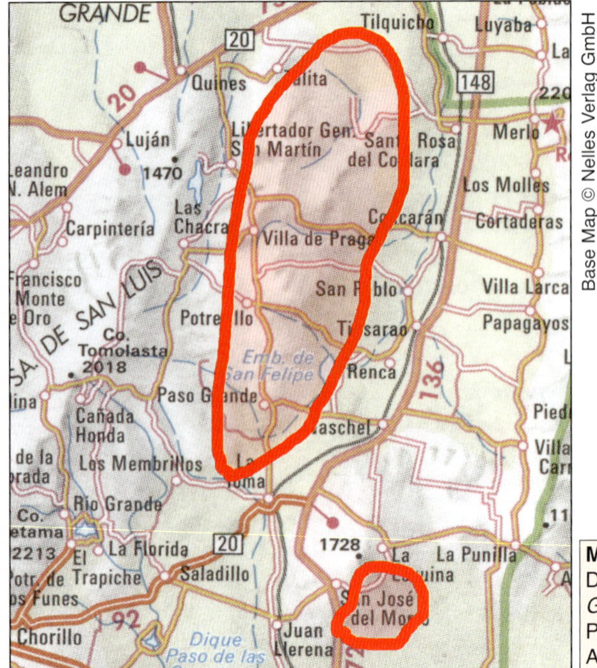

Map 16 Distribution of *G. berchtii* in Prov. San Luis, Argentina

Boca de la Quebrada to Quines, near Los Chañares, a tiny settlement. It is a small plant with a grey body and pretty pale pink flowers.

Neuhuber (1997) says that *G. gibbosum borthii* also grows in the area and I found that subspecies a few kilometres to the west. *G. berchtii* can be distinguished by its small grey-black bodies, T-arranged black spines and the disposition of the stamens.

Cultivation

This is a very good miniature species to grow. It has a grey body which in culture develops a waxy coating in good light. It is still quite uncommon in collections but it is easy to grow and flowers freely when small.

Fig.121-123 *G. berchtii, Neuhuber* 89 158-430 from Los Chañares, the type locality on Route 5, Prov. San Luis, Argentina

Gymnocalycium bruchii (Spegazzini) Hosseus

First Description (Spegazzini)

Breves Notas Cactológicas - Anales de la Sociedad Cientifica de Argentina 96: pp.73-75 (1923)

Frailea bruchii Speg. (n. sp.)

Body 10-20mm diameter and tall, more or less sunken in the crown, strongly offsetting, clumping or forming a cushion, up to 10-30 heads, almost globular, up to 100mm high, 150mm diameter, each with 8-12 ribs or more, running longitudinally, separated into tubercles, colour dull green.

Spines small, ashen, interwoven with the spines from neighbouring areoles. Areoles narrow elliptical, armed with 13-17 spines, all radiating, no centrals, 6-8 each side, 1-3 pointing down, all spines are white, thin, 2-5mm long, 0.15-0.30mm thick, lying flat or curved towards the body, relatively stiff, sharp, white; under the microscope the spines are covered with numerous papilla.

Flowers are odourless, coming from the edge of the central depression, numbering 1-3, bell shaped, 15-20mm long and diameter when open, tube very short, covered with green scales, which have a reddish mid-stripe, tufts of hair and a few bristles coming from the axils; petals number more or less 25, oblanceolate, 15-18mm long, 2-4mm wide, generally with a small soft point, pink with a deep violet central stripe, entirely without hairs; stamens inserted in the perianth tube, filaments nearly white, anthers yellow, style straight, white, longer than the stamens, carrying a spray of 5-8 stigma lobes, which are 2.5-3mm long, 0.5mm diameter, round, yellow. In spite of the many flowers in La Plata, one is unable to obtain fruits.

Fig.124 *G. bruchii*, GC390.01, Sierra. Chica, west of Alta Gracia, Prov. Córdoba, Argentina, near the type locality

Habitat: In the mountains of Sierra de Córdoba, near to Alta Gracia, aest. 1918 (lgt. et comm. Dr. C. Bruch).

Lectotype: Illustration in Spegazzini - *Anales de la Sociedad Cientifica de Argentina* 96: p.74 (1923). Designated by Metzing & al. in Allionia 33, pp.194-195 (1985)

Fig.125 The habitat of *G. bruchii*, GC1001.01 in the Sierra Chica, east of La Cumbre, Prov. Córdoba, Argentina

Fig.126 *G. bruchii,* GC1001.01 in the Sierra Chica, east of La Cumbre, Prov. Córdoba, Argentina

Etymology

Named after Dr Carlos Bruch, cactus collector in Córdoba, Argentina at the time of Carlos Spegazzini.

Other Names

Frailea bruchii Speg. in *Breves Notas Cactológicas - Anales de la Sociedad Cientifica de Argentina* (1923)

Synonyms

G. albispinum Backeberg in Backeberg and Knuth: *Kaktus-ABC*: p.285, 416 (1936)

First Description of *Gymnocalycium albispinum* Bckbg.

Body small globular, often offsetting, dull green. Ribs c. 14, c. 3mm high and 4mm wide, with transverse grooves. Areoles nearly submerged in the grooves, round, often only 1mm apart, at first with dense white hair. Spines c. 25, bristly, white, 10mm long, sideways interwoven, centrals are brown, spreading. Flower lilac-pink, wheel shaped, c. 3cm diameter. Fruit elongated-globular. – Argentina: Córdoba.

Closely related to *G. bruchii*. A larger body and longer spines, the flowers are similar.

G. lafaldense Vaupel in *Zeitschrift für Sukkulentenkunde* 1(14): p.192 (1924)

***Gymnocalycium lafaldense* Vpl. Spec. Nov.**

The habitat is the mountain La Falda in the Sierra de Córdoba in Argentina at a height of 1500-2500m. They grow in grass; this detail is the reason that, although this region has been well searched, until now, one hardly expected to find a new species.

Body small, only a few cms. high and wide, flattened on top, dark green, freely offsetting, crown sunken, full of flowers. Ribs (on my plant) 12, relatively high, humps formed by fairly deep cross cuts. Humps under the areoles, not chin-like, areoles close together, elliptical, 2-3mm long, with short white woolly felt, later losing some felt. Spines bristle-like thin, white with a brown base; radial spines c. 12-14 curved back to the body, up to 6mm long, coming from the sides and the bottom of the areoles; central spines 1, upright, more brown, frequently missing.

Flowers coming from near the centre of the crown, 3cm long, 3.5cm wide, opening for 3 days, closing at night. Ovary short, 5mm long, with 4 wide scales, 1 at the bottom, 3 on the top half. Ovules in thick bunches, attached singularly. Petals, many, inner petals pointed, 3-4mm wide, a delicate violet-pink, with a darker central stripe, outer petals blunt, greenish-brown, only lighter at the edge. Many stamens, much shorter than the petals, the longest at the bottom of the flower tube, not in 2 groups, filaments white, anthers pale yellow; style with 8 stigma lobes, all pale green, the highest stamens not overhanging the stigma lobes. Fruit not yet seen.

This described plant is not grafted. It originates from a high mountain habitat.

Lectotype: Vaupel, *Zeitschrift für Sukkulentenkunde* I(14): p.192 (1924) coloured illustration in Vaupel (l.c.). Type locality:

Fig.127 *G. bruchii* (albispinum), UCBG 68-514, said to be a propagation from an original plant from Backeberg.

Argentina, Córdoba, Sierra de Córdoba, La Falda. Fig.1. Designated by Metzing & al. in Allionia 33, p.207 (1985)

G. bruchii subsp. *lafaldense* (Vaupel) Neuhuber in *Gymnocalycium* 16(2): p.502 (2003)

Commentary on *G. lafaldense*

This description was made by Vaupel from plants sent to Schwebs of Dresden in 1923, probably in ignorance of Spegazzini's publication of *Frailea bruchii* a year earlier. Osten (1926) says that he found the plant in 1923 and that the habitat is a low hill behind the hotel in La Falda at about 1000m.

G. lafaldense was made a subspecies of *G. bruchii* by Neuhuber in 2003 although most authors had considered it to be just a synonym of that species. It is found on the northern Sierra Chica and Neuhuber (2003) tells us that, in contrast to the type, its small offsets are morphologically different from the mature stems.

G. bruchii subsp. *pawlovskyi* Neuhuber in *Gymnocalycium* 16(2): p.503 (2003)

In 1993 Neuhuber and Amerhauser found this population of plants with pectinate spines north of Dean Funes growing in meadow with sharp granite rocks between La Esperanza and El Camarón. The

Fig.128 The lectotype illustration of *G. lafaldense* Vaupel from *Zeitschrift für Sukkulentenkunde* (1924)

Fig.129 *G. bruchii, Piltz* 173 from Copina - El Condor, Prov. Córdoba, Argentina

plants differ from those from further south by not forming large clusters of stems, instead they become more elongated. The flowers are often dioecious.

G. bruchii subsp. *susannae* Neuhuber in *Gymnocalycium* 16(2): p.509 (2003)

This plant comes from near Copina just to the east of the Sierra Grande. It is said to differ from the nearby type by its larger body, pale epidermis, sunken crown and smaller flowers with a different structure.

Distribution (Map 17)

G. bruchii is found throughout the Sierra Chica in Prov. Córdoba. The variety *brigittae* was described from the western side of the Sierra Grande, presumably from a disjunct habitat location. The species grows in shallow soil associated with rock outcrops amongst grass.

Conservation Status

Least Concern. This species is widespread and not subject to any obvious threat.

History

First described by Spegazzini as a *Frailea* in 1923, it was transferred to *Gymnocalycium* by Hosseus in 1926. The plants had been found by Dr.

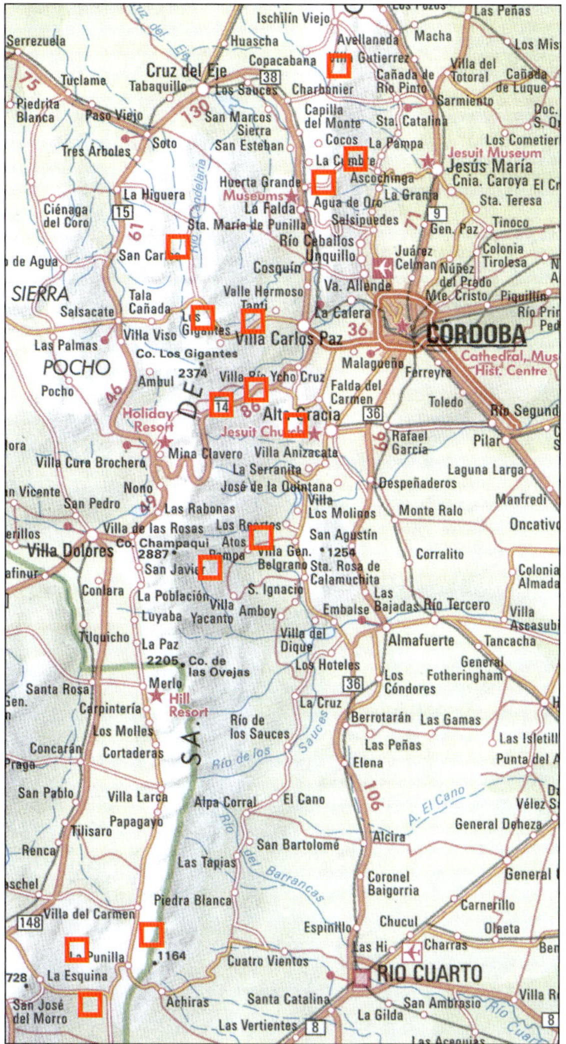

Map 17 Localities of *G. bruchii*, in Prov. Córdoba and Prov San Luis, Argentina

Karl Bruch in the autumn of 1918 above 1600m near Alta Gracia, Prov. Córdoba, Argentina.

The original description of the flower tube referred to "tufts of hair and a few bristles coming

Fig.130 *G. bruchii*, GC1001.01 in the Sierra Chica, east of La Cumbre, Prov. Córdoba, Argentina

Fig.131 *G. bruchii* (brigittae), *Piltz* 214 from Taninga, Prov. Córdoba, Argentina

Fig.132 *G. bruchii, Neuhuber* 91 321-1092 from the Observatory, Prov. Córdoba, Argentina

from the axils". This created a question as to whether it was really a *Gymnocalycium* but Neuhuber (2003) explains that this phenomenon can occur in *Gymnocalycium* when the weather encourages a flower to develop from an areole that was in the process of producing an offset.

Commentary

This small species is very variable with respect to body size, spine colour and flower colour. Its variability has resulted in many infraspecific names for the various populations, none of which are accepted here.

Piltz (1987) described the variety *brigittae* for a form with pink flowers from the western slopes of the Sierra Grande. It has subsequently become common in cultivation. Near to Ongamira, one can find a northern form of *G. bruchii* which Rausch called var. *niveum* although plants with white spines are also found elsewhere.

Fig.134 *G. bruchii, Piltz* 200 from east of Copina, Prov. Córdoba, Argentina

Cultivation

G. bruchii is an easily grown plant that flowers freely and is one of the first species of the genus to bloom in spring. The flower colour can be pink, white or even pale yellow, often with a central darker stripe. Since it freely forms clumps of stems, it is easy to propagate by removing the offsets that often already have their own roots. In this way clones can be passed around to ensure that those with desirable characteristics, or of habitat origin, are maintained in cultivation. White-spined forms of *G. bruchii* are particularly attractive and these are reported from various localities. Rausch (1989) described his *Rausch* 727 as var. *niveum*.

Illustrations

Osten, C. in *Zeitschrift für Sukkulentenkunde* II(9): p.146 (1925-26). B&W photograph and colour plate under the name *Gymnocalycium lafaldense*.

Fig.133 *G. bruchii, Neuhuber* 90 232-693 from Las Truchas, Prov. Córdoba, Argentina

Fig.135 *G. bruchii* flowering in early spring in the author's collection in England

Gymnocalycium calochlorum (Bödeker) Ito

Fig.136 The original illustration of *E. calochlorus* from *Monatsschrift der DKG* (1932)

First Description (Bödeker)

Monatsschrift der DKG 4(11): p.260-262 (1932)

Echinocactus calochlorus Böd. sp. n.

Among the imports from Herr Stümer to Herr Hahn during the summer of 1930 I found among the examples at hand, a not too large, pretty, shiny light green species that reminded me of one of the different forms of *Ects. denudatus* v. *paraguayense* Hort. As Herr Stümer mainly collected in N.W. Argentina, one comes to the conclusion that here we have a denudatum form, as they usually appear in S. Brazil, Uruguay and Paraguay. A short time ago I received a flower from Herr W. Andreae - Bensheim, from Herr Hahn I received a fruit, the seed was entirely different from that of G. denudatus L. & O. and varieties. Therefore in view of the light green colour of the plant I give the following: sp. n. with the above name.

Body somewhat flattened-globular, sometimes offsetting from above the ground, seldom from the base. From the examples at hand, up to 4cm high, 6cm diameter, body colour a beautiful light leaf green, later a little darker, crown flat, slightly sunken, ± in young plants, no spines in the crown, Ribs 11 separated by sharp vertical grooves, humps separated by sharp cross-cuts, at the base of the plant 1.5cm wide, at the top of the plant rounded, strongly chin shaped. Areoles c. 1.5mm with short white wool sitting on the top part of the hump, soon becoming bald. Spines up to 9, only radial, sideways and downwards pointing, to 9mm long, thin, rough, grey-white, the youngest (when damp) slightly pink in colour, all slightly curved, often unevenly so, lying close to the body.

Flowers few, near the crown, flowers 5-6cm long when open, ± the same length. Ovary, with reference to the tube, 3cm long, 1cm wide at the top, 0.5cm wide at the bottom, shining leaf green, covered with 4mm wide, white coloured scales, rounded with a point, set in a spiral on the tube, 1cm apart, becoming 1-2cm long, 7mm wide, oblong. Pink coloured with grey-green tips, sharp edged petals;

Fig.137 A large clump of *G.calochlorum* GC943.02 in fruit near Ojo de Agua, Prov. Cordoba, Argentina

Fig.138 *G. calochlorum* GC993.01 in habitat south of Ischilín, Prov. Córdoba, Argentina

Fig.139 *G. calochlorum, Lambert* 109 from San Sebastian, Prov. Córdoba, Argentina

inner petals long lanceolate, 7mm wide, 3-4cm long, the central ones a little shorter, ± slightly pointed, sharp edged, pale pink, with a darker central stripe running into a grey-pink point, throat carmine-pink. Style short and thick, pink coloured with c. 10 short yellow radiating lobes.

The fruit at hand is flattened, more egg-shaped, 6-12mm. Seed c. 1mm diameter, round or hat shaped, black, semi matt, with a long mouth-like white lipped hilum.

Habitat is unknown to me, surely it is from N.W. Argentina.

According to Dr. Rose belonging to *Gymnocalycium*. We place this plant according to Schumann, in the sub-genus *Hybocactus* K. Schumann, directly following *Ects. denudatus* Lk. et Otto as it has more ribs and spines and the flowers are pale pink among other things.

Etymology

From the Greek 'kalos' meaning beautiful and 'chloros' meaning green.

Other Names

Echinocactus calochlorus Böd. in *Monatsschrift der DKG* 4(11): pp.260-262 (1932)

Synonyms

G. proliferum (Backeberg) Backeberg in Backeberg and Knuth: *Kaktus-ABC*: pp.294-295 (1936). Basionym: *Echinocactus prolifer* Backeberg (1932)

Original description of *Echinocactus prolifer* Bckbg. n. sp. in *Der Kakteen-Freund* 1(12): pp.132-133 (1932)

Plants small, 2-6cm diameter, fairly flat, dark green, offsetting from the areoles and roots, soon forming a large clump. Ribs, up to 11, wide and low, separated into humps having chin-like protrusions below the areoles, which are separated from each other by strong cross-cuts. Areoles up to 15mm apart, elongated, at first covered with strong yellowish felt. Radial spines mostly 9, four pairs spreading sideways, pressed onto the body, 7-10mm long, one spine pointing straight down, c. 6mm long, central spines always missing, dirty grey in colour.

Flowers large, up to 5.5cm long, bud and tube bluish-green porcelain texture. Scales wide and short, white edged. Outer petals c. 25mm long, light brownish-white with an olive-green central stripe; inner petals light brownish-white, all are red at the base, slightly rounded at the top. Style and stamens yellowish-white. Prov. Córdoba, Argentina.

Commentary on *G. proliferum*

Backeberg got his plants from Stümer in 1932 and eventually accepted his species as just a variety of *G. calochlorum*.

G. amoenum (H. Till) Lambert in *Succulenta* 81(5): p.230 (2002). Basionym: *G. parvulum* var. *amoenum* H. Till (1994)

Original description of *Gymnocalycium parvulum* (Speg.) Speg. var. *amoenum* Till var. nov. in *Gymnocalycium* 7(2): p.125 (1994)

Body globular to flattened-globular, with strong occasionally ramified turnip-like or stake-like roots. Crown filled with young spines and areole wool. Flowering size plants, at first mostly single, offsetting later, c. 30mm diam., 28mm high. Older plants (35 years), up to 50mm diam., 35mm high, profusely offsetting, occasionally from the centre of the body. Epidermis blue-green to blue-grey-green. Ribs 9-11, straight, slightly raised, in the area of the crown flat round, quite flat at the base, divided into fine wart-like humps with chin-like points by horizontal furrows, these are also on older plants very distinct.

Areoles oval, 1.2-2mm x 2.5mm, 5-7mm apart, in and around the crown with white wool, later fading and disappearing. Spines 9-11, fine, bristly, mostly curved sickle-like, lying onto the body, light brown to pink. The top pair ± arranged horizontally, the remainder radiating sideways and downwards, 5-7mm long, mostly the top 2 or 3 pairs are the longest. The top third of the areole is spineless.

Flowers rising from the centre of the crown 40-65mm long, 30-40mm diam. Pericarp large, (in proportion to the small body) elongated barrel-shaped, somewhat conical at the base and at the top, lightly waisted where the outer petals begin, because the ovary is slightly swollen, 25-27mm long, 7-8mm diam. at the base, 12mm in the middle and 10mm waisted part), dark olive green with half round, white edged scales, which mostly exhibit a red tip. Outer petals lanceolate, grey-green, occasionally with a delicate pink tip, white edged, 13-15mm long, 5mm wide. Inner petals lanceolate, 20-22mm long, 5mm wide, a dirty white (grey-white), with a dark grey- green middle stripe. Filaments inserted in 2 series, the lower row 4-5mm long lying against the central inserted style, therefore enclosing the 2-3mm deep, pink coloured nectar chamber: the upper series are inserted in the wall of the flower tube, the lowest small, the higher ones stronger curving in towards the style, their yellow anthers with yellow pollen enclosing the stigma.

Style standing in the middle, 14mm long without the stigma, 18mm long with the stigma, 1.5mm diam., pale pink at the base,

Fig.140 *G. calochlorum* (proliferum), *Neuhuber* 90 212-588 from the Sierra Chica, Prov. Córdoba, Argentina

light green towards the top. Stigma with 7-8 yellow lobes. Ovary longish, heart-shaped, 18-20mm long, 7-8mm wide, white walled with bundles of white funiculae. Fruit spindle to club-shaped, 27-29mm long and up to 18mm diam. at the thickest part, bluish-green with a slight bloom, splitting lengthwise. Seeds globular, 1.2mm diam. with a small, somewhat sunken hilum, without arilus. Subgenus *Gymnocalycium*.

Habitat: Argentina, Prov. Córdoba, near Las Palmas.

Type: Argentina, Prov. Córdoba, near the town Las Palmas, at 1100-1200m. 16-10-1988, leg Hans Till HT88-199/1736. (Holotype in WU).

Commentary on *G. amoenum*

Lambert (2002) did not agree with Till's application of the name *G. parvulum* to plants from the Salsacate area, believing that he had found a better match for Spegazzini's name from further north. He therefore elevated Till's *G. parvulum* var. *amoenum* to species rank in order to apply it to Till's plants, although they are clearly forms of *G. calochlorum*. His illustration on page 227 of his article looks much more like *G. taningaense* which also grows in the same area, so one wonders if Lambert was talking about the same plants as Till.

Distribution (Map 18)

The west side of the Sierra Grande, Cumbre de Achala and Sierra de Comechingones. It can also be found much further to the north in the hills surrounding Dean Funes. It is often buried in crumbling granite when it can be difficult to find if not in flower or fruit.

Conservation Status

Least Concern. It is very widespread and locally abundant without any immediate threat to its future survival.

History

First described by Bödeker (1932) from plants sent to Hahn by Stümer, the species was transferred to *Gymnocalycium* by Ito in 1952. The locality was not known, but assumed to be north-west Argentina. Later in the same year, Backeberg

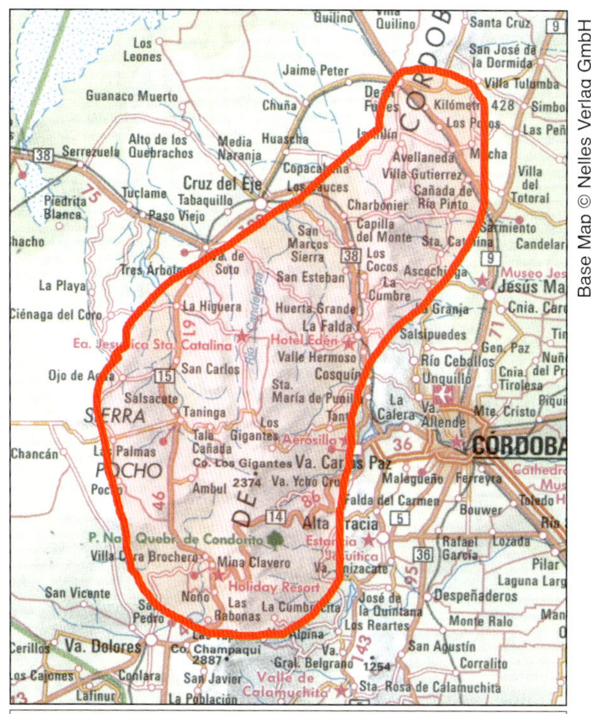

Map 18 Distribution of *G. calochlorum* in Prov. Córdoba, Argentina

(1932) published the name *Echinocactus prolifer* for the same taxon, then in his *Die Cactaceae* III (1959) made his new species a variety of *G. calochlorum*.

Till & Amerhauser (2007) have recently combined *G. calochlorum* as a variety of G. *quehlianum*. This is in line with their belief that *G. quehlianum* is a member of the subgenus *Gymnocalycium*, a view which is not accepted here, see page 221.

There has also been speculation linking forms of this taxon with Spegazzini's *Gymnocalycium parvulum*, a name rejected here as indeterminate.

Commentary

This taxon is similar to *G. capillaense* and has the same type of seeds. It differs by its smaller bodies and usually more and finer spines. Meregalli (pers. com.) reports that he has seen the two species growing together near Ongamira. Till (1994a) also reports them growing together on the eastern side of the Sierra Chica and the Sierra de Ischilín.

Cultivation

This easily grown plant soon makes clusters of neatly-spined stems. The spines can be light brown but there are also pretty white-spined clones. The large flowers are freely produced.

Gymnocalycium capillaense (Schick) Hosseus

First Description (Schick)

Neue Kakteen aus der Sierra de Córdoba. - *Möllers Deutsche Gärtner-Zeitung* 38(26): pp.201-203 (1923)

Echinocactus capillensis Schick n. sp.

Body flattened-globular, smoky green, crown slightly sunken with warts, almost unarmed, but with a little wool, 3.5cm high, 6cm diameter. Ribs 9. Areoles 4mm wide, 2cm apart. Spines only 5, all radials, 15mm long, light horn colour, round, straight, sharp. Flowers from the topmost spined areoles, when open 8cm long, 6mm diameter, outside of flower naked, but with half-round, greenish, white edged scales. Petals somewhat fleshy, spatulate, ivory colour, with a greenish mid-stripe on the back, inner petals lanceolate, ivory colour with a light pink central stripe, wine red in the throat. Stamens and anthers yellowish, Style with 10 stigma lobes yellowish-white. Berry spindle shaped, 4cm long, 1.5cm wide, bluish-green with whitish scales.

Habitat: Argentina, on the dry hills near Capilla del Monte.

Neotype: leg. H. Fechser 1961, ex col. H. Till, HT54 (BA) Designated by Till in *Gymnocalycium* 16(1): p.497 with an illustration. [It is a pity that Till chose to designate a plant without a precise habitat locality for the neotype of this name]

Etymology

Occurs near to Capilla del Monte in Prov. Córdoba, Argentina. The spelling was corrected by the addition of the letter 'a' to *capillaensis* by Oehme (1938) at the request of Schick.

Fig.141 *G. capillaense, Piltz* 82 from Capilla del Monte, prov. Córdoba, Argentina

Fig.142 The original illustration of *Echinocactus Sigelianus* from *Möllers Deutsche Gärtner-Zeitung* (1923)

Fig.143 The original illustration of *Echinocactus Sutterianus* from *Möllers Deutsche Gärtner-Zeitung* (1923)

Other Names

Echinocactus capillensis Schick in.'Neue Kakteen aus der Sierra de Córdoba' - *Möllers Deutsche Gärtner-Zeitung* (1923

Synonyms

G. deeszianum Dölz in *Kakteenkunde* 11(3): pp.54-55 (1943)

First description of *Gymnocalycium deeszianum* Dölz. spec. nov.

Body flattened-globular, 6.5cm wide, 4.5cm high, deep dull green, new growth in the crown shiny olive green; crown unarmed, with a little felt between the humps. Ribs 7-8, wide, flat at the bottom. Areoles oval, c. 1.5cm apart, with dirty white felt, at first strong, later disappearing. Spines usually 7, all radials, 1 pointing straight down, 2 pointing diagonally downwards, 1 pair pointing diagonally upwards, the top pair the shortest, c. 2.5cm long, dirty pale yellow to whitish, at the base ± brownish, sometimes blackish, rough, robust, 2/3mm thick at the base, sharp pointed, ± bent or twisted, therefore irregular, in age ± lying close to the body, in youth more porrect and interwoven with the spines from the neigh-

Fig.144 *G. capillaense,* GC948.01 near to Capilla del Monte. Prov. Córdoba, Argentina. In December, the plants had ripe fruits and open flowers.

Fig.145 The habitat of *G. capillaense* GC948.01 near to Capilla del Monte. Prov. Córdoba, Argentina.

bouring areoles, occasionally one or both the top pairs are missing. (Sometimes a stronger central spine is present, Andreae).

Flowers from the young areoles in the crown, c. 4-5cm long and diameter. Pericarp 6-7mm diameter, 8-9mm long, light green. Ovary 3.5-4mm long and diameter, almost square. Scales semicircular, whitish, yellow to brown in the centre; outer petals white, pale pink at the base, with a grey-brown central zone on the back, and a shiny delicate lilac-pink mid-stripe on the inside, inner petals white (cream-white) , pale pink at the base, running in to a point, especially the inner petals, that are tooth-like and irregular; inside of flower tube pink, outside light brownish-red. Style strong, white, 1.8cm long (with stigma lobes), 8 stigma lobes, very pale yellow, that reach over the highest stamens, filaments white, anthers yellow. Seeds longish, hat shaped, c. 1.2mm long, 0.8mm wide, dull black, with a sunken hilum

Commentary on *G. deeszianum*

A form with 5 long spines, said by Till (2003a) to be from the east side of the Sierra Chica. It was found by Andreae in 1930 while on a visit to the nursery of De Laet in Contich, Belgium, among an importation of plants identified as *G. sigelianum* but without a habitat locality.

G. sigelianum (Schick) Hosseus in *Apuntes sobre las Cactaceas. - Revista del Centro Estudianteo de Farmacia* 2(6): p.16 (1926). Basionym: *Echinocactus sigelianus* Schick (1923)

First description of *Echinocactus sigelianus* n. sp. in *Möller's Deutsche Gärtn.-Zeitung* 38(26): p.201 (1923)

Body simple, flattened globular, light grey-green, 8cm diameter, 4cm high, crown sunken, roots thick, turnip-like. Ribs 11, blunt, separated by sharp grooves, clear cross-cuts form chin-like humps, with areoles sitting on the lower part of the humps. Areoles c. 2mm apart, 7mm wide, with short yellowish-white wool, later becoming grey, finally bald. Radial spines 3, one pair radiating horizontally, the third pointing down, straight or curved, sickle-like, lying close to the body of the plant, 10-12mm long, robust, in new growth grey horn coloured, later grey, round, stiff.

Flowers quite long, coming from near the crown, just before opening 8cm long, when open 6cm diameter. Ovary 12mm thick, with half round, light green, reddish-white edged scales. Petals flesh pink, with a green central stripe; sepals pink with a darker central stripe. Many stamens, shorter than the flower tube; filaments and anthers yellow. Style with 12 stigma lobes yellowish-white.

Habitat: Argentina. Capilla del Monte, Sierra Córdoba.

I name this plant after Herrn Kaufmann Carlos Sigel in Capilla del Monte, for his help and the plants he sent from Córdoba.

Commentary on *G. sigelianum* (Fig.142)

After examining plants of *G. capillaense* in habitat, Till (2003a) reports finding plants matching the description of *G. sigelianum* from the eastern part of the distribution, between Ascochinga and Río Cabello.

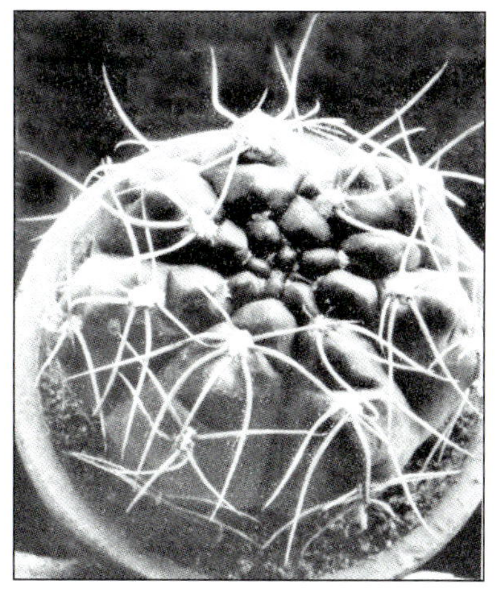

Fig.146-147 The original illustrations of *G. deeszianum* from Kakteenkunde (1943)

Fig.148 *G. capillaense,* a propagation from a plant collected by Fechser in the collection of H. Till.

G. sutterianum Schick (Hosseus) in *Apuntes sobre las Cactaceas. - Revista del Centro Estudianteo de Farmacia* 2(6): p.16 (1926). Basionym: *Echinocactus sutterianus* Schick (1923)

First description of *Echinocactus sutterianus* Schick n. sp. in *Möller's Deutsche Gärtn.-Zeitung* 38(26): p.201 (1923)

Body semi-globular with a sunken spineless crown, grey-green, 8cm diameter, 4.5cm high. Ribs 9, running vertically, having humps formed by cross-cuts, the humps are chin-like under the areoles. Areoles 2-2.5cm apart. 6mm long, 4mm wide, in youth with yellowish-white felt, soon becoming bald. Spines, 5 radials, the top pair 2.5cm long, angled upwards, the lower pair spreading horizontally, 17mm long, the bottom spine pointing down, also 17mm long, not lying directly onto the body, all spines robust, grey horn colour, in new growth dark honey yellow.

Flowers from near to the edge of the crown, just before opening nearly 10cm long, when fully open 6cm diameter. Ovary light green, with semi-circular white edged scales, each has a pale red fleck on the upper edge. Flowers funnel shaped, outer petals oblong, almost fleshy, whitish to whitish-pink, with a light green back stripe, inner petals narrow spatulate, pale pink, with a darker mid-stripe, wine red at the base. Filaments white, numerous; anthers yellow. Style and the 9 stigma lobes yellow. Berry spindle-shape, 5cm long, 5.5cm wide, Grey-green with whitish scales.

Habitat: Argentina, Sierra de Córdoba.

Commentary on *G. sutterianum* (Fig.143)

Till & Amerhauser (2008) have applied this name to plants from a disjunct population near the border between Prov. Córdoba and Prov. San Luis. This border region is the southern end of the Sierra de Comechingones, which can be regarded as part of the Sierra de Córdoba. They indicate that Sigel had travelled to this region when looking for a place for German emigrants to settle.

The plants from there have a longer and narrower tube than the plants from the north, clearly shown by Till's illustrations 15 & 16 on page 789 of the 2008 article. The provisional name *G. sanluisensis* n.n. for this plant has appeared in seed lists.

Distribution (Map 19)

G. capillaense was originally said to come from Capilla del Monte, a town to the east of the Sierra Chica in Prov. Córdoba. More recent explorations

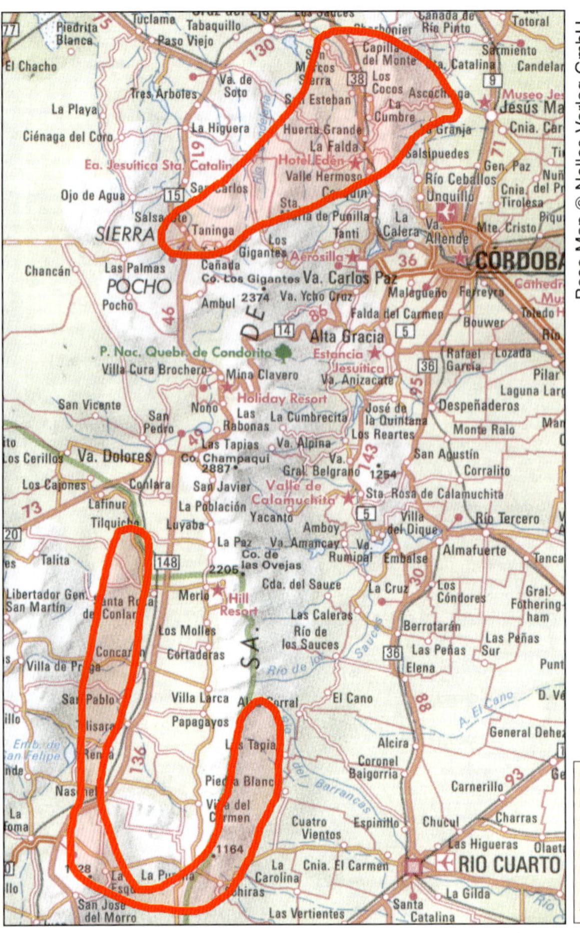

Map 19 Distribution of *G. capillaense* in Prov. Córdoba and Prov. San Luis, Argentina

have revealed its distribution to stretch from Ascochinga in the east to Salsacate in the west. There is also the disjunct habitat of the *'sutterianum'* form at the southern end of the Sierra de Comechingones.

Conservation Status

Least Concern. The plant is locally common on the dry hills near to the town and much further afield.

History

Carl Schick published this species with two others, *Echinocactus sigelianus* and *E. sutterianus*, in the September 1923 issue of Möllers Deutsche Gärtner-Zeitung. They were described from plants sent by Carlos Sigel who lived at Capilla del Monte, Prov. Córdoba. It is believed that Schick had wanted to publish his new species in the DKG's 'Zeitschrift für Sukkulentenkunde' but its editor, Dr. Vaupel, declined the article, believing that there could not be more new cacti from the region because Spegazzini had already done such a thorough exploration!

Till (2003a) considered the form of *G. capillaense* from the west of its range near to Salsacate to be the taxon that was described as *G. mucidum* by Oehme in 1937. He made it a variety of *G. capillaense* on page 498 of his article. This is just one of three attempts to associate Oehme's name with different

Fig.149 *G. capillaense* growing on a large ant hill near Route 17 crossing the Sierra Chica, west of Ongamira, Prov. Córdoba, Argentina

Fig.150 *G. capillaense* (sigelianum) GC1002a photographed in the rain near to Las Tres Cascades, west of Ascochinga, Prov. Córdoba, Argentina

plants known in habitat today, hence the name is here abandoned as unattributable. See page 262

Commentary

The three taxa described by Schick were all said to be from the area surrounding the town of Capilla del Monte and the Sierra Córdoba. They have been treated by most authors as descriptions of individuals within the variation of a single species. This was the opinion of Hosseus (1926a) who made the combinations in *Gymnocalycium*. However, Till (2003a) tells us that only plants matching the description of *G. capillaense* can be found near the town, the other two being from elsewhere. He maintains that plants from the eastern part of the distribution, east of the Sierra Chica correspond with *G sigelianum* and *G. deeszianum*.

Till & Amerhauser (2008) have applied the name *G. sutterianum* to plants from a southern population near the border between the Provinces of Córdoba and San Luis in the Sierra de Comechingones, a southern extension of the Sierra de Córdoba.

Cultivation

It is easy to grow, enjoying a good supply of water in the growing season when it will produce its large pinkish flowers and slowly grow into a cluster of stems.

Fig.151 *G. capillaense,* GC3.01 from Capilla del Monte, Prov. Córdoba, Argentina

Fig.152 *G. capillaense* (sigelianum), GC393.01 from the Sierra Chica, east of La Falda, Prov. Córdoba, Argentina

Fig.153 *G. capillaense* (sutterianum), *Papsch* 75-99, La Punilla, Prov. San Luis, Argentina

Gymnocalycium carolinense (Neuhuber) Neuhuber

First Description (Neuhuber)

Gymnocalycium 7(3): pp.127-130

Gymnocalycium andreae subspecies *carolinense* Neuhuber, subspecies nov.

Body simple, flattened-globular, the part above ground 12mm high, 20mm diameter, in cultivation 35mm high, 58mm diameter, with a longer stake-like, ramified root. Older plants occasionally offset. Crown slightly sunken, epidermis blue-green, dull. Ribs flat mostly (9-)11(-12), vertical, flatter towards the base, separated into flat humps by shallow cross-cuts, somewhat wider at the areoles. Areoles round, 3mm diameter, to oval, 3 x 5m, developing dense spines, with yellowish-white wool, later white, never becoming bald in cultivation, 10mm apart. Spines round, seldom flattened and awl shaped, curved, with a tendency to be hook shaped, 7-9 radial spines, the one pointing down the shortest, 10(-15)mm long, light brown, tips lighter, rough, lying onto the body sideways and downwards, some times with 2 spines pointing upwards, only 5mm long, on older plants 1-2 true central spines. Only sometimes observed in habitat, one spine at the top and one at the bottom, standing away from the body, 10-13mm long.

Flowers from the younger areoles in the crown, short funnel shape, 24-35mm long, 30-43mm diameter, white, sometimes light pink, with a sweet scent, in cultivation, opening at midday. Pericarp short conical, 4-6mm long, 7-9mm diameter, dark green, silky-matt, with a few wide, rounded scales, with whitish to pink edges. Outer petals spatulate, up to 19mm long, 5-8mm wide, white to pink, with a wide green middle stripe. Inner petals lanceolate, 15-16mm long, 2-4mm wide, from white with a yellowish tip, to light pink with a darker middle stripe. Relatively few stamens inserted in the receptacle, grouped around the stigma, not lying against the style. Filaments white, thin, straight, 7-8mm long. Anthers light yellow oval, 0.7 x 0.4mm long. Style white, 10-13mm long, 1mm diameter. Stigma cream-white, 5-6mm long, chalky, with 7-8 lobes, the highest anthers overhang the stigma. Nectar chamber whitish-yellow to delicate pink, 0.3mm long, 1.9mm diameter, not closed. Ovary urn shape, white walled, 2-4mm long, 3-5mm diameter. Fruit pear shape, 13-16mm high, 6-13mm diameter, dark green, olive green when ripe, splitting lengthwise in 1 or 2 places. Seeds mostly 1.3mm long, 1.1mm diameter. Testa dull black, with small warts, often with a light brown arillus. Hilum-micropyle region large, nearly round, scarcely sunken. Subgenus: *Gymnocalycium*.

Photo: Ludwig Bercht

Fig.155 The habitat of *G. carolinense, Bercht* 3221 near Carolina, Prov. San Luis, Argentina

Photo: Ludwig Bercht

Fig.154 *G. carolinense, Bercht* 3221 in habitat near Carolina, Prov. San Luis, Argentina

Habitat: Argentina, prov. San Luis, in the southern part of the Sierra de San Luis, 1400-1700m.

Type: Argentina, Province of San Luis, Sierra de San Luis, 1400-1600m, leg. *G. Neuhuber* GN88-31/52, 8th April 1988 (Holotype WU)

Other Names

Gymnocalycium andreae subsp. *carolinensis* Neuhuber in *Gymnocalycium* 7(3): pp.127-130 (1994)

Etymology

Named after its finding place, Carolina, originally a gold mining area in Prov. San Luis.

Distribution (Map 20)

Only known from the type locality and within a radius of 20km from there. Neuhuber (1994) tells us that the habitat is an undulating high plain with rocky outcrops which he illustrates on page 127 of

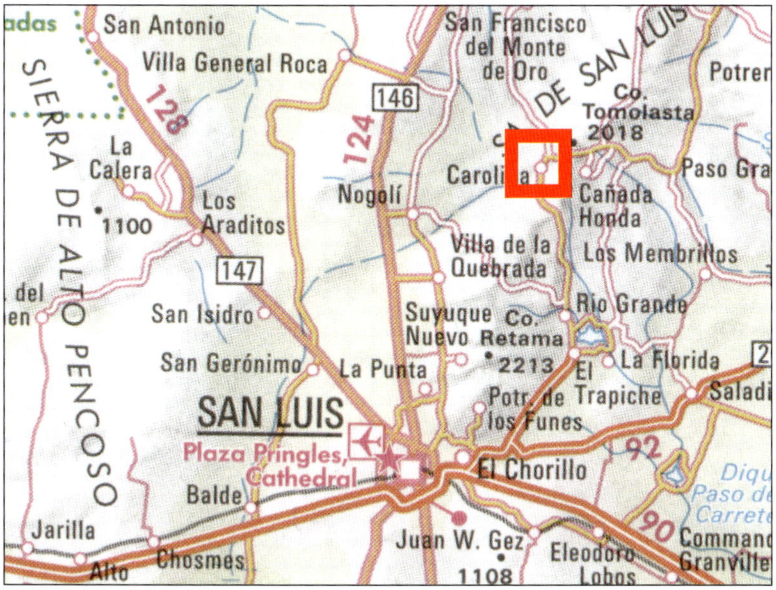

Map 20 Location of *G. carolinense* in Prov. San Luis, Argentina

Fig.156 *G. carolinense, Neuhuber* 90 273 881 from Cto. Pajoso, north of Carolina, Prov. San Luis, Argentina

Fig.157 *G. carolinense, Neuhuber* 90 273 880 from Cto. Pajoso, north of Carolina, Prov. San Luis, Argentina

Fig.158 *G. carolinense, Meregalli* 753 in habitat near Carolina, Prov. San Luis, Argentina

his article. The rock is magma, gneiss and quartz where no shrubs grow with the cacti (Fig.155).

Conservation Status

Least Concern. The author states that he found a number of populations within a radius of 20km suggesting that the plant is plentiful and not under any immediate threat.

History

Neuhuber found this plant in April 1988 when visiting the east side of the Sierra San Luis. He first described it as a subspecies of *G. andreae* in 1994 then he (2005c) elevated it to the rank of a separate species. Further populations which exhibited minor differences were found on subsequent visits to the area.

It is unclear why Neuhuber considered it to be related to *G. andreae* since it appears to have more in common with *G. bruchii* where it was synonymised in the *New Cactus Lexicon* (2006). It is accepted here because of its solitary habit and more widely-spaced areoles separated by prominent humps. Bercht (pers. com.) believes that its closest relative is *G. bruchii* var. *brigittae*.

Commentary

This is a very interesting plant because, as far as we know, it is far away from similar taxa. Its location and distinct characters are the reasons for its acceptance here as a separate species. The relationships of the many similar species in the subgenus *Gymnocalycium* from Prov. San Luis are poorly understood and it is likely that some can be rationalised when better known.

Cultivation

This is a pretty plant that is easy to grow and flowers when small, if somewhat later than *G. andreae* with which it was associated. Unfortunately, it is not yet readily available to growers, so hopefully some enterprising nurseryman will soon propagate it for sale.

Illustrations

New Cactus Lexicon Atlas (2006) Illn. 270.4

Gymnocalycium erinaceum Lambert

First Description (Lambert)

Succulenta 64(3): pp.64-66 (1985)

Plant solitary, flattened-globular to globular, 50mm high, 55mm diameter, grey-green, testa velvet-like, crown sunken, with a little white wool, covered with young spines, root turnip-shaped.

Ribs 12, straight, lost in humps, each with a chin under the areole, c. 12mm wide, only half as high, horizontal furrows are short.

Areoles round to oval, 3mm wide, 3-4mm long, at first with white wool, later bare, 7-8mm apart.

Radial spines 7-9, 3-4 pairs, displayed comb-like, 1 pointing downwards, sharp, round in section, blackish-brown in the new growth, later becoming grey with a brown point, dark red towards the base, 6-8mm long. Central spines 1-2, fine, standing erect, also round in section, the same as the radial spines, up to 1cm long

Flower from the young areoles near the crown, funnel shaped, c. 55mm long, 48mm diameter. Pericarpel short, c. 13mm long, 7mm diameter. grey-green with a whitish bloom, carrying a few semi-circular scales, 3.5mm wide, 2mm long, faintly pink at the tip.

Outer petals spatulate, somewhat tapering, 27mm long, 7mm wide, dull yellow to greyish-white with a very pale mid green mid stripe, on the outside greenish. Middle petals lanceolate, 25mm long, 7mm wide, with a fine grey line on the outside. Inner petals lanceolate, shorter, white. Receptacle carmine on the inside.

Primary stamens are inserted in a ring around the style, secondary stamens inserted in the receptacle wall, the topmost stamens curving over the stigma, all stamens white, pale pink at the base. Anthers yellow. Style greenish-white, with stigma 17mm long. Stigma white, 11 lobes.

Fig.159 *G. erinaceum,* GC994.02 in habitat east of Sauce Punco, Prov. Córdoba, Argentina

Fruit spindle shape with a bluish bloom. Scales few, whitish, 16mm long, 13mm wide.

Seeds c. 1.3mm long, 1mm wide, testa covered with small warts, black, covered with a pale brownish arilus, hilum pear to rhomboid shape, flat, blackish. Type **Ovatisemineum* (Schütz). Baldiana (Buxbaum).

This species is distinguished from the other **Ovatisemineum* by its quite typical spination, which somehow reminds me of a hedgehog; hence the epithet erinaceum.

* Now subgenus *Gymnocalycium*

Fig.160 The habitat of *G. erinaceum,* GC994.02 east of Sauce Punco, Prov. Córdoba, Argentina

The central spines develop on small plants of 1.2-2cm. Plants with a diameter of 3.5cm are ready to flower. This species remains small.

Type: *Lambert* 40 from near Sauce Punco, Sierra de Tulumba, Province of Córdoba, Argentina at 1050m (Herb. Univ. Rheno-Traiecti, Holotype)

Etymology

From the Latin for hedgehog.

Synonyms

G. gaponii Neuhuber in *Gymnocalycium* 14(3): p.410 (2001)

First description of *Gymnocalycium gaponii* Neuhuber spec. nov.

Body up to 85mm high, of which, only about a quarter of the body is above the soil, 50-60(-80)mm diameter, always simple, inverted skittle shape. Additional side roots would sometimes be found on the main root. Crown mostly bald or slightly woolly, slightly sunken, armed. Epidermis matt; green to light olive green. Ribs 8-9(-11), running straight down, very quickly thickening, up to 30mm wide at the base. The ribs are divided into flat humps, through shallow, slightly curved horizontal grooves, which have small chins under the raised, white woolled, soon balding, 3.5mm round areoles, which are 11mm apart. Spines 5-7, 13-16mm long, 0.5-0.6mm diameter, awl-shaped, stiff, slightly elastic, straight or slightly recurved to the body, a dirty white colour, distinctly brown at the base, older spines becoming dark brown; one or two horizontal pairs, or the top pair pointing slightly upwards, the lowest pair directed sideways to below, the lowest spine is the same length, often curved to the side, their points often bent, only the larger plants have central spines.

Flowers 57-64(-74)mm high, 42-45(-51)mm diameter, white to pale pink, coming from young areoles in the crown, hardly fragrant. Outer petals white, pale pink at the base and the tip, also with a pale pink middle stripe, lanceolate to broad lanceolate, up to 29mm long, and 8mm wide, inner petals white to pale pink, pale pink at the base, the tip lanceolate, up to 22.5mm long and 6mm wide. Receptacle funnel shaped, pale pink to pink, always darker than the petals, 15-20mm high, 9.2-11.3mm diameter. Nectar chamber light orange, 1.2-4.3mm high 2.7-3.8mm diameter. Many filaments, mostly white, seldom pale pink, thin, the lowest row 10mm long, the upper rows 7-10.5mm long, irregularly distributed in the receptacle wall. Anthers yellow, 0.5x0.4mm, one or two of the rows of stamens are higher than the stigma. Style pale green, 12-13mm high, 1.3-1.5mm diameter, not free standing. Stigma yellow, 2.7-5mm long, with 9-10 lobes. Ovary very small, pale pink or white, 10.3-13.5mm high, 4.9-5.4mm diameter, filled with white ovules. Pericarp 15-21mm high, 9.5-11.3mm diameter, light grey-green, with mostly small crescent shaped, pale pink to olive colour, lighter edged scales. Fruit at first light green, elongated egg shaped to club shaped, later egg shaped, at the base light green, resolving to dark green at the top; olive-green when ripe, 16.5-22mm high, 8-13mm diameter, splitting vertically at one or two places; spermatozoon light brown, seldom pink, occasionally sticky. Seed 1.1mm-1.6mm long, 1.1-1.5mm diameter, when ripe partially or completely covered with a dry brown skin, which is only lacking when the seed is not ripe, Hilum-micropyle-region small, roundish, and at the funiculus side pointed, sunken, Testa black, (subgenus *Gymnocalycium*).

Type: Argentina, Province Córdoba, near Villa Rafael Benegas, at 900m, 7th December 1991; collected by G. Neuhuber 91-362/1222; (in alcohol, as well as dry in Herbarium), Holotype: BA, Isotype: WU; at the same habitat, G. Neuhuber 96-850/2783 + 2782 (in alcohol), Paratype: BA, WU.

G. gaponii subsp. *geyeri* Neuhuber et Gapon in *Gymnocalycium* 21(1): p751-754

G. papschii H. Till in *Gymnocalycium* 14(3): pp.405-408 (2001)

***Gymnocalycium papschii* H. Till, spec. nov.**

Body simple (only after damage do 1-3 offsets appear with out a thickened root-throat, coming from a relatively thin stake-like root with several short secondary roots. Elongated globular to short columnar, 40-95mm long, 30-65mm diameter in age, mostly wrapped in spines. In its natural habitat the plant often appears to be flattened globular, while often more than 2/3 of the body is sunken into the soil. Ribs narrow, 12-14(-17), separated by deep vertical furrows, with 10-13(-20) chin-like humps, which are arranged one above the other. Areoles in the area of the crown are round with short, yellowish-white felt, later oval, longish, up to 5mm long, the felt becoming grey or missing.

Fig.161 *G. erinaceum,* STO 3-2 from Rio Pinto - Palo Cortado, Prov. Córdoba, Argentina

Spines in a semi-circle, directed towards the base, 5 pairs, of which the two topmost are up to 4mm long, both the middle pairs up to 7mm long, the lowest pair up to 4mm long, and the shorter lower spine 3mm long and usually a 4-5mm longer, curved or twisted central spine. All spines in the crown are white or pale yellow, older spines light grey to pale brown and mostly darker at the base.

Flowers funnel-shaped, according to the age of the plant, 45-60mm high, 45-50mm diameter, ivory white with a weak pink to pale pink sheen and a deep pink throat. Outer petals spatulate, 9-22mm long, 5mm broad, 3mm wide at the base, olive green with a wide paler edge, inner petals lanceolate, up to 24mm long, 5mm broad, slightly curved outwards at the tip, white to pink, dark pink at the base, innermost petals only 9-12mm long, 3.5-4mm wide. Flower throat deep pink (the intensity of the colours can vary). Filaments thin 3mm long, yellowish-white, one or two rows lying against the style, the rest thickly arranged over each other on the inner wall of the flower tube overhanging the style. Anthers oval, 2x1mm, white, pollen yellow. Style with stigma 12mm long, 2mm thick, stigma globular (the stigma lobes do not radiate), reaching up to the centre of the petals, yellowish-white. Ovary 16-29mm long, 8-9mm thick, olive green, with many half- round, 5x3mm olive green scales, each with a wide light (white) edge, resolving into the outer petals. Ovary 10mm high, 5mm wide, inner-wall white, filled with spermatic cords and ovules. Seeds globular, 1.3-1.5mm diameter, with a layer of skin, which is partially loose, hilum tapered, black.

Distribution and Ecology: On the western slopes of the Sierra de Comechingones, near San Javier, Luyaba to Loma Bola, where, until now, plants of this species have not been observed. (WP 83/111, Be 517, HA 521, HT 2668) near Luyaba and Loma Bola the species grows on stony ground. The flowers of these forms have a squat pericarp c. 45mm high, which gives it the appearance of being densely scaled.

Type: Argentina, Province of Córdoba, San Javier, Cerro Champaqui, 1200m, collected by W. Papsch WP89-83/111, November 1989 (CORD, Holotype; WU, Isotype)

Commentary on *G. papschii*

This population was discovered by Papsch and Hold in 1999 when they climbed the Cerro Champaqui in the Sierra Comechingones, Prov. Córdoba. Till (2001) compared the plants with *G. bruchii* and his concept of *G. parvulum*.

Base Map © Nelles Verlag GmbH

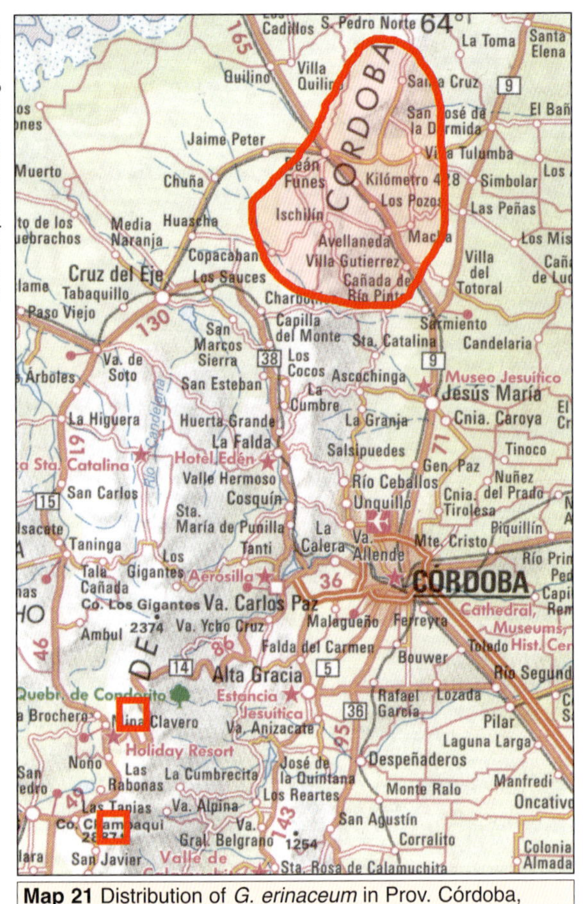

Map 21 Distribution of *G. erinaceum* in Prov. Córdoba, Argentina

Distribution (Map 21)

The original location is in the mountains east of Dean Funes, to the north of Prov. Córdoba, but the synonyms extend the distribution far to the south of that province. Such a disjunct distribution suggests that either there are more populations to be found or other species recognised here, such as *G. amerhauseri*, are really part of this complex.

Conservation Status

Least Concern. It has a number of known habitat locations with no reported threat to them.

History

This species was discovered by Jacques Lambert on 5th December 1981, near Sauce Punco, Sierra de Tulumba, Province of Córdoba, Argentina, at a height of 1050m. The plants were given the field number JL 40. They were found in a biotype of open bushes, together with *G. calochlorum*, while *G. monvillei* occurred further down the slope.

Piltz (1994) described a larger form of this species from Ongamira in the Sierra Chica as var. *paucisquamosum*, field number *Piltz* 400. He found it growing with *G. capillaense* which he says is very abundant in the area among sparse shrubs.

Commentary

This easily recognised species is probably most closely related to *G. baldianum* and *G. amerhauseri*.

Cultivation

This plant presents no difficulty in cultivation but should be kept moist in summer and protected from strong sun to prevent dehydration. It can occasionally offset in culture, making a cluster of small dark-bodied stems. It flowers early in the year and when still only small.

Most plants in collections are descendants of the type collection, although other enthusiasts have subsequently found and propagated it. The variety *paucisquamosum* is seen under the type number *Piltz* 400 as well as collections by others.

Fig.162 *G. erinaceum* (paucisquamosum), *Bercht* 1130, Las Palmas, south of Ischilín, Prov. Córdoba, Argentina

Photo: Massimo Meregalli

Fig.163 *G. erinaceum* (paucisquamosum), STO391/1 from Ongamira, Prov. Córdoba, Argentina

G. fischeri Halda, Kupcák, Lukasik & Sladkovsky

First Description (Halda et al.)

Acta musei Richnoviensis 9(1): p.60 (2002)

Gymnocalycium fischeri spec. nov.

Solitary or caespitose, 80-150mm diameter, flat; ribs 7-10 lost in tubercles 5-8mm long, 5-8mm broad, 4-10mm high formed by horizontal grooves; areoles on top of the tubercles, 2-3mm long and wide, 15-20mm from each other, covered with abundant white wool. Spines 3-7, stiff, needle-like, radiating, 10-35mm long, white or brown.

Flowers white or pink, 70-120mm long, inner petals c. 15, 8-12mm wide, 30-40mm long, with a magenta mid-stripe, mucronate and ends dentate; outer petals scale-like, reddish or brownish, lighter at the edges, 3-5mm long, 4-6mm wide. Pericarp 25-35mm long, 8-12mm diameter, top part glaucous, base paler, ovary 8mm diameter, style c. 25mm long, stigma with 6 lobes, white. Fruit 40mm long, 15mm diameter, smooth, glaucous, paler at the base. Seeds 1 x 0.8mm, testa black, tuberculate, hilum sunken.

Holotype: no. 11.702; leg. V. Sorma 1.1.1997, no. VS1. Habitat in Prov. San Luis, Argentina, El Volcán, near Campo "La Sierra" in dry meadows at 1200m.

Etymology

Named for Ladislav Fischer, Czech cactus collector.

Synonyms

Gymnocalycium fischeri subsp. *suyuquense* Berger in *Gymnocalycium* 16(4): pp.539-540 (2003)

Gymnocalycium miltii in *Act. Mus. Rich. Natur.* 9 (1): pp.64-65 (2002)

First description of *Gymnocalycium miltii* Halda, Kupcak, Lukasik et Sladkovsky spec. nov.

Stem singular, c. 9 ribs, taproot more or less napiform, stem globose, top flattened, 50-70mm across, partly hidden, humps c. 5mm long, 8mm wide at base, angled and carinate, green or yellowish-green. Areoles grey tomentose, spines 4-7, needle-like, up to 20mm long, grey, irregularly reflexed.

Flowers 30-40mm across, 50-60mm long, interior segments c. 20, 8-11mm wide and 20-25mm long, milky white or pale pink, the outside with a brownish tint, top mucronate; exterior segments scale-like, reddish, with pale margin, 2-5mm long, 2.5-4mm wide. Pericarp c. 20mm long, 4mm across, green or glaucous, with whitish bottom; ovary c. 4mm across and 10mm long. Style c. 20mm long, excerted up to 5mm; stigma lobes yellowish-white. Fruit up to 20mm long and 8mm across, glabrous, glaucous, bottom whitish, opening with one or two slits. Seed 1mm long and 0.9mm thick, testa blackish-brown, tuberculate, hilum sunken. Named after Ivan Milt.

Holotype: PR no 11.706; J. Procházka 10.11.1994, no. JPR95/219

Habitat: Argentina, Prov. San Luis, Quebrada de los Cóndores, 1000m

Fig.164 *G. fischeri, Neuhuber* 90 272 873 from Tamboreo, Prov. San Luis, Argentina

G. poeschlii Neuhuber in *Gymnocalycium* 12(3): pp.295-300 (1999)

First Description of *Gymnocalycium poeschlii* Neuhuber spec. nov.

Body simple, flat, 20(-30)mm high and 80(-140)mm diameter above the surface of the soil, conical c. 57mm below the surface in a stake-like root from 100(-250)mm long, which seldom divides, and with many thread-like subsidiary roots coming from the end of the stake root. Epidermis dull blue-grey. Ribs flat, mostly 8-12, vertical, wavy by the areoles and humps, quickly widening towards the bottom, then 20-28mm wide, on older plants the ribs appear to be lost in the humps. The flat-round humps on younger plants, are formed by light cross-grooves, on older plants the humps are at first very angular around the crown, then usually separated by deeper cross-cuts. Areoles on older plants oval, 2mm long, 1mm wide, or round, 13-15mm apart, slightly sunken, seldom with white

Fig.165 *G.fischeri* (suyuquense), *Sorma* 8 from Suyuque Nuevo, Prov. San Luis, Argentina

Fig.166 *G. fischeri* , GC931.01 near La Petra, Prov. San Luis, Argentina

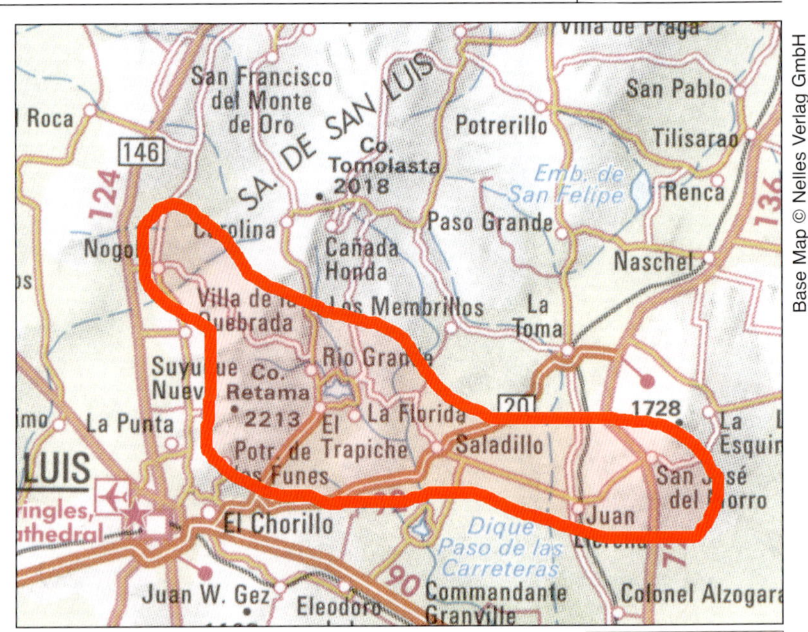

Map 22 Distribuation of *G. fischeri* in Prov. San Luis, Argentina

areole-wool, soon becoming bald. Spines 5(-7), arranged in 2-3 pairs, 7.5-13mm long, round, later somewhat flattened, red-brown, sometime becoming lighter towards the point, the lower 3 spines radiating or claw-like, mainly lying onto the body, the lowest spine not always the longest, the topmost or highest pair are sometimes missing, are shorter and arranged pointing sideways, central spines absent.

Flowers coming from the areoles in the crown, light lilac-pink. Throat lilac-pink to carmine-red, with a silky sheen, funnel shaped, 45-65(-80)mm long, 40-70mm diameter when fully open with recurved petals. The flowering time is in the spring when no other plant of this seed group from this region flowers. Outer petals lanceolate, often dentate, 20-30mm long, 6-8mm wide, pale lilac-pink with a paler mid-stripe and darker at the base, also the outside has an olive point. Inner petals tip lanceolate, often dentate, pale lilac-pink, with a paler mid-stripe and a darker base, 16-23mm long, 4-7mm wide, 3 rows of petals are not uncommon. Receptacle funnel-shaped, light lilac-pink, 10-14mm high, 8-10mm diameter. Nectar chamber lilac-pink, 1.3mm high, 2.4-3mm diameter. Filaments mostly set in 5 rows, white to lilac-pink, the lowest 6mm long, mostly lighter than those above, which are inserted at a distance and are 9.5mm long, white to yellow, slightly leaning towards the centre, the top row standing straight. Anthers yellow, 1.5mm long, the top ones mostly overhanging the stigma, 1-3 rows inserted higher than the stigma. Style pale pink, 10-15mm long, 2mm diameter, never standing in the ovary. Stigma yellow, 8-12 lobes, 4-9.5mm long. Ovary very pale pink or white, always with a domed roof, 14-15mm high, 3-4mm diameter, according to its sex, thickly filled or almost empty of white ovules. Pericarp grey-green to olive-green, occasionally with a light grey bloom, 18-15mm long, 8mm diameter, with half round to triangular, seldom pointed, dark green scales, with lighter or light pink edges, which resolve into the outer petals, which are mostly olive-green edged light pink.

Fruit club-shaped, to spindle-shaped, grey-green, sometimes with a grey-blue bloom, when ripe 22-50mm high, 12-22mm diameter, split perpendicularly in one or two places, producing 500-700 seeds. Seeds mainly 1.4mm long, 1.25mm diameter, partly or completely covered with a brown skin, black underneath, hilum-micropyle region small, short tear-drop shaped, strongly sunken, mostly black, centre slightly raised. Subgenus *Gymnocalycium*. It occurs as male and female, which determines the stigma size, the style, the chamber and the ovary.

Type: Argentina, Province San Luis, south of La Toma by the Rio Quinto, 980m. collected by G. Neuhuber 88-24/35, 27th April 1988. (BA, holotype), GN88-24/36, 27th April 1988. (WU, isotype).

Commentary on *G. poeschlii*

Discovered in 1988, it was not described until 1999 and since then has remained rare in cultivation. It is one of a number of recently described taxa from Prov. San Luis which belong to the seed group *Gymnocalycium*. It is similar to *G. berchtii* in that it has a grey body but it grows larger and comes from

Photo: Massimo Meregalli

Fig.167 *G. fischeri, Meregalli* 989 in habitat south of El Volcan, Prov. San Luis, Argentina

Fig.168 The habitat of *G. fischeri* (poeschlii) GC932.01 near Saladillo, Prov. San Luis, Argentina

Fig.169 The habitat of *G. fischeri* (poeschlii) GC932.01 near Saladillo, Prov. San Luis, Argentina

much further south. It also flowers about four weeks earlier.

Neuhuber (1999b) suggests that plants from nearby are a good match for *G. terweemeanum,* although he maintains that they are not the same as *G. poeschlii.*

Distribution (Map 22)

Known from many localities around the southern end of the Sierra San Luis, it grows under spiny bushes on the slopes of the low hills.

Conservation Status

Least Concern. It is quite widespread and safe from immediate threats to its survival.

History

This plant has been known since the mid 1980's when it was considered by some as a southern relative of *G. capillaense*. Kiesling thought it was Spegazzini's *G. stuckertii* although it does not agree with Spegazzini's illustration. The name *G. sutterianum* has also been applied to this plant although the original description of that taxon states that it was from the Sierra de Córdoba. Berger (2003) maintains that *G. sutterianum* is a different plant, which he says grows in San Luis Province, but further to the east. See page 80.

G. fischeri was also found by Piltz and distributed under his field numbers *Piltz* 106 and 106a. Berger (2003) described a minor variant from the western side of the Sierra San Luis as subsp. *suyuquense*. It grows at Suyuque Nuevo, at the same locality as *G. neuhuberi* but this taxon is lower on the hill. Berger claims that it occurs further north than the type.

Commentary

This is one of a number of related taxa from San Luis province which may prove to be one variable species. *G. poeschlii* is similar, growing further to the east and is here treated as a synonym. *G. fischeri* has similarities with *G. capillaense* (sutterianum) which is from the same seed group but grows further north and east within Prov. San Luis but mainly in Córdoba. Wolfgang Papsch (pers. com.) believes that *G. fischeri* is no more than a subspecies of *G. capillaense*.

Cultivation

G. fischeri is easy to grow and produces its large flowers in a range of colours from cream to pink. Most older plants in culture originate from Piltz seed under the numbers P106 or 106a.

Fig.170 *G.fischeri, Piltz* 106a from El Trapiche, Prov. San Luis, Argentina

Gymnocalycium gibbosum subsp. *gibbosum* (Haworth) Pfeiffer ex Mittler

First Description (Haworth)

Synopsis Plantarum Succulentarum: p.173 (1812)

Cactus gibbosus Haworth

"Flattened globular, with less than 16 prominent deep ribs; apex sunken, without spines; ribs have prominent humps with clusters of spines; spines black.

Habitat unknown; Flower unknown. In cultivation before 1808.

I have not seen this nor any other species with so small a number of ribs as ten. Described from Mr. Vere's collection.

I presume that this has not been specifically named by any other author."

Since this description is brief, it is useful to refer to a later one assumed to be of the same species. A few years after this first description of *G. gibbosum* by Haworth, an English author Edwards published a description with a good illustration, (*Botanical Register* 2. pl.137 [1816]). This illustration shows a plant that conforms to the Haworth's description.

Cactus gibbosus

Div. Echinomelocactus, subrotundi.

C. gibbosus, short columnar, with 16 ribs. Spines set in rows of tubercled ribs, the longish tubercles alternate with bare, pressed together humps.

Elongated-globular, a dirty green colour, c. 3" (76mm) high, 2.5" (63.5mm) diameter, with angled ribs. Ribs with swollen humps; humps having a

Fig.171 The illustration of *Cactus gibbosus* Plate 1524 from Loddiges *Botanical Cabinet* (1829)

Fig.172 The lectotype illustration of *Cactus gibbosus*, the Edwards *Botanical Register* illustration, Plate137 (1816).

depressed, round, brown felted areole, which alternates with thinner, naked projecting warts, which are all pressed together. Areoles carry up to 7 very strong, pointed spines. Flowers 3" (76mm) long, come from the topmost areoles with spines; the bottom of the flower-tube is green, towards the petals becoming dirtier and paler, internally white, odourless. Calyx cylindrical, carrying blunt, adpressed, whitish scales, irregularly inserted. Numerous petals, pointed spatulate, narrow at the base, inner petals longer but hardly wider. Ovary 1" (25mm) long. Spines are at first short, erinaceus, the older spines lose their colour.

This description is followed by these observations:

This species is not yet classified in the plant kingdom, nor till now are the flowers known to botanists. It has been cultivated in the greenhouse of Mr. Vere, at Kensington Gore, from before 1808, and flowered in June last year for the first time. This gentleman's gardener informs us, that it was raised from seed received from Jamaica by Messers Lee and Kennedy, of the Hammersmith nursery.

It belongs to the section of the genus Echinomelocactus, or Melon-Thistle, from the form and armature of the species of which it consists. Some of the strange looking plants are said to

Fig.173 Plate 5 of *Echinocactus gibbosus* from Lemaire *Iconographie des Cactées* (1843)

Fig.174 Plate 55 of *Echinocactus gibbosus* from Vol.5 of Gürke *Blühende Kakteen* (1905)

exceed two yards in girth in their native places, and are composed of a succulent green flesh of one consistence throughout. In times of drought they are known to be sought after by the cattle, who after stripping off the spiny covering with their horns, devour them greedily. The entire genus, with the exception of *C. opuntia*, common to both Europe and America, is indigenous to the West Indian Islands, and the warmer parts of the American continent; where it is numerous and multiform species are said to grow from fissures in the sides of the steepest rocks. The fruit (or fig or pear, as it is sometimes called from its shape) is succulent in most of them.

The present specimen, the only one we have seen of this species, was not much more then three inches high, of an oblong cylindrical form, depressed at the summit, somewhat narrowed towards the base, and had 16 angles or ribs.

Each rib or angle consists of a vertical rank of tubercles of two different forms, alternating one with the other, one sort having a depressed tomentose crown armed with a divergent fascicle of about 7 or 8 horny acicular thorns, the other soft and narrow, free from all pubescence and armature, and projecting much beyond the other. The flowers appear from the axils of the uppermost thorny fascicles, which terminate each rib at the outer edge of the depressed thornless tessellated area of the summit of the plant; these in this instant were two, nearly 3 inches in length, externally of a reddish or liver-coloured green, internally white, scentless. Corolla composed of numerous obcuneatly ligulate petals, arranged in several imbricating ranks. Germen about one inch long enclosed within the green cylindrical outwardly loose-scaled persistent calyx.

Lectotype: The Edwards *Botanical Register* illustration; plate 137 (1816). Designated by Kiesling, *Flora of Patagonia*, p.226 1988.(see Fig.172)

Etymology

From the Latin 'gibbous' meaning hump-like swellings referring to the tuberculate plant body.

Other Names

Cactus gibbosus Haworth (1912)
Echinocactus gibbosus (Haworth) DeCandolle (1826)
Cereus gibbosus (Haworth) Pfeiffer (1837)

Synonyms

G. brachypetalum Spegazzini in *Nuevas Notas Cactológicas'Anales de la Sociedad Cientifica de Argentina* 99: pp.135-137 (1925)

First description of *Gymnocalycium brachypetalum* Speg. (n. sp.)

Body globular to cylindrical, dark greenish-blue. Ribs 13 with swollen rounded humps, separated by deep sharp zig-zag grooves. Tubercles three sided, the fronts flattened to rotund. Areoles elliptical, sunken with a little greyish felt. Spines 5-7, round, thin, rigid, more or less spreading, radiating, curved upwards, yellowish when young, later brownish-grey. Flowers longish, narrow, funnel shaped, outside greenish-blue, with scales; ovary narrow cylindrical; petals broad ovate, sharp pointed, pure white; many stamens arranged in two sets, filaments white, anthers ochre; style greenish-white with 12 stigma lobes of the same colour, standing above the anthers.

Habitat:- In the hills and gullies of the Rio Negro, and in the region of Carmen de Patagones.

Observations: Intermediate species or variety between *G. gibbosum* and *G. chubutense*, in its characteristics. Style large with the stigma level with or above the anthers, the outstanding length

Photo: Pat & Sally Craven

Fig.175 *G. gibbosum* in habitat at Peninsula Valdes, growing in sand

of the ovary and perianth tube giving the appearance of being almost pedicel-like. The body is almost cylindrical, 8-10cm high, 6-7cm diameter colour dark bluish-green. Ribs 13, separated by shallow but sharp wavy furrows. Tubercles three sided, 10-12mm long, 15mm diameter, at first flattened, later tooth or humped shape. Areoles elliptical, 6mm long, 3mm wide, somewhat sunken, covered with a very short ash coloured felt.

Spines 5-7 to each areole, all radiating, 5-25mm long, round, rigid, sharp, curved outwards, never bulbous at the base, yellow in youth, ash grey in age. Flowers come from near the shoulder of the plant, 55mm long, straight, outside bluish-green with c. 12 semi-circular scales, distributed in three spiralled rows, gradually increasing in size towards the top. Ovary swollen cylindrical, 20mm long, 9mm diameter.

Perianth tube 20mm long and diameter, interior violet, petals crowded, like slates on a roof, long and wide, 20mm long, 12mm wide, tips pointed, white. Stamens formed into two groups, the lower separated from the top group by a 4mm gap, all filaments white, the fine anthers yellow. Style 22mm long, cylindrical, lower part slightly swollen, green and smooth, white at the top, covered with small papillae,12 white stigma lobes reach up to or above the highest anthers.

Photo: Pat & Sally Craven

Fig.176 *G. gibbosum* in habitat at Peninsula Valdes, growing in sand

G. chubutense (Spegazzini) Spegazzini in *Anales de la Sociedad Cientifica de Argentina* 99: pp.137-138 (1925). Basionym: *Echinocactus gibbosus* var. *chubutensis* Spegazzini (1902)

First description of *Echinocactus gibbosus* D.C. var. *chubutensis* Speg. in *Anales del Museo Nacional de Buenos Aires* 7(Ser. 2, IV): p.285 (1902)

Habitat: frequently found on the dry banks of the Rio Chubut. Summer, 1899-1900.

Observations: Variety of the type, always low growing, colour greyish-blue, fewer spines, flowers remarkably large.

Body often solitary, rarely offsetting, hardly raised above the soil, the underground part is almost conical, (5-10cm long), dingy brown, spineless, with transverse wrinkles. The roots are stringy, (15-30cm long, 5-10mm thick), whitish ochre, ± branched. Above the ground greenish-brown to purplish-blue, flattened on top, almost bare in the centre of the crown. Ribs 12-13, blunt, separated by shallow, not very sharp grooves. Tubercles at first small, sharp, (to medium round), without spines, later becoming larger and more flattened, ± running into each other. Areoles not quite round, 4mm long, 3mm wide, with short greyish felt, often sunken. Spines 5-6 radials, appressed, sometimes one finds a straight central on the higher areoles, at first grey, scaly, becoming darker, 5-20mm long, 0.6-1.0mm thick, bottom of spines not thickened.

Flowers from the older areoles often solitary, 85mm long, 40mm diameter. Ovary club-shaped, 24mm long, 12-14mm

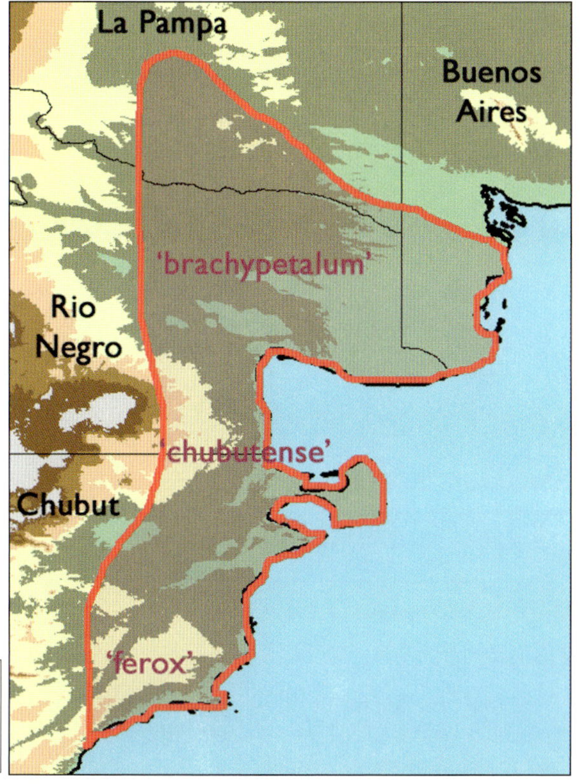

Map 23 Distribution of *G. gibbosum* in southern Argentina

diameter, inside wall violet, outside bare, dark bluish-green; scales c. 12, the lower ones semi-circular, innermost spatulate with a sharp point, higher up becoming spatulate resolving into petals, scales fleshy with transparent pink edges. Petals 6-7 rows, outermost slightly fleshy, dingy reddish-green, innermost spatulate with a sharp point, all petals slightly fleshy, 45mm long, 15mm diameter. Stamens in two rows; filaments 12mm long, white to greenish-white; anthers ochre; style round 25mm long, 3mm thick. Stigma lobes fine, white, c. 10, 5-6mm long. Fruit unknown.

Gymnocalycium gibbosum subsp. *ferox* (Labouret ex Rümpler) Papsch in *Gymnocalycium* 9(4): p.201 (1996)

Gymnocalycium gibbosum subsp. *radekii* Halda & Milt subsp. nov in *Acta musei Richnoviensis* 9(1): pp.63-64 (2002)

Distribution (Map 23)

This is the most southerly species, on the edge of the overall distribution area of the genus *Gymnocalycium*. As a result, it has not been visited by as many enthusiasts and collectors as those habitats further north where several species can be found in close proximity.

Fig.177 *G. gibbosum* in habitat at Caleta Valdes, Peninsula Valdes, Prov. Chubut, Argentina

Photo: Guillermo Rivera

Fig.178 *G. gibbosum* (brachyanthum), *Papsch* 121-170 from Grl. Conesa, Boca de la Trevesia, Prov. Rio Negro, Argentina

G. gibbosum likes the sandy or gravelly alluvial soil along the Río Negro and Río Colorado where it grows under bushes and other plants. This form was described as *G. brachypetalum*. The region has gentle hills, temporary salt lakes and bush-covered slopes with typical Monte vegetation. The distribution area extends westward into Prov. Neuquen and northwards into La Pampa.

Further to the south, the form known as *G. chubutense* can be found. Here, south of San Antonio del Oeste, the ground becomes stony with broken volcanic rock formations and gentle hills like that in Patagonia. In this area can be found a form which grows on basalt corresponding to the variety *ferox*. The most southerly habitats are believed to be north of Comodoro Rivadavia.

Conservation Status

Least Concern. The plant is locally common and has a wide distribution with no immediate threats.

History

This is the first description of a cactus species which would eventually be placed in the genus *Gymnocalycium*. In fact, it is now considered to be the type of the genus and hence the subgenus previously known as *Ovatisemineum* must now be called *Gymnocalycium*.

Haworth (1912) published the name *Cactus gibbosus* and in the same book *C? nobilis* said to be from Mexico. This name was illegitimate since there was a prior use of the name by Wildenow. This taxon was later validly named as *Cactus reductus* by Link in 1822. See page 108.

Four years later, Edwards published a more complete description and fine plate of *Cactus gibbosus* in his *Botanical Register* 2. Plate 137 (1816). This plate was designated as the lectotype by Kiesling in *Flora Patagonia* 5: p.226 (1988). Edwards tells us that the plant was in the collection

Fig.179 *G.gibbosum* (ferox), *Papsch* 49-62 from Dolavon, Prov. Chubut, Argentina

of Mr. Vere at Kensington Gore and was grown from seed received from Jamaica by Messrs. Lee and Kennedy of the Hammersmith nursery. It is possible that this is the same plant that Haworth described in Mr Vere's collection before it flowered for the first time in June 1815.

It is likely that this plant was brought to England from the coastal region near the mouth of the Río Negro or the Río Colorado, now in the far south of Prov. Buenos Aires. It was here, at the end of the 18th Century, that ships would stop to take shelter and collect fresh water. Their journey from Europe would first take them to the Caribbean before following the coast south to Buenos Aires and beyond. Their return journey, the same way, often resulted in plants from Argentina being reported as being from a place en-route, such as Jamaica.

A most useful review of this species was written by Papsch (1996a,b) in which he provides a detailed description and suggests that the type locality is probably near the estuary of the Río Negro at

Fig.180 *G.gibbosum* (brachypetalum), *Piltz* 101 from Choele Choel, Prov. Río Negro, Argentina

Fig.181 *G.gibbosum* (brachypetalum), *Papsch* 37-47 from Barranca de Gualicho, Prov. Río Negro, Argentina

Carmen de Patagones where plants agreeing with the first description can still be found.

Commentary

Many varieties of this species have been described, some of which are actually forms of *G. reductum,* a species which has frequently been confused with this taxon following Salm-Dyck's treatment of it as a synonym of *Echinocactus gibbosus*.

Plants from further north near the Sierra de la Ventana, Prov. Buenos Aires, which here are regarded as *Gymnocalycium reductum,* have been confused with *G. gibbosum* for many years. Belonging to the same seed group, *G. reductum* can be distinguished by its spination, shorter flower tube and body shape, which becomes short cylindric in age.

Cultivation

G. gibbosum is an impressive plant to grow with its dark body and often black spines. The large white flowers are also magnificent and are easily produced on quite small plants. It presents no difficulty in cultivation and quickly grows into an impressive plant. It is able to withstand low temperatures in winter.

It should be noted that many plants in collections labelled *G. gibbosum* are in fact *G. reductum*, for instance seedlings of *Piltz* 94 were distributed as *G. gibbosum* but are really *G. reductum*.

Gymnocalycium gibbosum subsp. *borthii* (Koop ex H.Till) Charles

First Description (Koop ex H. Till)

Gymnocalycium borthii Koop nom. inval. KuaS 27(2): pp.25-27 (1976). Validated by H.Till in KuaS 38(8): p.191 (1987)

Gymnocalycium borthii Koop spec. nov.

Body simple flattened to elongated globular, grey-green to violet-brown, up to 90mm diameter, and 100mm high, with a very tough branch-like root. Ribs (9-) 16, running vertically downwards, yet they often appear to run into uneven humps, up to 8mm high and 15mm wide. Humps and protruding chins mostly rounded, very rarely with a sharp edge. Areoles slightly sunken, to c. 20mm apart, oval, up to 5mm long, 3mm wide, covered with yellow-grey felt. Radial spines; mostly 5 (-7, very rarely 9), central spines absent, elastic, flexible to stiff, awl shaped, straight, (Very rarely curved), radiating ± diagonally away from the body, one pointing down and one pointing up, not always but sometimes longer than the rest, all spines variable in colour, from white to yellow to grey, at the base dark grey to brown.

Flowers come from the periphery of the crown, funnel shaped, 43-46mm high, 48-52mm in diameter, white with a pink throat, outer petals up to 20mm long, 5mm wide, lanceolate, with a brownish tint; inner petals up to 26mm long and 6mm wide, lanceolate, sharp pointed, white to shiny white, all pink at the base, (the form of the outer petals that also have pale edges is particularly prominent on the buds just before opening, they appear to be slender and pointed).

Pericarpel c. 20mm long, 8mm max. diameter, outside dark grey-green to violet brown, covered with many light ± wide round scales. The tube of the receptacle that is lighter in colour, carries large, less blunt scales and is up to 12mm long, the inner wall is magenta-pink. The stamens are inserted over the whole of the receptacle, the double ring of primary stamens lean in towards the style, the length of the stamens range from, the primary 6mm, and up to 9mm for the rest, filaments cotton-like, whitish to yellow, pale pink at the base, anthers intensive yellow, oval, c. 1mm long, 0.5-0.6mm wide. Style without the stigma c. 12mm long, 1.6-1.8mm diameter at the base, 1.2-1.4mm at the top, greenish at the base, yellowish at the top, stigma lobes 10-13, light yellow, cylindrical, the blunt ended lobes reach over the top of the anthers. Nectar chamber, wide funnel shaped, up to 1.6mm high, 2.5mm diameter, at the top. Ovary elongated, max. height 12mm, 4.5mm diameter, completely but loosely filled with the seed structures. Fruit spindle to elongated to barrel shaped, max. diameter 15mm, the colour has

Fig.182 *G. gibbosum borthii,* GC938.02 from Los Chañares, east of Quines, Prov. San Luis, Argentina

Fig.183 The habitat of *G. gibbosum borthii,* GC938.02 from Los Chañares, east of Quines, Prov. San Luis, Argentina

Fig.184 *G. gibbosum borthii,* GC938.02 from Los Chañares, east of Quines, Prov. San Luis, Argentina

a strong similarity to the plant body, but rather darker, the fruit splits length-ways, fruit flesh whitish, rather dry.

Seeds: 0.9-1.1mm long, the same width, egg shape, with a mud-black testa, testa with fine warts. Hilum; oval, pointed, separated sharply from the testa but without a swelling, sunken. (Seed from the type is similar to the Baldianum sub-type of Buxbaum and Frank). Published in KuaS. 2/1971/ p. 28 (often with arillus).

Habitat: Near Quines, Argentina. 600-800m.

According to the material found, about 40 plants, one must say that *G. borthii*, especially in some details is very variable, also in culture, yet in spite of this, the hardly different characteristics, colouring and the general habitat impression, even to the novice, appears unusually singular, *G. borthii* is surely related to *G. gibbosum*, yet because of its wide distribution in habitat and its different strong characteristics, I believe this to be a good species. I name this very attractive species *G. borthii* after the Vienna collector Hans Borth, who keeps this species under the collection number BO55.

Photo: Massimo Meregalli

Fig.185 *G. gibbosum borthii* (nogolense), *Meregalli* 629, Rte.147 near Mina la Calera, prov. San Luis, Argentina

Fig.186 *G.gibbosum borthii, Neuhuber* 89 157-413 from Bajo de los Pejes, Prov. San Luis, Argentina

The Holotype was presented to the protection and type collection of the Botanical Garden Linz. Austria.

Etymology

Named for Hans Borth, Austrian plant collector.

Other Names

Gymnocalycium borthii Koop ex Till in KuaS 38(8): p.191 (1987)

Synonym

Gymnocalycium borthii subsp. *nogolense* Neuhuber in *Gymnocalycium* 20(1): p.701 (2007)

Distribution (Map 24)

Reported localities of this plant show the distribution area as both sides of the Sierra San Luis from Quines in the Sierra de Varela in the south. The type locality is believed to lie to the east of Quines, near to Los Chañares.

From the west of the distribution area, and west of the Sierra San Luis, in relatively flat land, Neuhuber (2007) published his subsp. *nogalense* quoting only minor differences. He points out that

Photo: Massimo Meregalli

Fig.187 *G. gibbosum borthii* (nogolense), *Meregalli* 982 in habitat at Macho Muerto, Prov. San Luis, Argentina

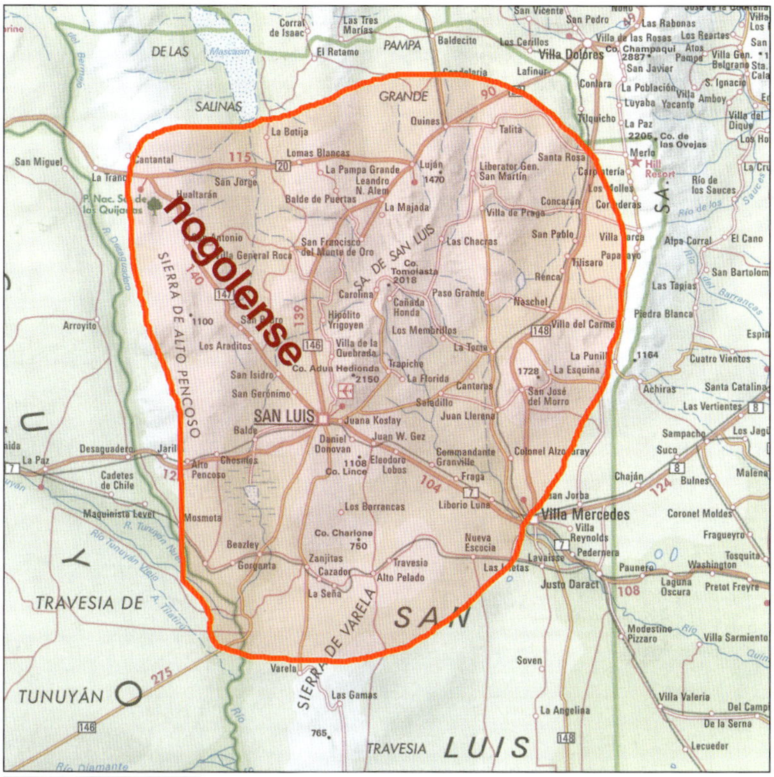

Map 24
Distribution of *G. gibbosum borthii* in Prov. San Luis, Argentina

plants of this form occur in the flat land at the foot of the Sierra de las Quijadas and the Sierra del Gigante from where Meregalli and Guglielmonte (2006) described their *G. striglianum* subsp. *aeneum*. He makes the point that these authors included illustrations of his *G. borthii* subsp. *nogalense* in their article (Figs. 10 to 13) considering them, in error, to be their new subspecies of *G. striglianum*.

Conservation Status

Least Concern. Although the known populations are small, reports suggest that this taxon is widespread and not under immediate threat.

History

The name first appeared with the description reproduced above in 1976 but it was invalid because no type was preserved. The situation was corrected by Till (1987) when he preserved a plant of BO55 as the type and validated the name.

I made it a subspecies of *G. gibbosum* in CSI 20: p.18 (2005) from which it differs by its smaller size, more tuberculate ribs and a disjunct distribution, separated by some 380km from the type.

Commentary

This taxon has remained little known since its description. Field number lists of *Gymnocalycium* specialists have various localities for this taxon including some from much further south than the type locality. It is related to *G. striglianum* which occurs further to the west.

The representatives of the subgenus *Gymnocalycium* in San Luis Province of Argentina

Fig.188 *G.gibbosum borthii, Bercht* 301 from San José de Morro, Prov. San Luis, Argentina

Fig.189 *G.gibbosum borthii, Papsch* 89 63 78 from Zanjitas, Prov. San Luis, Argentina

remain confusing. In his article about this taxon, Neuhuber (2007) endeavours to distinguish it from other similar plants in the region such as *G. striglianum* and *G. taningaense* forms.

Cultivation

In culture, this plant is like a small version of *G. gibbosum*. It is easy to grow and flower but remains rare in collections. This may be because, until recently, seed or plants have rarely been offered for sale.

Illustrations

Cactus & Co. X(2): pp.77-78 (2006) Figs 10 -14

New Cactus Lexicon Atlas (2006) Illn. 272.3

Fig.190 The habitat of *G. kieslingii* GC984.03 at the Cuesta de la Cebila, on the border of Prov. La Rioja and Catamarca, Argentina.

First Description (Ferrari)

CSJ(US) 57(4): pp.244-246 (1985)

Gymnocalycium kieslingii sp. nov.

Central root napiform. Stems globose-depressed, grey-green, with the underground portion obconic, simple, 6-9cm in diameter, and 2cm high from the soil level, with the apex sunken. Ribs 12 (9-13), low, with 5-6 tubercles of 0.7-0.8cm in height and 1.4-1.8cm in width at the base, separated completely or incompletely by a transverse furrow, with a prominent chin in a lower half. Areoles at the upper edge of the tubercles, more or less circular, large, nearly 3mm wide and high when young, with much caducous wool, becoming smaller with age. Spines 5-7 (-9), adpressed, all radial, slightly arching to the stem, 0.5-0.8cm long, whitish with pink base.

Flowers campanulate-funnelform, 5.5-6.5cm long and 5-6cm in diameter. Pericarpel conical, 2.5-3cm long and 0.7-1.2cm diameter, pale glaucous green with about 15-17 subcircular, whitish or pink scales of nearly 5mm diameter. Receptacle conical, wide, 1.5cm long, 1.2-1.7cm in diameter, with about 8 scales, larger than those of the pericarp, with transition to the perianth segments. The receptacle may be pink to carmine red inside. Perianth white, slightly tinged green only on the larger segments; tepals 25-30, the outer ones largest, spatulate, nearly 2cm long and 0.7cm wide, grading to the inner ones which are spatulate-lanceolate, 1.1cm long and 0.4cm wide. Stamens in 3 series: the external ones around the border of the receptacle with a single row nearly 1cm long, the middle ones numerous, spirally covering the internal wall of the receptacle, 0.6-1cm long, and the inner ones composing a single row curved in towards the style, 0.8cm long; filaments and anthers yellow-cream. Style cylindrical, 1cm long and 1.5-2mm in diameter, green with a pink base, ending in a stigma composed of 7-8 subcylindric, 3mm long, yellow lobes. Ovary truncate-obconic, 1.5-1.7cm long and 3-5mm in diameter.

Fruits fusiform, narrow at the base, about 3.5cm long and 1cm in diameter, with persistent dry perianth. Seeds truncated globose with verrucose surface, cuticle particularly rugose and non-sticky, brown in appearance; hilum large, sub-rhomboid, somewhat sunken.

Habitat: in sandy soil, under bushes with a predominance of Larrea spp (creosote bush), along with *Trichocereus cabrerae*, *T. terscheckii*, *T. strigosus*, *Opuntia sulphurea* and *Gymnocalycium mazanense*. Type locality and distribution: Argentina, La Rioja, Dept. Arauco, Cuesta de la Cebila, El Alto, O. *Ferrari 19/1980* (SI holotype). Also Dpto. Arauco. Sanagasta, Cuesta de Huaco, O. *Ferrari 34/1982* (SI, isotype).

Etymology

Named for Dr. Roberto Kiesling, Argentinian botanist and cactus specialist.

Synonyms

Gymnocalycium frankianum Rausch ex H.Till & Amerhauser in *Gymnocalycium* 20(3): p.730 (2007)

First Description of *Gymnocalycium frankianum*

Body flattened-globular with a slightly sunken crown, dark grey-green to dark blue-green; roots strongly branched. Ribs 9-11(13), straight, running straight down, few raised and slightly angled, during the rainy season almost flat. Spines 5-7, radiating, lying against the body, fine bristle-like, unequal, 2-5.5mm long, dirty white, occasionally grey at the base.

Flowers coming from near the crown, thin, with pointed buds, 50-55mm long, at full anthesis 40-45mm diam. first, at full anthesis they are fully open with ± a dark magenta throat. Ovary narrow funnel-shaped, somewhat shorter than the petals in the closed condition,. Petals 2-3mm wide at the base, becoming thicker

Fig.191 *G.kieslingii* GC984.03 at the Cuesta de la Cebila, on the border of Prov. La Rioja and Catamarca, Argentina

Fig.192 *G.kieslingii* GC984.03 growing with *G. hossei* (centre) at the Cuesta de la Cebila, on the border of Prov. La Rioja and Catamarca, Argentina

towards the top, 5-6mm wide, mostly dark green with a light edge, resolving into the outer petals, these are pointed lanceolate , 13-16mm long, up to 5mm wide, greenish-white with a darker point, the innermost petals, mostly half as long as the outers, are white, often long covering over the dark throat. Stamens thin, white, with a dark base. The primary stamens surround the style enclosing the nectar chamber, the secondary stamens are regularly inserted on the inner wall of the tube overhanging the style. Style thin, greenish-white, with a small yellowish stigma.

Fruit short spindle-shaped, splitting lengthwise. seeds globular, about 1.5mm diam, only with a thin layer of skin.

Type: Argentina, Prov. Santiago del Estero, Sierra de Guasayan, leg. W. Rausch R 722, CORD (Holotypus).

Commentary on *G. frankianum*

The description is based on a collection of Rausch but the accompanying illustrations show such diversity that one wonders if they really are all the same taxon. It comes from a region that is rarely visited by cactus enthusiasts, further to the east than the well-known localities.

G. rosae H. Till in *Gymnocalycium* 8(3): pp.159-163 (1995)

First description of *Gymnocalycium rosae*

Body flattened-globular, with a slightly sunken crown, 42-68mm high, 55-80mm diameter, 1/3 - 1/2 of the body is under the soil. Tap roots mostly strong, stake-like with many secondary roots. Epidermis dark olive green, finely granulated, not shiny. Ribs 7-8, straight, wide, formed by deep vertical furrows, (especially in the upper part of the body), rounded and separated into wide chin-like humps by ± deep cross-cuts. The humps are very prominent, and on

Fig.193 *G.kieslingii, Muhr* B63 , Prov Catamarca, Argentina

older plants reach down to the base, where they are so enormous that they almost reach the areole below. However on younger plants the humps tend to flatten near the base of the plant, they also become wider, 22-30mm wide. Areoles oval to round, 4x3 - 3x3mm diameter, sunken into shallow indents, in the crown covered with yellowish-white felt, soon becoming grey, and then disappearing. More than a third of the areoles are spineless.

Spines 5-7, radiating fan-like, all thin, in youth flexible, also often brittle, 2-3 pairs pointing sideways, 8-10mm long, one pointing downwards, always shorter, 5-6mm long, yellow, red or brown at the base. In age damaged, therefore older plants are often spineless.

Flowers funnel-shaped, 60mm long, 52mm diameter at full anthesis, white with a red throat. Pericarp strong, conical, 20mm long, 6mm diameter near the top, 13mm above, dark olive green, with white edged, pink at the tip, rounded scales, which towards the top, gradually resolve into the outer petals. Outer petals spatulate, 20-31mm long, 9mm wide, the lowest olive green with a white edge, the middle ones are greenish-white, the top ones, which are the longest, are white, only at the tips are they olive-green. Inner petals wide lanceolate, 28mm long, 9mm wide, innermost petals becoming smaller, c. 21mm long, 6mm wide, pure white to ivory white. Filaments pale green, inserted in one group over the pink wall of the flower tube, slightly curving over the style, the lowest and those that lie onto the style 5mm long, the higher ones 7mm long. Anthers yellow, round 1.5mm diameter, pollen yellow. Style middle standing (i.e. it reaches to the middle of the anthers), without the stigma 11mm long, 2mm thick, green at the base, becoming white towards the top. Stigma, in proportion to the style, small, 5mm long with 10 yellowish lobes. Ovary longish heart shaped, 14mm long, 6mm wide at the top, white walled.

Fruit elliptical, 25mm long, 11mm diameter, splitting lengthwise. Seeds elongated globular, 1.5mm high, 1.2mm thick, with ample dried covering. Testa weakly humped, black. Hilum-micropyle region small, round with a slightly swollen rim, which is leaning in, without elaiosome, micropyle elevated.

Photo: Massimo Meregalli

Fig.194 *G. kieslingii* (frankianum), *Meregalli* 912 in the Sierra de Guasayan, Prov. Santiago del Estero, Argentina

Fig.195 The habitat of *G.kieslingii* (castaneum) GC976.02 north of Huaco, Prov. La Rioja, Argentina

Type: Argentina, Province Salta, Department La Candelaria, bush covered hills near El Brete, 850m, 22nd Feb. 1993, collected H.Till 93-639 (Holotype WU).

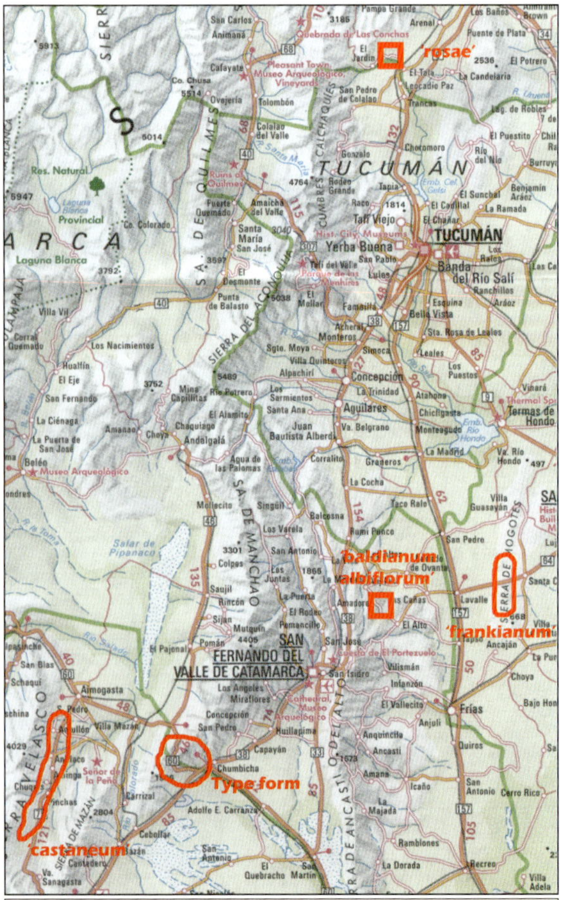

Map 25
Locations of the forms of *G. kieslingii* in Argentina

Commentary on *G. rosae*

Till (1995) found this plant while looking for the place where a herbarium specimen with red flowers, identified as *G. baldianum,* had been collected by Schreiter in 1931. The location was surprisingly far north for this taxon, which is known from 200km further south, and would represent a significant northerly extension of the subgenus *Gymnocalycium*. At a lower altitude, and to the east of Schreiter's locality, Till found plants of subgenus *Gymnocalycium* but later in cultivation they flowered with a white flower. He described them as a new species, *G. rosae*, which because of their overall similarity, is here referred to *G. kieslingii*.

Distribution (Map 25)

G. kieslingii was originally known from just a few localities in Prov. La Rioja, Argentina. The type

Fig.196 *G. kieslingii* (baldianum albiflorum), *Neuhuber* 91 451-1440 Sierra Ancasti, Prov. Catamarca Argentina

Fig.197 *G. kieslingii* (frankianum), *Neuhuber* 91 445-1425, Santa Catalina, Sierra de Guasayan, Prov. Santiago del Estero, Argentina

Fig.198 *G. kieslingii* (castaneum), GC976.02 in habitat north of Huaco, Prov. La Rioja, Argentina

Fig.199 *G. kieslingii* (castaneum), GC215.02 from the Cuesta de Huaco, Prov. La Rioja, Argentina

locality on the Cuesta de la Cebila is a pass through the Sierra Ambato. The forms *castaneum* and *albiareolatum* were described from the east side of the Sierra Velasco, at the foot of the mountains north of Villa Sanagasta. The plants there grow on gentle slopes of crumbling granite usually under bushes which are often of the genus *Larrea*.

Till (1990a) also reports seeing larger plants which might be this species in the Sierra Los Colorados further to the south.

Further east, on the Sierra de Ancasti one can find plants greatly resembling *G. kieslingii*, including Bercht's *G. baldianum* var. *albiflorum*. Then the furthest north locality occurs just in Prov. Salta from where Till (1995) described *G. rosae*, a form which grows amongst trees and has a greener body.

Neuhuber & Till (1999) consider that *G. rosae* is a synonym of *G. baldianum* var. *albiflorum*, both of which are treated as forms of *G. kieslingii* here. Neuhuber described *Gymnocalycium* X *hiediae* for hybrids between this and *G. baldianum*.

The last range of hills to the east of this region is the Sierra de Guasayan in Prov. Santiago del Estero, the type locality for the recently described *G. frankianum*. (shown as Sierra de Mogotes on the map opposite)

Conservation Status

Least Concern. The populations are large and under no major threat. Road improvement and agriculture has destroyed many plants but overall their future looks secure.

History

Although this plant grows near to main roads, it was not described until 1985 in the US journal. The first plants I saw in cultivation were sent, before it was described, by Frau Muhr under her number B63 collected in Prov. Catamarca. From their appearance, they would appear to have been collected in the vicinity of the type locality where I have only recently been able to find it.

Commentary

A distinct species which has an uncomplicated history. The unfortunately named fa. *albiareolatum* has led to confusion with *Gymnocalycium albiareolatum*, a different species that belongs to the seed group Microsemineum. The two taxa look alike and can both be found near Villa Sanagasta. There are no reports of them growing together, rather that there is an altitude gap with *G. albiareolatum* below 1000m and *G. kieslingii* forms above 1500m.

The type form of *G. kieslingii* grows larger than the fa. *castaneum* and has a grey, rather than bronze epidermis. The latter has also been confused with *G. albiareolatum* Rausch, a plant growing in the same area from the subgenus Microsemineum. This confusion is compounded by the description of *G. kieslingii* fa. *albiareolatum* Ferrari (1985) from the same area.

The type locality of the species is on the Cuesta de la Cebila [sometimes spelt Sebila] which is a pass through the Sierra de Ambato. The plant is not common there and can be difficult to find. Its two forms, described at the same time as the type are really very different and may warrant a higher rank.

Its nearest relative is probably *G. robustum* from north-east Prov. Córdoba. There are plants in that region and in nearby Prov. Santiago del Estero which have characteristics intermediate between the two species and are difficult to assign with certainty.

Cultivation

Although described more than 20 years ago, this species is not often seen in collections. The forma *castaneum* is probably more common than the type. All the forms are quite slow growing so a large well-grown plant should do well at shows, particularly in a restricted pot size class.

Gymnocalycium kroenleinii Kiesling, Rausch & Ferrari

First Description (Kiesling et al.)

KuaS 51(12): pp.315-318 (2000)

Gymnocalycium kroenleinii spec. nov.

Simple, flattened-globular, with a turnip-like root, light grey-green with a brownish-pink tint. Ribs 8-10, up to 6mm thick, separated into 7mm long chin-like humps by horizontal furrows. Areoles round, 3-4mm diameter, white felted. Radial spines in (1-)3-4 pairs and one pointing downwards, radiating to porrect, up to 17mm long, central spines 0-1, these up to 20mm long, all spines awl-shaped and irregularly curved, brownish-grey with a brown base.

Flowers coming from near the crown, 3cm long and wide, pericarpel skittle-shaped, 7mm long, 5mm broad. Receptacle green or pink, with wide, rounded, whitish-pink scales. Outer petals round, whitish-pink with a brown-green mid-stripe, inner petals spatulate, pink with a darker central stripe, or whitish-cream with a pink base. Throat pink, stamens pale pink, style and the stigma lobes (7) white.

Fruit, globular, flattened above and below, 12mm thick, violet or pinkish-grey with light pink scales, splitting vertically. Seeds pot-shaped with a large hilum, dull black and with a loose cuticular covering, 1.2-1.4mm long, 1-1.2mm thick.

Distribution area: Argentina, Prov. La Rioja, in the mountains of Sierra Malanzán, 1300m.

Holotype: Argentina, Prov. La Rioja; Sierra de Malanzán, 1300m, XI-1987, R. Kiesling, O. Ferrari & W. Rausch, ex cult. La Plata (KF 92) (prep. XI-1999) sub R. Kiesling & O. Ferrari 9304 (SI; offset and flower).

Photo: Massimo Meregalli

Fig.200 The habitat of *G. kroenleinii, Meregalli* 977 in the Sierra Quintana, Q. Cóndores, near Malanzán, Prov. La Rioja, Argentina.

Paratype: Argentina, Prov. La Rioja. Sierra de Malanzán, 1300m, 25-XI-1987, R. Kiesling, O. Ferrari and W. Rausch 6786 (SI; only fruit and seeds).

This is a very characteristic species, which is different from all others of this subgenus, distinct by its prominent humps and round fruit, also there are the pink coloured flowers, seldom seen in this group. Only *Gymnocalycium bruchii* (Speg.) Hoss., and the near related *Gymnocalycium andreae* have round plump or flattened round fruit. By way of these characteristics, this species is certainly different.

Chin-like humps are typical of many *Gymnocalycium* species, but in the foregoing case, the humps are very characteristic and distinct; no other species of this sub-genus resembles *G. kroenleinii* in this characteristic.

The flowers are mostly white or whitish, sometimes red or yellow. One also finds deep pink-

Photo: Massimo Meregalli

Fig.201 *G. kroenleinii ,Meregalli* 977 in the Sierra Quintara, Q. Cóndores, near Malanzán, Prov. La Rioja, Argentina

Photo: Massimo Meregalli

Fig.202 *G. kroenleinii ,Meregalli* 977 in the Sierra Quintara, Q. Cóndores, near Malanzán, Prov. La Rioja, Argentina

Base Map © Nelles Verlag GmbH

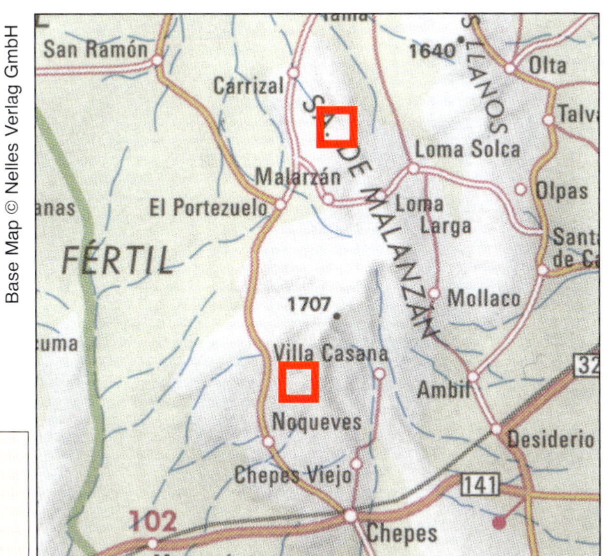

Map 26 Locations of *G. kroenleinii* in Prov. La Rioja, Argentina

(to cream-) coloured flowers on some species within this sub-genus; i.e. on *Gymnocalycium parvulum* (Speg.) Speg. and *Gymnocalycium uebelmannianum* Rausch.

Etymology

Named for Marcel Kroenlein (1928-1994), former director of the Jardin Exotique, Monaco.

Distribution (Map 26)

The type locality is in the south of the Sierra Malanzán but Meregalli (pers. com.) has found a population further north. It grows in Monte vegetation, among granitic rocks with a pinkish colour, outcrops of which occur in this region.

Conservation Status

Data Deficient. The extent of this species in the wild is not known.

History

Discovered in November 1987, this taxon was first thought by the authors to belong to the *Trichomosemineum* seed group, perhaps related to *G. intertextum*. However, examination of the seed showed that it was actually part of the subgenus *Gymnocalycium*.

Photo: Detlev Metzing

Fig.204 *G. kroenleinii, Metzing s.n.* from near the type locality the Sierra de Malanzán, Prov. La Rioja, Argentina

Commentary

This is an interesting recent find because it is the first report of a member of the subgenus *Gymnocalycium* from this mountain range. It could be widespread in these mountains but there are few roads at the right altitude so exploration of large areas would be difficult to undertake. It may have eluded discovery before because it grows only on a particular localised rock type, the pink granite mentioned in the description.

Cultivation

There are few plants in cultivation yet but three seedlings given to me by Omar Ferrari have not presented any difficulty in cultivation and they flowered when only 20mm in diameter. The flowers were a dirty white, unfortunately not glorious pink like one plant pictured in the original article.

Illustrations

New Cactus Lexicon Atlas (2006) Illn. 273.1

Fig.203 *G. kroenleinii ,Ferrari s.n.* from the type locality in the Sierra de Malanzán, Prov. La Rioja, Argentina

Fig.205 *G. kroenleinii ,Ferrari s.n.* from the type locality in the Sierra de Malanzán, Prov. La Rioja, Argentina

Gymnocalycium neuhuberi H. Till & W. Till

First Description (H. Till & W. Till)

Gymnocalycium 5(1): pp.59-60 (1992)

Gymnocalycium neuhuberi H. Till & W. Till, spec. nov.

Body simple, flattened globular, up to 48mm high, 70mm diameter, mostly with an armed, slightly sunken crown. Roots ramified. Epidermis grey-green, not smooth, granular. Ribs 9-12, formed by wavy vertical furrows, separated into 5 sided wart-like humps with blunt chins by deep cross-cuts, humps at the base of the plant flattened, often wider than high (15-20mm wide). Areoles round to oval, c. 6-7mm in diameter, slightly sunken, 15-20mm apart, bald, only in the crown have the areoles wool-felt. Spines awl-shaped, stiff, mostly thicker at the base, radiating, straight or slightly curved, mostly 9 radial spines, the lowest spines 15-20mm long, the upper two pairs shorter, 11-12mm long, often 1-3 bristle-like spines are found on older plants at the upper edge of the areoles. Young plants have 1 central spine, older plants have 4 central spines, forming a cross, somewhat longer than the radial spines, spines mostly yellow, the same colour at the base, occasionally darker (orange or brown). Buds appear at the upper edges of the areoles where the spines are pointing down and are irregular.

Flowers from the areoles near the crown, funnel shaped, 27-36mm long, 26-34mm diameter, sometimes larger, lilac coloured, opening (in cultivation) at midday. Pericarpel short weak conical shaped, 6-7mm long, 9-10mm in diameter, light green, shiny, with little wide scales, rounded with a little point, that are light edged with brownish tips, the upper scales a little longer, becoming the outer petals. Outer petals spatulate to long lanceolate, up to 21mm long, 6mm wide, lilac, with a relatively wide green mid-stripe, inner petals lanceolate towards the centre becoming smaller (from 20mm long, 5mm wide to 13mm long, 4mm wide), lilac, with a darker mid-stripe. Stamens in 2 groups, the 2 bottom rows lean onto the style, the upper stamens are inserted in the wall of the tube; Filaments thin, white, 5-6mm long; anthers yellow round, c. 0.8mm in diameter, pollen light yellow. Style thin, white, 12-14mm long, 1.5mm in diameter, Stigma claw-like, with 9, cream-white, thin 3-4mm long stigma lobes, overhanging the higher anthers. Nectar chamber 3.5mm wide, 1-2mm high, enclosed by the lower rows of anthers. Ovary urn-shaped, c. 5x5mm, white walled, filled

Fig.206 *G.neuhuberi, Neuhuber* 77-1181 from the type locality at Suyuque Nuevo, Prov. San Luis, Argentina

Fig.207 The type locality of *G. neuhuberi* at Suyuque Nuevo on the western slopes of the Sierra San Luis, Prov. San Luis, Argentina

Fig.208 *G.neuhuberi, Neuhuber* 77-1181 from the type locality at Suyuque Nuevo, Prov. San Luis, Argentina

with funiculae. Fruit inverted egg-shaped, 15-18mm high, 10-20mm in diameter, dark green, with flower remains, dehiscent lengthwise in one or two places. Seeds 1.2mm high, c. 1.0mm in diameter, pot-shaped, with small raised testa cells, the large flat almost round hilum-micropyle region black, the seeds have a dry loose skin.

Habitat: Argentina, Prov. San Luis, in the South of the Sierra de San Luis, 1300m.

Holotype: Argentina, Province de San Luis, in the south of the Sierra de San Luis, 1300m leg. G. Neuhuber GN88-77/179, December 1988(WU)

Isotype: GN88-77/189, flowers and fruits (WU)

Etymology

Named after Gert Neuhuber, Austrian *Gymnocalycium* specialist, co-founder and former president of the Austrian specialist group 'Arbeitsgruppe Gymnocalycium'.

Distribution (Map 27)

The type locality, a small village Suyuque Nuevo on the south-west of the Sierra San Luis was not mentioned in the original description but has

Fig.209 *G. neuhuberi, Schädlich* 191b. A young plant at the type locality, Suyuque Nuevo, Prov. San Luis, Argentina.

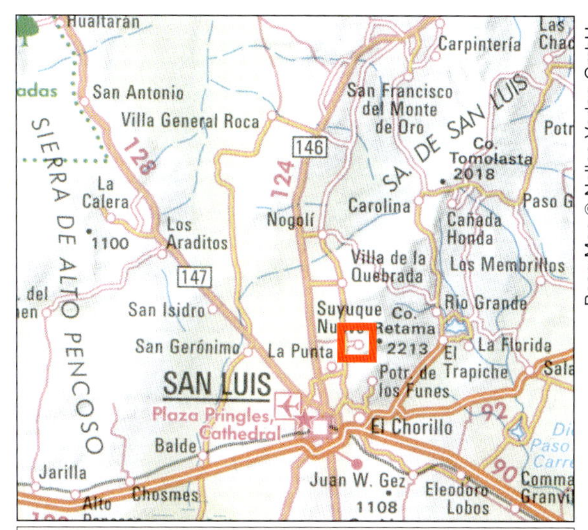

Map 27
Location of *G. neuhuberi* in Prov. San Luis, Argentina

become well known and is often visited by *Gymnocalycium* enthusiasts.

This species is known only from the type locality which would make it one of the most localised *Gymnocalycium* species. The place is described as a patch of barren granite and scree above the tree line. It may occur at a similar altitude elsewhere in the Sierra San Luis but access is difficult.

Conservation Status

Vulnerable. The original locality is described as just 3 by 2km, although the number of plants in this area is not known. Regrettably, it is likely that plants continue to be illegally collected from the population even though the plant can easily be grown from seed.

History

Discovered in 1988, the first description appeared in 1992. Since then little has been published about this remarkable plant.

Commentary

When *G. neuhuberi* was first found, there was speculation that it was related to the widespread *G. monvillei,* perhaps because of its pink flowers. However, its *Gymnocalycium* seed type rules out this suggestion. In fact, it grows close to the local form of *G. monvillei* and sympatrically with a form of *G. fischeri,* also from subgenus *Gymnocalycium*, which was named as *G. fischeri* subsp. *suyuquense* by Berger (2003).

Cultivation

G. neuhuberi is probably the best new *Gymnocalycium* species from the point of view of the grower. It flourishes in cultivation and flowers while still small. It has dense spination and spectacular pink, perfumed flowers which are produced in profusion early in the year. The spination varies from light to dark brown.

Gymnocalycium reductum subsp. *reductum* (Link) Pfeiffer ex Mittler

First Description (Link)

Cactus reductus Link in *Enumeratio plantarum horti regii botanici berolinensis altera* 2:21 (1822)

This first valid description of the name refers to Haworth's illegitimate name *Cactus nobilis* with just a brief Latin description:

"Short straight spines"

A more complete description is given by Pfeiffer (1837) *Beschreibung und Synonymik der in deutschen Gärten lebend vorkommenden Cacteen* under the synonym *Cereus reductus* DC:

Mexico, Guatamala.[an error, perhaps reflecting where the ship had last called on its way to Europe]

Stem thick, brownish-green, naked on the crown, only humps. Ribs flattened, with cone-shaped blunt tubercles. Areoles between the tubercles, wide, with felt. Spines awl-shaped, very stiff, brown, tips horn-coloured, radials 7-9, radiating, (the top spines the shortest), centrals, 1, upright, seldom absent.

In the Berlin Garden, more than a foot high, 4" (101mm) diameter. Areoles c. 4-5" (101-127mm) apart. Central spines 1 1/4" (30mm) long, the rest somewhat shorter.

Etymology

From the Latin meaning 'remote'

Other Names

Cactus reductus Link
Cereus reductus (Link) DeCandolle (1828)
Echinocactus reductus Schumann (1895)

Synonyms

Cactus nobilis Haworth (1812) non Miller ex Linné (1767) nom. illeg.
Echinocactus gibbosus var. *nobilis* Lemaire (1839)

G. mackieanum (Hooker) Metzing, Meregalli & Kiesling in *Allionia* 33: p.209 (1995). Basionym: *Echinocactus mackieanus* Hooker (1837)

First description of *Echinocactus mackieanus* Hooker in *Curtis's Botanical Magazine* plate 3561

General characteristics

Sepals numerous, overlapping, elliptical, attached to the flower tube at the base, enveloping inner petals. Stamens numerous. Style thread-like, apex divided into numerous lobes. Fruit carrying flower remains with many scales. Cotyledon none? Fruit fleshy, ovate or globular. Body ribbed, without leaves. Ribs, as if formed by tubercles. Dorsal spines growing in clusters. Cephalium or spadix absent. Flowers coming from the clusters of spines near the apex of the ribs. Flowers waxy, hardly with a tube.

Specific characteristics

ECHINOCACTUS *mackieanus*: Obovate, mammillose. Tubercles large, flattened-conical, arranged in 16-17 almost regular ribs, apex woolly. Spines 8-10, long, thin spreading, brown at the tips. Flowers white, pink at the tips

Fig.210 Plate 4443 of *Cactus reductus* from *Curtis's Botanical Magazine* (1849)

Fig.211 Plate 12 of *Gymnocalycium reductum* from Pfeiffer *Abbildung und Beschreibung Blühender Cacteen* Vol.II (1846-50)

Fig.212 Plate 3561 of *Echinocactus mackieanus* from *Curtis's Botanical Magazine* (1837)

Commentary on E. mackieanus (Fig.212)

This taxon was described from a plant in the collection of Messrs. Mackie of the Lakenham Nursery which was said to originate in Mr. Hichin's collection where it was derived from Kew. The suggestion that it came from Chili [Chile] made authors reticent to accept it as a species of *Gymnocalycium* but it was placed there by Metzing et al. (1995) who designated the original illustration as the lectotype.

Papsch (2000) tells us that plants matching the description can be found on the west side of the Sierra de la Ventana, Argentina, and he equates it to *G. reductum* subsp. *reductum*.

Gymnocalycium gibbosum subsp. *ferdinandii* Halda, Malina & Milt subsp. nov in *Acta musei Richnoviensis* 9(1): p.61 (2002)

Gymnocalycium gibbosum subsp. *gastonii* Halda & Milt subsp. nov in *Acta musei Richnoviensis* 9(1): p.62 (2002)

G. sibalii Halda & Milt in *Acta musei Richnoviensis Sect natur*. 13(1): 7 (2006)

First description of *Gymnocalycium sibalii*

Taproot napiform; stem single, dwarf, discoid, partly submerged, 50-70mm diam.; ribs 10-12, obtuse, mostly more or less crenate; tubercles to 10mm long and 15mm wide, angular, glaucous or purplish; areoles oblong, to 10mm long, white woolly (later greyish); spines 7-9 rigid, needle-like, to 20mm long, greyish to brownish, irregularly recurved; flowers white with a reddish centre, to 10mm diam, 75mm long; c. 20 petals to 12mm wide and 75mm long, mucronate, with a pale brown central stripe; scales brownish-olive with a paler margin; pericarp up to 25mm long; style to 30mm long; stigma yellowish; fruit spindle-shaped, brownish-pink to brownish-violet, 35mm long, 12mm diam, glabrous, scales pinkish, thick, flexible, white edged. Seeds subglobose, bottom truncate, 0.7mm long, 0.6mm across, testa brownish-yellow, shallowly tuberculate, hylum sunken.

Distribution: On the low schist hills in the Sierra Chica, 1000m ASL.

Type: Herbarium Haldianum no. 061135; Hortus miltianus e seminibus enatus JPR77/173; leg, I. Milt 20. 5. 2006.

Commentary on G *sibalii*

Neuhuber (2008) helps us find the Sierra Chica, Prov. La Pampa. It is one of a number of isolated hill in the pampa near to the National Park of Lihuel Calel, one of a number of such hills where relatives of *G. gibbosum* can be found. The Sierra Chica is only 430m high so the 1,000m given in the description should probably be feet. He explains that while the nearest population of *G. reductum* is some 220km away, the form of the flower makes it clear that these plants belong to that species. This is strange because *G. gibbosum* is the usual species found on these isolated hills.

G. reductum subsp. *sibalii* (Halda & Milt) Neuhuber in *Gymnocalycium* 21(1): pp.758-759 (2008)

Distribution (Map 28)

Plants of this taxon can be found growing on rocky places in the Sierra de la Ventana and nearby hills in the south of Prov. Buenos Aires, Argentina. The distribution is extended to the west by subsp. *sibalii* from the Sierra Chica and Sierra de Lihuel Calel in Prov. La Pampa, Argentina.

Conservation Status

Least Concern. The extensive habitat range is little affected by human activity so the plant appears to be secure.

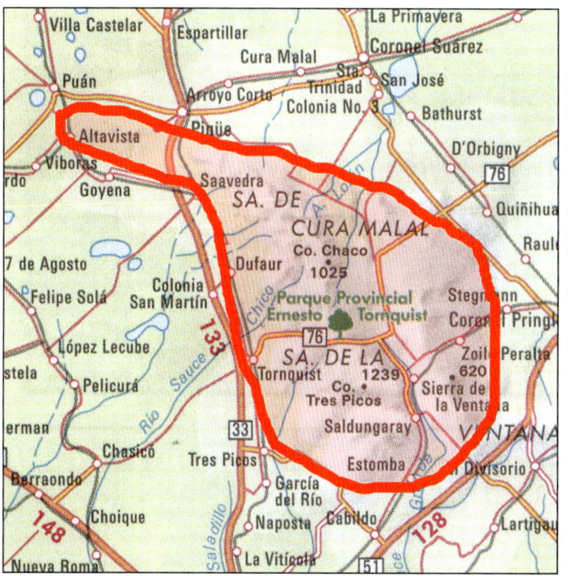

Map 28 Main distribution of *G. reductum reductum,* Prov. Buenos Aires, Argentina

Fig.213 The habitat of *G.reductum reductum* GC929.02 in the Sierra de la Ventana, Prov. Buenos Aires, Argentina

History

This plant has first been published as *Cactus nobilis* by Haworth in 1812. However, there was already a *Cactus nobilis* Willd, a synonym of the older *Cactus recurvus* Miller. It was Link who gave Haworth's plant a new valid name when he called it *Cactus reductus*.

It was treated by many authors as the same taxon as *G. gibbosum* after Salm-Dyck (1844) made it a synonym, a mistake which perpetuated for a very long time, except for the view of Pfeiffer.

In his *Abbildung und Beschreibung Blühender Cacteen* II (1846-50), Pfeiffer published a fine engraving of a plant growing in the collection of Mr. Fennel of Cassel as plate XII. He tells us that he had visited Kew in the spring of 1845 and had seen there a plant of this species which was four feet high. He disagrees with Salm-Dyck and states that he considers *G. reductum* to be distinct from *G. gibbosum*.

G. reductum was one of the original three species which Pfeiffer included in his new genus *Gymnocalycium* when he proposed the name in the catalogue of Schelhas in 1843-44.

Fig.214 *G. reductum reductum* GC929.02 in the Sierra de la Ventana, Prov. Buenos Aires, Argentina

Fig.215 *G. reductum reductum* GC929.02 in the Sierra de la Ventana, Prov. Buenos Aires, Argentina

Fig.216 The habitat of *G.reductum reductum* GC928.01 in the Sierra de la Tuna, Prov. Buenos Aires, Argentina

Fig.218 *G. reductum reductum, Papsch* 89 23-26 from the Sierra Cura Malal, Prov. Buenos Aires, Argentina

Another splendid illustration of this species appeared as Tab. 4443 in Curtis's Botanical Magazine of 1849 where it was described as the 'Dingy Cereus' (Fig.10). It had been in cultivation at Kew for some time so perhaps it was the same plant that Pfeiffer had seen.

Spegazzini visited the Sierra de la Ventana and published his findings in 1896. He mis-identified *G. reductum* as *Echinocactus ottonis* Lehm. Later, in 1902, he revised his opinion and named the plant *Echinocactus gibbosus* var. *ventanicola*.

Fric also visited the same Sierra but says he failed to find *G. platense* but rather saw a plant resembling *G. gibbosum* but almost always in many headed clumps. He gave it the provisional name *G. caespitosum* which was later made an invalid variety of *G. gibbosum* in 1964 by Fleischer to whom Fric had given a cutting.

Commentary

G. reductum reductum differs from *G. gibbosum* by its less-applanate body, spiny crown, and larger flowers. The long standing confusion of *G. reductum* with *G. gibbosum* was not followed by Kiesling who described the plants from the Sierra de la Ventana in 1982 as *G. platense* var. *ventanicola*, so breaking the link with *G. gibbosum*. The question was further studied by Piltz (1995) who perpetuated the view that *G. reductum* was a synonym of *G. gibbosum* and so proposed *Echinocactus mackieanus* as the oldest name for these plants. Papsch (2001b) wrote an informative review of the species.

Cultivation

G. reductum is easy to grow and produces large pure white flowers when about 8cm in diameter. It is usually seen as a large single head but can occasionally offset in culture. Large specimens are particularly imposing with a dark epidermis and strong spination.

Illustrations

Schumann, K. *Gesamtbescheibung der Kakteen* (1899) Fig.71 on page 407 as *Echinocactus gibbosus*.
Maasz, H. *Die Schönheit unserer Kakteen* p.45 (1925) as *Echinocactus gibbosus*
Deess, H. *Sukkulentenkunde* 5(1): p.6 (1942) as *Gymnocalycium gibbosum* var. *nobile*
Neumann, H. *Sukkulentenkunde* 5(3): p.65 (1942)

Fig.217 *G..reductum reductum, Piltz* 94 from the Abra de la Ventana, Prov. Buenos Aires, Argentina

Fig.219 *G..reductum reductum, Papsch* 89 15-15,Sierra de Tuna - Sierra Pillahuinco, Prov. Buenos Aires, Argentina

Gymnocalycium reductum subsp. *leeanum* (Hooker) Papsch

First Description (Hooker)

Curtis's Botanical Magazine 71: plate 4184 (1845)

Echinocactus leeanus Hooker

Raised by Messrs. Lee of Hammersmith nursery, from seed sent from the Argentinean Province of Buenos Aires by Mr. Tweedie in 1840.

The specimen here figured blossomed in May, in the cactus house of the Royal Botanical Gardens of Kew. I do not meet with its description in any book to which I have access, and therefore venture though not without hesitation to publish it as new. It may rank near our Multiflorous (supra, Tab. 4181), but is really very different.

Description: A small species, globose but depressed at the top. Tubercles which compose the surface, rather large, hemispherical, but having about six obtuse angles, of a rather glaucous green colour, not arranged in distinct lines or series so as to form ridges with their corresponding furrows, but placed with a good deal of irregularity, becoming, below especially, confluent and obsolete, at the top small and very numerous. Areoles oval, downy, or rather woolly, producing about eleven rather slender spines, of which one, the central one stands forward and is quite straight; the other ten are slightly recurved and spread horizontally (especially on the older tubercles), most of these are nearly equal in size and about half an inch long.

Flowers from the summit or depressed portion above, one or two moderately large, Tube short, covered with green roundish to oblong obtuse scales, the upper ones larger, with pale edges and tips, gradually passing into the pale sulphur or almost cream coloured petals.

Fig.221 *G.reductum leeanum,* GC926.01 in habitat north of Tandil, Prov. Buenos Aires, Argentina

Fig.220 Plate 4184 of *Echinocactus leeanus* from *Curtis's Botanical Magazine* (1845)

Etymology

Named for John Lee, a nineteenth century nurseryman in Hammersmith, London.

Other Names

Echinocactus leeanus Hooker in *Curtis's Botanical Magazine* 71: plate 4184 (1845)

Synonym

G. schatzlianum Strigl & W. Till in KuaS 36(12): pp.250-253 (1985)

First description of *Gymnocalycium schatzlianum* Strigl and W. Till spec. nov.

Body mostly flattened-globular, becoming very large. The plants at hand are up to 12cm diameter, and 8cm high, in habitat plants are much larger (Rausch). Body colour dark grey-green, crown sunken bare. During the dry season the plants appear to be completely covered with spines. Ribs 16-20 straight, at the base c.

Fig.222 *G.reductum leeanum,* GC926.01 in habitat north of Tandil, Prov. Buenos Aires, Argentina

Fig.223 The habitat of *G.reductum leeanum*, GC926.01 north of Tandil, Prov. Buenos Aires, Argentina

20mm wide, formed by weak angular humps. Areoles c. 20mm apart, not sunken, rounded-oval, with short white felt. Spines radiating, thin, strong, rigid, sharp, a light horn colour with a darker, mostly red-brown base; radial spines 9-11, the side and lower spines being the longest (up to 25mm long), central spines (1)4-7, one of these usually longer, up to 30mm long, all strong, the same as the radials, rarely the spines are less than 1cm long.

Flowers coming from the crown, wide funnel shaped, 6-7.5cm long, 7-8cm in diameter, tube short, only 15mm long, 10mm diameter, dark olive-green with light edged, wide round scales. Outer petals pointed, c. 35mm long, 7mm wide, yellowish to white with a greenish central stripe, inner petals somewhat shorter and narrower, pointed, lanceolate, lighter in colour. Many stamens inserted in the wall of the flower tube, the lower stamens lying against the style. Filaments thin, white, anthers yellow. Style 28mm long (with the stigma 34mm), 2-2.5mm thick, greenish-white. Stigma lobes 12-14, slightly curved, 6mm long, white. Seeds hat shaped, 1.5mm high and wide, 1mm thick, matt black, with some arillus remains. Hilum large oval, without a protruding edge. Belonging to the sub-genus Ovatisemineae, in the Schütz system.

Type: Argentina, Province Buenos Aires, near Balcarce, 1972 collected by W. Rausch 541, cultivated in collection of H. Till, no.1326 ex coll. H. Till praep. in F.P.A. July 1985 (WU)

Map 29 Locations of *G. reductum leeanum* in Prov. Buenos Aires, Argentina

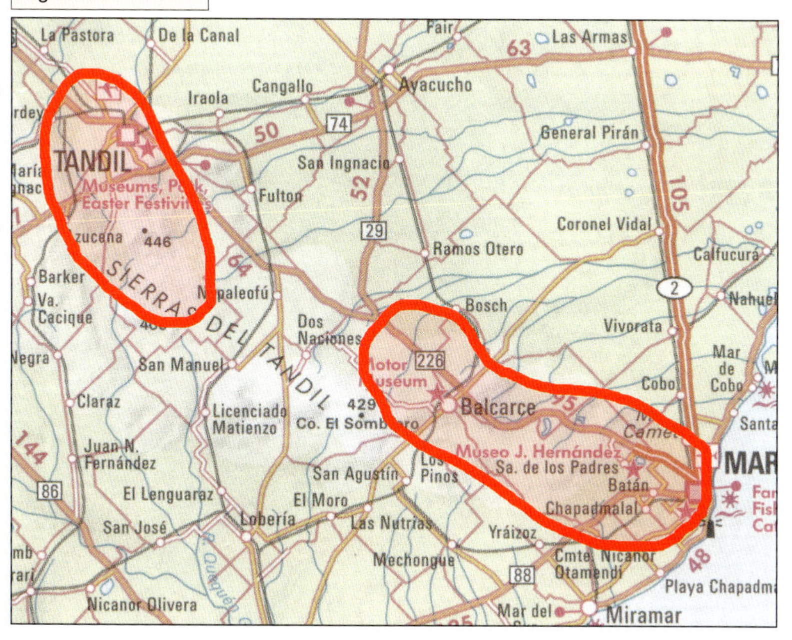

Commentary on *G. schatzlianum*

Rausch found this plant in 1972, believing it to be related to *G. gibbosum*. The authors considered whether their plant was actually *G. hyptiacanthum* but listed the characters and concluded that they were different. The authors also describe a curious characteristic of the species which they call heterostyly, some flowers having a short style and others a long one. They claim that the most efficient pollination is achieved when the two types are crossed.

Distribution (Map 29)

The type form grows in organic material among rocks in the Sierra de Tandil, a small range of hills rising to 500m near the town of Tandil in the south of Province Buenos Aires, Argentina. Plants are often in the shelter of bushes and occasionally a plant up to 30cm high can be found.

To the south-east, there is a series of individual hills that extends past Balcarce as far as the sea at Mar del Plata. Here can be found the form of *G. reductum* subsp. *leeanum* which was described as *G. schatzlianum*. The small isolated habitats get more humidity and receive more rainfall as one gets nearer the ocean, so the plants grow with moss and ferns. The type location of *G. schatzlianum* is on the Sierra Barrosa, just to the south of the town of Balcarce where it grows on the flat tops of the hills.

Conservation Status

Vulnerable. The plant grows on hills in an area where there is increasing human activity (such as animal grazing) so putting pressure on the populations which are already small. Papsch (2000) describes how populations have been destroyed by land clearance, grazing and fires. He also reports the deterioration of populations by the sea near to Mar del Plata where, following heavy rain, the plants had been eaten by insect larvae and rodents.

Fig.224 *G.reductum leeanum, Papsch* 89 5-7 from Tandil, Sierra Tandil, Prov. Buenos Aires, Argentina

Fig.226 *G.reductum leeanum* (schatzlianum), *Papsch* 89 10-10 from Balcarce, Sierra. Cinco Cerros, Prov. Buenos. Aires. Argentina

History

This taxon was first described by Hooker as *Echinocactus leeanus* in *Curtis's Botanical Magazine* of 1845, Tab 4184. The plant had been raised by Mr Lee of Hammersmith nursery from seed sent from Province Buenos Aires, Argentina by Mr. Tweedie in 1840.

John Tweedie was a landscape gardener who lived in Argentina from 1825 until his death in 1862, aged 87. He lived in Buenos Aires but made long trips looking for plants. One of these trips took him to the Sierra del Tandil in 1837 where he probably collected the seeds he sent to Hooker.

Even though the source of the seed was stated in the first description, the species was confused with yellow-flowering plants from Uruguay by Britton and Rose. This interpretation was then followed by later authors and, as a result, Backeberg made the combination *G. leeanum* var. *netrelianum* in *Kaktus-ABC* (1936).

Metzing (1991) came to the conclusion that *G. leeanum* was in fact a plant related to *G. reductum* and *G. schatzlianum* from Prov. Buenos Aires.

Commentary

The application of the name accepted here is based on the research of Papsch (2000). He put forward a persuasive argument that plants from the Sierra Tandil are a good match for *G. leeanum*.

Cultivation

Capable of becoming a large solitary plant, this beautiful taxon easily grows into a flowering specimen. The flowers are large, and a pretty shade of pale creamy-yellow.

It is the form known as *G. schatzlianum* that is most often seen in collections, especially plants grown from *Piltz* 93 seed collected from plants originating from the Sierra Barrosa. Its flowers are said to have a strong smell of carbolic, a characteristic which can also be detected in other forms of *G. reductum*.

Fig.225 *G.reductum leeanum* (the type collection of *G. schatzlianum*), *Rausch* 541, Sierra Barrosa, Balcarce, Prov. Buenos Aires. Argentina

Fig.227 *G.reductum leeanum* (schatzlianum), *Papsch* 92 118-162 Sierra Barrosa, Balcarce, Prov. Buenos Aires, Argentina

Gymnocalycium robustum Kiesling, Ferrari & Metzing

First Description (Kiesling et al.)

CSJ(US) 74(1): pp.4-8 (2002)

Gymnocalycium robustum R. Kiesling, O. Ferrari & D. Metzing nov. spec.

Roots thick, obconical, c. 5cm long, 2cm maximum diameter. Stems simple, globose-depressed, 3-5cm high, 8-11cm diameter, epidermis grey or grey-green or slightly mauve, dull. Ribs 9-11, straight, 1.5cm high, 2-3.5cm wide, obtuse, with 3-5 areoles and below them, a prominent chin (podarium) 0.5cm high and a conspicuous transverse furrow.

Areoles elliptical, 6 x 3mm, with white hairs, a basal spine and 2-3 pairs of lateral spines; all adpressed irregularly against the podaria, rigid, chalk-white when dry, yellowish with a spot of brown at base and apex when wet; centrals 0.

Flowers c. 6cm long and 6cm diameter when completely open (7cm x 3cm just before opening). Receptacle funnel-shaped, 4.5cm long, 0.8-1.4cm diameter, scales nearly sub-circular, wider than long; 0.4cm long, 0.6cm wide, succulent, pure grey or tinged reddish. Pericarpel slightly conical, 3cm long, 0.5-0.8cm diameter. Ovary chamber conical, 2cm long, 0.5cm wide, ovules many. Style cylindrical, very short, just reaching the base of the first

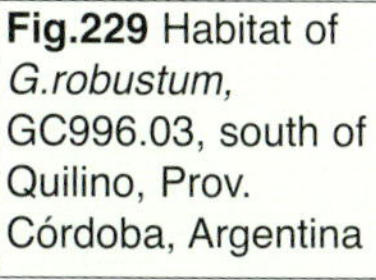

Fig.229 Habitat of *G.robustum*, GC996.03, south of Quilino, Prov. Córdoba, Argentina

Fig.228 *G.robustum*, GC996.03, south of the town of Quilino in the north of Prov. Córdoba, Argentina

stamens, 1.3cm long, 2mm diameter, white, with small tubercles. Stigma 8-9(-14!)- lobed, clear yellow, nearly 3mm long. Nectary chamber pink, 0.4mm long. Stamens many, the basal ones straight, shorter than the stigma and parallel to them, filaments 1.1cm long, tuberculate at the base, anthers 2mm long; the other filaments inserted in a continuous spiral, longer than the stigma, curved along the centre, white, 1.5cm long, anthers yellow, 2mm long, 1mm wide. Perianth rotate when completely open, with the outer segments similar to

Fig.230 *G.robustum, Meregalli* 823 in habitat south of Quilino, Prov. Córdoba, Argentina

Fig.231 *G.robustum* (kuehhasii), *Meregalli* 906 in habitat north-east of Villa de Maria, Prov. Córdoba, Argentina

the scales but longer and with the transparent marginal area becoming gradually wider; inner segments white, pink at base, thin lanceolate, acute, the innermost shorter than the intermediate ones.

Fruits clavate, rarely fusiform, grey, 4-4.5cm long, 1.5-1.8cm maximum diameter, 1cm diameter at the base; with pure grey scales. Seeds globose-truncate, symmetric, 1.2mm long and thick, black, with the cuticle almost completely separate from the testa and then appearing brown, prominently wrinkled; testa cells convex, all similar; hylum-micropylar region oval, umbilicate, 0.9mm long, 0.6mm wide.

Holotype: Argentina, Prov. Córdoba, Dep, Ischilín, Quilino, cultivated by Omar Ferrari under number FK 120; prepared for herbarium December 2000, *R. Kiesling et O. Ferrari* 9883 (SI).

Paratype: Argentina, Prov. Córdoba, Dept. Ischilín, Quilino, 19-XI-1987, *R. Kiesling et O. Ferrari* 6787 (SI). Note: The paratype was collected and preserved in the field and consists of two complete stems, roots and several flowers. Specimens of the same collection were cultivated and one of them was used to prepare the holotype, which consists of a single plant with one fruit.

Synonym

G. kuehhasii Neuhuber & Sperling in *Gymnocalycium* 21(1): pp.747-750 (2008)

First Description of *Gynocalycium kuehhasii*

Body solitary, flattened, semi globular, not offsetting, up to 40mm high, and 90mm diam. Stake-like root short, mostly does not appear to be strongly attached. Epidermis in cultivation grey-green to dark grey-green, matt. Crown sunken, armed. Ribs up to 12, straight down, wide at the bottom, close together, divided by humps with strong projecting rounded chins, separated from each other by deep notches, areoles round sunken, with strong white wool about 9mm apart. Spines (3-)4-5(-7), thin, occasionally flattened, slightly projecting or lying onto the body horizontally, light horn-coloured, almost white at the tip, darker at the base, becoming black in age, no central spine.

Flowers coming from the crown, few in number, 65-70mm long, 40-45mm diam., white to dirty white, bell shaped, not fully opening, outer petals white with a light pink or light green mid stripe, on the back sometimes with a wide light grey- green stripe, light pink at the base, average 23.5mm long, 6.5mm wide, lanceolate, inner petals white, 19mm long, 5mm wide, narrow lanceolate and pointed. Flower throat with a thick tissue, pale pink to pink, 15mm high and 9mm diam., Nectar chamber narrow, light green to white at the top light pink to light orange, up to 8mm high, 2-3mm diam. Filaments up to 6.8mm long, mostly light yellow, one primary row, 4mm long, without the small anthers, the anthers lying on, in the middle or just below the stigma. Filaments white, light green at the base, primaries clearly at a distance, 4(-6) rigid rows inserted in the inner wall of the flower tube, their anthers hanging over the stigma like a roof. Anthers 1.3mm long, 0.6mm wide, yellow. Style light green, 11mm long, 1.2mm diam., very strong, standing free of the ovary, the partition wall to the ovary white and often curved under. Stigma light yellow, 2.5mm long, 9-11 stigma lobes, always below the anthers of the secondary rows. Ovary white, 16.5mm high, 4mm diam. Pericarp 22mm high, 10mm diam., conical, dark green with a light grey bloom. Scales wide semi-globular, with no or only a trace of light green to light pink edging. Fruit dehydrated, about 23mm long, 12mm diam., green to blue-green, elliptical to spindle-shaped, splitting lengthwise, the seeds pushing out of the fruit.

Seeds 1.15-1.2mm long, to 1.1mm diam., HMR wide tear drop shape, deep, brown, sharp edged, testa black with mostly strong rounded structures, with some dried brown skin. (subgenus *Gymnocalycium*).

Habitat: Argentina, Province Córdoba, district of Villa Maria, 550-650m altitude, always growing in shade under bushes.

Type: Argentina, Province de Córdoba, near the village of San Miguel, 550 -650 m.s.m., 16. January 1993, leg. F. Kühhas KF 93-128/472, Holotype CORD (ex WU 1060, flower in alcohol), isotype: WU (Nr. 889 et Nr. 1212, flower in alcohol; Nr. 2508, dry fruit); l. c. F. Kühhas KF93-128/472a, Paratype: WU (Nr. 2172, flower in alcohol).

Fig.232 *G.robustum* (kuehhasii), *Meregalli* 907 in habitat near Va. Ojo de Agua, Prov Santiago del Estero, Argentina

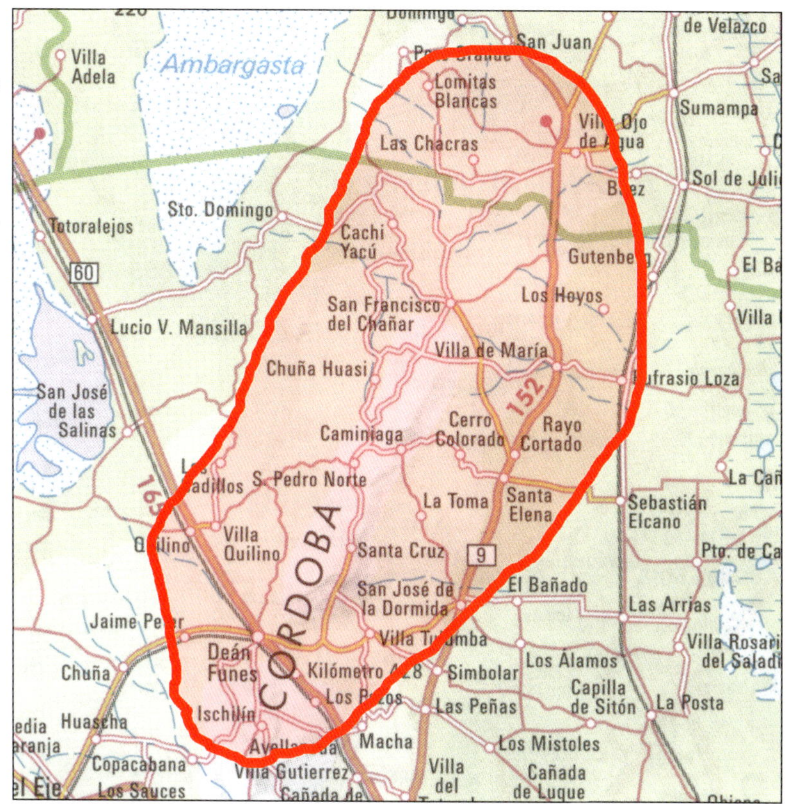

Map 30
Distribution of *G. robustum* in Prov. Córdoba, Argentina

Fig.233 *G.robustum, Meregalli* 823 from south of Quilino, Prov. Córdoba, Argentina

Fig.234 *G.robustum, Meregalli* 823 from south of Quilino, Prov. Córdoba, Argentina

Commentary on *G kuehhasii*

A interesting plant somewhat intermediate between *G. kieslingii* and *G. robustum*, its habitat being closer to the latter.

Distribution (Map 30)

It was said to grow only near Quilino in the north of Prov. Córdoba, so if this is true, it would be a very localised species. At the type locality, *G. robustum* grows on a series of low hills comprising crumbling granitic rock. Also here one can find the local form of the widespread *G. quehlianum* (*Trichomosemineum*) from which *G. robustum* can be distinguished by its pronounced humps between the areoles, although the two have often been confused. Also reported from the same habitat is *G. prochazkianum*, a member of the subgenus *Microsemineum*, also with a very limited distribution.

More recent reports suggest that *G. robustum* grows north and east of Quilino, but there are also similar forms belonging to the the same subgenus (*Gymnocalycium*) in the area which may be considered to belong here, or are undescribed. One of these has recently been given the name *G. kuehhasii* and is included here at least until it is better understood.

Conservation Status

Vulnerable. If one accepts that the plant occurs only at and near the type location, then this species is vulnerable to habitat destruction in a region where agriculture may expand.

History

The publication of *G. robustum* in 2002 reignited the discussion about which seed group the original *G. quehlianum* belongs. Till (1993) had claimed that *G. quehlianum* belonged to the seed group *Gymnocalycium*, not *Trichomosemineum* where it has historically been placed. Till (2002) repeated his claim, stating that *G. robustum* is a redescription of the original *G. quehlianum*. His argument is unconvincing and is rejected here. See the full story under *G quehlianum* on page 218.

Commentary

When this species was described, Till (2002) tried to prove that it was redescription of *G. quehlianum* to which it has some resemblance, especially in culture.

Its closest relative is probably *G. kieslingii* from further north and there are plants in north-east of Prov. Córdoba which have intermediate characteristics.

Cultivation

This species is rare in cultivation and I have not yet grown it, but there is no reason to think it will provide any difficulties.

Illustrations

Till in *Gymnocalycium* 6(2): p.91 (1993). Illustrations 3 & 4 as *G. quehlianum*.

Gymnocalycium schroederianum Osten subsp. *schroederianum*

First Description (Osten)

Gymnocalycium schroederianum n.sp.

Notas Sobre Cactáceas - Anales del Museo de Historia Natural de Montivideo 2nd series, Vol V, No.1; p.60 (1941)

Number 16.873 comes from the area north of the Río Negro, the muddy banks of the Río Uruguay, on muddy pampas soil, endless grass-land with no cover except for groups of shrubs and a number of small trees.

I have only one adult example that is very old, 7cm high, 14cm diameter, semi-globular, the colour is clear dark grey-green, Crown is flat to ± sunken. Humps in crown unarmed. Ribs 24 with deep cross-cuts, which form strong bold humps with conical chins that on this specimen number 4-6, c. 1.5cm diameter. Areoles c. 2cm apart. Spines, usually 7, yellow in youth, soon becoming grey, reddish-purple at the base, the spines resemble a dragonfly at rest, the top pair being the shortest, the next two pairs pointing sideways (the wings) with the longest spines pointing straight down (the abdomen of the dragonfly), in some areoles a number of the spines abort.

This example flowered abundantly from November 1922 through to February 1923, but it only produced one fruit. Flowers c. 7cm long, 5.5cm diameter, when open. Ovary 20mm long, 7mm diameter, exterior colour dark green, hairless but provided with scales. Scales 4mm wide, 2.5mm high, olive-green with a white edge, flower tube narrow at the base, in the form of a funnel, colour lighter than the ovary, the scales gradually change form becoming the outer petals, also clear olive-green, wider towards the top, edges white, at the tip a clear chestnut tone, the scales on the tube 5mm high and wide. Outer petals lanceolate with a point, 25mm long, 6mm wide, central petals a pale greenish colour with a thin green back stripe, inner petals usually smaller, lanceolate, dentate, pale green, clear green below. Filaments are inserted very untidily, the highest reaching to half way up the inner petals, yellowish-green with cream anthers. Style pale green, 14mm long on a male flower, (anthers well developed) with stigma lobes still closed, clear yellow, the style reaches only half way up the tallest stamens. Unfortunately I have made no observations on female flowers.

Fig.236 *G.schroederianum, Herm* 42 from Nuevo Berlin, Río Negro, Uruguay

In mid March 1923, after many flowers this plant produced one ripe fruit, inverted elongated pear-shape, 25mm long, 5mm diameter at the widest part, light grey-green. scales wide, 2mm high, 4mm wide, with a 1mm white margin. Dried flower remains sit on the crown of the fruit, 25mm long, the fruit splits vertically. Seeds spherical, truncate below, 1.2mm diameter.

Type: J. Schröder 04.1922, Uruguay, Distr. Río Negro, Nueva Melhem, Herb. C. Osten 16-873.

Photo: Ludwig Bercht

Fig.235 *G.schroederianum, Bercht* 2271 in habitat at Río Gualeguaychú, Prov. Entre Rios, Argentina

Photo: Ludwig Bercht

Fig.237 *G. schroederianum* (paucicostatum), *Bercht* 2273 in habitat south of Pucherta, Prov. Corrientes, Argentina

Synonyms

Gymnocalycium schroederianum subsp. *paucicostatum* Kiesling

First Description of *Gymnocalycium schroederianum* subsp. *paucicostatum* in CSJ(US) 59(2): p.49 (1987)

Differs from the typical variety by the smaller number of ribs, only 9-11, wider and very angulate. Areoles 2-3 per rib, nearly 1.7mm apart, elliptic, 0.6-0.7cm long. Spines normally 3, rarely 5, disposed in a "Y", the lower longer, 2-4cm, strong, subulate, the other 2 laterals, 1-1.5cm long, all suberect and arching outwards, horn-colour, reddish-brown at the base. Flowers pure white with red throat, style pale green.

Holotype of the subspecies: Argentina, Corrientes, Dpto. Curuzu Cuatia, Arroyo Mocoreta, 12-XI-1981, A. Schinini 21678, cum S. Cacers et C. Quarin (SI); "crece en suelos degradados, entre Sellaginella sellowii, asociada con Dyckia sp. y Portulaca sp. Flor externamente verdosa e internamente blanca. Florece a medio dia".

Paratype: Corrientes, Dpto. Paso de los Libres, Río Mirinay, 16-II-1979; A. Shinini 17,288, cum E. Cabral et R. Vanni (SI), Entre Rios, Dpto. Federacion, Ea. Buena Esperanza, Dpto. Federacion, Ea. Buena Esperanza, 25-X-1961, T. M. Pedersen 6274 (SI).

Etymology

Named for Dr. L. Schröder, a local medical doctor and cactus grower who found the plant.

Distribution (Map 31)

The type locality is Nueve Melhem, a ranch north of Fray Bentos near to Berlin, close to the border with Argentina.

Kiesling (1987) lists a number of specimens he had seen which suggest that this plant can be found in localities all over the Argentinian Province of Entre Rios and in the south of Province Corrientes. Some of these localities are not immediately next to the big rivers that border this region.

Having visited some of the reported localities, Kiesling says that he found dozens of plants. They grow in low, open forest in white or pinkish coloured clay that frequently becomes flooded.

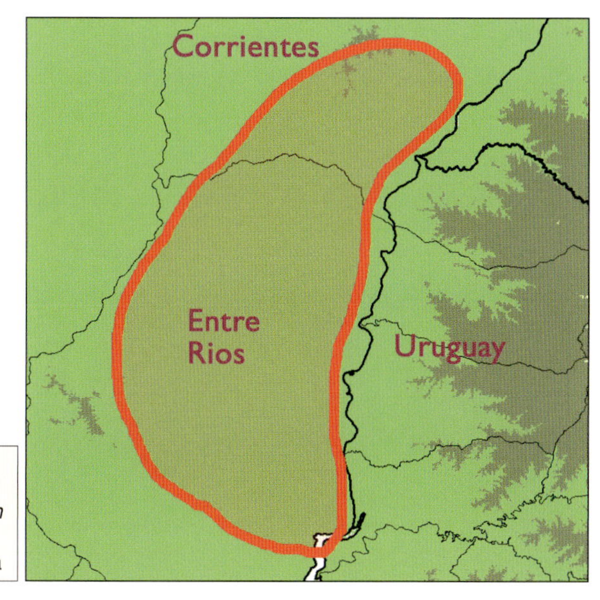

Map 31 Distribution of *G. schroederianum schroderianum* mainly in Argentina

Fig.238 *G.schroederianum* (paucicostatum), *Bercht* 960 from Curuzu Cuatia, Prov. Corrientes, Argentina

Conservation Status

Near Threatened. Although there are many populations containing numerous plants, Papsch (2001) tells of the threat of land clearance for grazing and the cultivation of oranges and mandarins. Even further north where the soil is less suitable for agriculture, the plants have been adversely affected by flooding in recent years.

History

First found by Dr. L. Schröder in May 1922, this species was described by Osten in 1941 from a single specimen, the publication appearing five years after his death. It was said to come from the muddy banks of the Río Uruguay and it remained little known in culture until its rediscovery in 1967 by Buining (1968). He found it growing in the wet muddy clay on the banks of the river. It is clearly difficult to find since Hugo Schlosser, a Uruguayan cactus specialist, took four trips before he eventually found it in 1979.

Kiesling (1987) gives us information from the records of Spegazzini which include two photographs of this plant. The first, of a plant collected near the Paraná River in Prov. Entre Rios, was reproduced as Fig.170 in Volume III of *The Cactaceae* by Britton & Rose (1922) and identified as *G. monvillei*. Kiesling (1997) described subsp. *paucicostatum,* a form with fewer ribs from Corrientes.

Commentary

The typical subspecies includes the northern populations in the Provinces of Entre Rios and Corrientes in north-east Argentina. It is similar to subsp. *bayense* but they are recognized because of the 500km distance between the habitat locations.

Cultivation

This species grows best when protected from strong sunshine and kept moist in the growing season. It remains quite rare in cultivation but is a pretty plant to grow, flowering when still quite small. Some clones of the form described as subsp. *paucicostatum* have spectacular spines.

Gymnocalycium schroederianum subsp. *bayense* Kiesling

First Description (Kiesling)

CSJ(US) 59(1): p.48 (1987)

Gymnocalycium schroederianum ssp. *bayense* Kiesling, ssp. nov.

syn.; *Gymnocalycium "bayensis"* Rausch nomen nudum

Differs from ssp. *schroederianum* by the smaller stems, 7(-10)cm diameter, pale grey epidermis, the occasional presence of 1 erect central spine in addition to the 5-7 radial spines, and shorter flowers, 4-5cm long, with obconic pericarpel.

Holotype of the subspecies: Argentina, Buenos Aires, Pdo. Olavarría, Sierras Bayas, July, 1981 (flowered in cultivation XII-1982), R. Kiesling et A. G. Lopez 4323 (SI).

This new subspecies grows on the top of these low mountains in the crevices of the granite rock or in organic material. The vegetation there is mostly grasses and other herbaceous plants. In cultivation it flowers sparsely and mostly in December which is the beginning of the summer here in the southern hemisphere.

Distribution (Map 32)

It grows near the tops of a range of hills by the town of Olavarría, the Sierra Bayas and other nearby Sierras in the south of Buenos Aires Province, Argentina.

Fig.239 *G.schroederianum bayense,Neuhuber* 90 289-970 from the Sierra Bayas, Prov. Buenos Aires, Argentina

Map 32 Locations of *G. schroederianum bayense* in Prov. Buenos Aires, Argentina

Fig.240 Habitat of *G. schroederianum bayense* GC927.01 Cerro de China, north of Olivarría, Prov. Buenos Aires, Argentina

Fig.241 *G. schroederianum bayense,* GC927.01, Cerro de China, north of Olivarría, Prov. Buenos Aires, Argentina

Conservation Status

Vulnerable. There are thought to be few plants left in habitat. Kiesling (1987) says there were few specimens when he visited in 1981-2. Observations by Papsch (2001a) describe a similar situation and my own recent observations support the view that the populations are suffering from human activity. Papsch also suggests that the climate is changing, resulting in fewer young plants becoming established.

History

Rausch found this plant when travelling with Ferrari in the Sierra Bayas, and gave it the provisional name *'Gymnocalycium bayensis'*. Kiesling (1997) validated the name as a subspecies of *G. schroederianum* in the American Journal.

The habitat is easily reached from Buenos Aires and must have been known for a long time. It would be surprising if plants from here were not described earlier. Papsch (2001) follows up this thought with detailed analysis of the possible contenders and reaches the conclusion that the oldest name for this taxon from the Sierra Bayas is *Echinocactus hyptiacanthus* and that Spegazzini's *Echinocactus platensis* is a later synonym.

Fig.242 *G. schroederianum bayense, Papsch* 92 112-149 from Olavarría, Sierra Bayas, Prov. Buenos Aires Argentina

Fig.243 *G. schroederianum bayense,* GC927.01, Cerro de China, north of Olivarría, Prov. Buenos Aires, Argentina

Commentary

Papsch thought it likely that the plants were first collected by John Tweedie who worked as a landscape gardener in Balcarce, Prov. Buenos Aires around 1837. He probably sent plants to England as well as to France and it was from these that *E. hyptiacanthus* was described by Lemaire (see page 45) as well as *E. leeanus* by Hooker. (see pages 45 & 110)

Papsch attempted to neotypify the name *Echinocactus hyptiacanthus* with a plant of this taxon from the Sierra Bayas. However, as pointed out by Kiesling et al. (2002), the name had already been neotypified by Kiesling (1999) with a different taxon which was valid, even though he had cited the specimen as a lectotype rather than a neotype. This means that the name *Gymnocalycium hyptiacanthum* has to be applied, in accordance with Kiesling's neotype, to the widespread yellow-flowered taxon from Uruguay that is better known as *G. uruguayense*. My personal view is that Papsch is probably right but there is no substantial evidence to challenge Kiesling's work. See Page 45 for more information.

The plant looks remarkably like the type subspecies, whose habitat is about 500 kilometres away to the north.

Cultivation

This is an easy plant to grow and is a handsome specimen in age. It flowers easily and should become more popular when plants become available in the trade.

Illustrations

New Cactus Lexicon Atlas (2006) Illn. 274.1

Gymnocalycium schroederianum subsp. *boessii* R. Kiesling, E. Marchesi & O. Ferrari

First Description (Kiesling et al.)

Gymnocalycium schroederianum subsp. *boessii* subsp. nov. in KuaS 53(9): p.226 (2002)

Roots fibrous. Body simple, depressed-globular, 6-7 (-8) cm diameter, up to 4.5cm high. Ribs c. 11 (-14), straight with horizontal furrows between the humps. Humps are 0.5cm high with the areole on the upper part with a more or less projecting chin below. Areoles elliptical, 5mm long, 1-2mm wide, with 5,7 or 9 spines in the lower half, one basal, the others in sideways pointing pairs, all lying close to the body, mostly only curved at the base but somewhat slightly curved, 0.5-1cm long, whitish with a reddish base, thin, flexible; central spines 0, very rarely 1, resembling the radial spines.

Flowers from the topmost areoles, only rarely from the side areoles, closed 6-9cm long, when open 5-8cm long and 4-6cm wide. Receptacle bright green, elongated, almost cylindrical, 3.5-4.5cm long, solid basal part 2.2cm long, 0.5cm in diameter at the base, 1cm diameter above, with 12-15 scales, these are bright green with the upper edge reddish. Ovary elongated, 1.5-2cm long, 0.4 cm broad, Style short, straight, reaching the lower anthers, 1.1cm long, 1.5mm diameter with c. 8-10 lobes, these 3-3.5mm long. Stamens 2 to 18mm long, inserted in a single spiralled row reaching up 12-14mm. Anthers yellow, 1.5mm long, 0.5mm wide. Outer petals white with a green central stripe, sometimes with a reddish tip, spatulate, 3 cm long, 1 cm wide, inner petals up to 3.5cm long and 1cm wide, more or less lanceolate, white with a pink base.

Fruit spindle-shaped, 3cm long, 1.3cm in diameter, bluish-green, splitting lengthwise, still with the remains of the 2.5-3cm long dried flower. Seeds urn-shaped, black that appears brown through the loose cuticle, 1.5mm long and 1mm diameter.

Holotype: Argentina. Santa Fé, Department Vera. Entrada a Gallareta. 17.09.1988. ex cult,. O. Ferrari 3/88 (prepared in December 2001) sub R. Kiesling 10020 (SI: body, flower and fruit).

Occurrence: Argentina. Santa Fé, Department Vera. Entrada a Gallareta, in the Monte Chaqueño (type habitat). The subspecies grows in a loamy, usually wet soil, in the north of Argentina, Prov. Santa Fé (Dept. Obligado, San Justo and Vera), it readily flowers and fruits between November and February.

Etymology

Named for Dr. Ernesto Boess (1900-1963), dentist, cactus specialist, amateur entomologist and apiarist in La Plata who, in the first half of the 20th century, worked with A. Castellanos and other Argentinian botanists.

Photo: Ludwig Bercht

Fig.244 *G.schroederianum boessii, Bercht* 2308 from Vera, Prov. Santa Fé, Argentina

Synonym

Gymnocalycium erolesii Neuhuber & Bercht in *Gymnocalycium* 15(4): pp.473-478 (2002)

First description of *Gymnocalycium erolesii* Neuhuber & C.A.L. Bercht, spec. nov.

Body simple, flat-globular, up to 92mm diameter and 48mm high (the holotype is 52mm diameter and 39mm high), the part below ground is short conical and with long, thin fibrous roots growing from two thick main roots. Crown sunken, ± armed, epidermis dull grey-green. Ribs 9-10, straight, 23mm wide and 4-6mm high, separated from each other by deep, wavy vertical furrows, in turn separated into humps by short horizontal grooves, which are slightly curved upwards. Humps rounded, on older plants with a projecting chin. Areoles sunken, small, oval, 4(-5.5mm long, 1(-2.2)mm wide, those in the crown covered in yellowish white wool, soon becoming bald. Spines 7(9) (11-13 only on very old plants), thin, soon becoming soft, slightly curved, radiating or also comb-like, lying onto the body, horn coloured, darker at the base and tip, not changing their colour in age, The topmost pair up to 8mm long, the 2 middle pair 8-10mm long, the rest 7-8.5mm long, shorter than the middle spines and pointing downwards and sideways, the lowest spine pointing towards the base 5-7mm long, shorter than two lowest pair, seldom a central spine, and then only one on older plants, (8-9mm long).

The flowers have a very light green pericarp while the plant body is dark grey-green, they are 70(-80)mm high, 46(-63)mm diameter, white, coming from the areoles in the crown, occasionally with the scent of lemon balm (*Mellisa officinalis* L.). Outer petals white, pale yellow towards the base, sometimes pale pink, with a creamy-white to pale yellowish-green middle stripe the tip often green to pink, wide lanceolate, 26(34)mm long, 9.5(7.5)mm wide, inner petals white, very pale yellow towards base, wide lanceolate (spoon shaped), 20(23)mm long, 8.2(9)mm wide. Interior of the receptacle thin bell-shaped, pale pink or pale pink with a white base, 18(19)mm high, 7(-10)mm diameter. Nectar chamber always

Photo: Volker Schädlich

Fig.245 Habitat of *G.schroederianum boessii, Schädlich* 85 north-east of Malabrigo, Prov. Santa Fé, Argentina

Fig.246 *G. schroederianum boessii, Schädlich* 84 in habitat north-east of Malabrigo, Prov. Santa Fé, Argentina

yellow, 2(-3.8)mm high, 1.5(-2.4mm) diameter, because it has the same diameter as the style, always closed. Filaments in 5-6 series, pale green to pale yellow-green, strong, the lowest curved towards the centre with the anthers touching the base of the style, the rest are curved towards the centre with the anthers forming a circle, the lowest series 4.3(-7)mm long the upper series 9(-12)mm long. Anthers white, 1 x 0.6mm. Style pale green, 10.2(-12)mm high, 1.5(-2.4)mm diameter, not free standing, (= not protruding into the ovary). Ovary wall below the base of the style mostly arched, seldom flat. Stigma creamy-white, 3.5-4.5mm long with 9-10 lobes. Ovary white, thin and obovate, 18-23.5mm high, always with a 4-11mm high basal area without ovules, the bottom 4-5mm diameter, is loosely filled with white ovules. Pericarp light green, always with a light grey bloom, shiny below, (25.5-32.3)-33.3mm high, 7(9)mm diameter. Scales half round to crescent-shape, thickened, light green pink at the top with a red tip.

Fruit pale green, spindle shaped, 29mm high, 11mm diameter, below the flower remains narrow, at the base stalk-like, elongated, splitting vertically, the dry fruit contains no solid or gluey pulp, the seeds come easily off the open fruit, 1.2mm long, 1.1mm diameter, less then 50% are covered with brown cuticle. Testa black, testa cells domed, the round hilum micropyle-region (HMR) round yet recognisably reduced on the funicle side, provided with a wide swollen edge, flat, and filled with elaiosome. (Subgenus *Gymnocalycium*). The dimensions in brackets are not those of the holotype but from other plants in the same collection.

Typus: Argentina, Provincia of Santa Fé, between Vera et Malabrigo, km 742 on ruta international No. 11, Department Vera, 58 m, 24th November 2000, collected by P. Eroles PE00-25; Holotype: BA [Body (part 1) and flower (part 2) in alcohol], Isotype: WU (flower in alcohol).

Base Map © Nelles Verlag GmbH

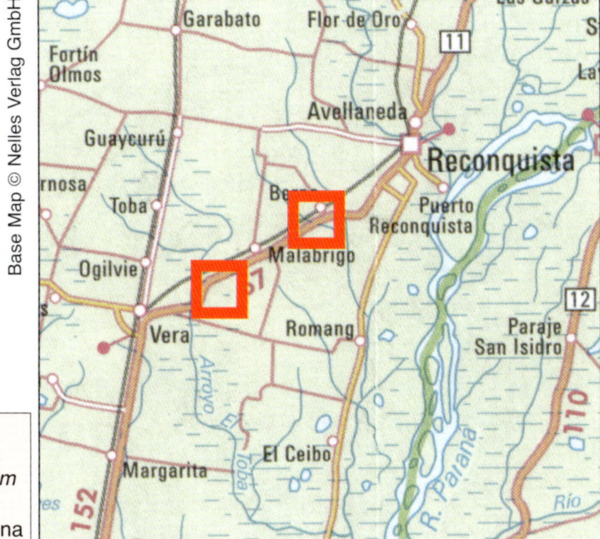

Map 33 Locations of *G. schroederianum boessii* in Prov. Santa Fé, Argentina

Distribution (Map 33)

This taxon is known from a number of places in Prov. Santa Fé, Argentina, Papsch (pers. com.) says that he has seen it at eight scattered localities over a range of 60km. It is said to grow in open flat land, at the edge of woodland, or under bushes, sometimes in grass and with moss.

Conservation Status

Least Concern. It is currently known from many locations without any recorded threats.

History

Kiesling (2002b) with other authors, created the name subsp. *boessii* for populations of *G. schroederianum* in the Argentinian Province of Santa Fé. Later in the same year, plants from these populations were named again by Neuhuber and Bercht (2002), this time as a separate species, *Gymnocalycium erolesii*, because the authors considered this taxon not to be closely related to *G. schroederianum*. They believe it to be associated with species from Córdoba and La Rioja, such as *G. erinaceum* and *G. kieslingii*. The subsp. *boessii* had been sent to Europe by Fechser as '*G. leptanthum*' some time earlier.

Commentary

This taxon may well not belong to *G. schroederianum* since it exhibits many differences especially the young seedlings. Meregalli (pers. com.) thinks this could be the original *G. leptanthum*, a name which has been associated with a number of taxa and hence is here rejected as unattributable. It certainly has long, thin flower tubes which are associated with *G. leptanthum*. Ludwig Bercht (pers. com.) still considers this to be a separate species for which *G. erolesii* would be the oldest name at species rank.

Cultivation

This taxon has not yet become widely available in cultivation and I have yet to grow it in my glasshouse. There is no reason why it should be any more difficult than other plants of *G. schroederianum*.

Gymnocalycium striglianum Jeggle ex H. Till

First Description (Jeggle)

Gymnocalycium striglianum Jeggle nom. inval. KuaS 24(12): pp.267-268 (1973). Validated by H.Till in KuaS 38(8): p.191 (1987)

Body simple, 30-50mm high, 40-80mm diameter. Epidermis blue-green to brown. Root short. Ribs 8-12, the grooves curved. Areoles longish, 2-3mm wide, up to 5mm long, 10-15mm apart, with grey-brown felt. Spines 3-5, up to 15mm long, blackish not becoming grey.

Flowers 40mm diameter and 50mm long, creamy white with a pink tint. Ovary longish, even so the flower has only a few, not pointed, white edged scales. Style with 10 yellowish lobes, not reaching up to the many stamens.

Fruit, grey-green to brownish, splitting lengthwise. Seeds 1mm diameter, 1.5mm long, black, rough. Hilum area oval with a light edge.

Habitat: Argentina, Prov. Mendoza.

Type: W. Rausch s.n. Argentina, Prov. Mendoza (WU) designated by H. Till (1987).

Etymology

Named for Franz Strigl, *Gymnocalycium* specialist and co-founder of the Austrian specialist group 'Arbeitsgruppe Gymnocalycium'

Fig.247 *G.striglianum,* GC219.03 from near Ugarteche, Prov. Mendoza, Argentina

Fig.248 The habitat of *G.striglianum* GC219.03 near Ugarteche, Prov. Mendoza, Argentina

Synonym

G. striglianum subsp. *aeneum* Meregalli & Guglielmone in Cactus & Co. X(2): pp.69-81 (2006)

Commentary on *G. striglianum* subsp. *aeneum*

Differentiated from the type by the bronze-coloured body with a thick waxy coating, the more prominent, stronger horn-coloured spines, and the flower with more slender pericarpel, almost completely white with very light-pink receptacle. It could probably be recognized as a good subspecies.

Distribution (Map 34)

Originally found near to Cacheuta in Prov. Mendoza, Argentina, and since also in the surrounding area to the south and east of Mendoza City. There is also a population growing in the Sierra de las Quijadas that was named subsp. *aeneum* by Meregalli & Guglielmone (2006).

The plants grow in sandy or gravelly loam soil under dense thorny bushes often among grass. Neuhuber (2007) says that the subsp. *aeneum* grows on limestone soils, a rare occurrence in South America

Map 34 Distribution of *G. striglianum* in Prov. Mendoza & San Luis, Argentina

Photo: Massimo Meregalli

Fig.249 *G.striglianum* (aeneum), *Meregalli* 627 in habitat at Mina la Calera, Prov. San Luis, Argentina

Photo: Massimo Meregalli

Fig.250 *G.striglianum, Meregalli* 763 in habitat at the Sierra Valera, Prov. San Luis, Argentina

Conservation Status

Vulnerable. The limited known populations and the effects of urban expansion and grazing are a cause for some concern.

History

The plants collected by Rausch on his first expedition included about 30 specimens of this species and Jeggle had the chance to see them when they arrived in Austria in February 1973. They had been collected near Cacheuta, Prov. Mendoza, Argentina. He took them away and gave some to Strigl and other *Gymnocalycium* enthusiasts.

Following observation of the flowers and fruits, he described *G. striglianum* but without preserving a type, so it was left to Till (1987) to preserve one of the original Rausch plants and designate it as the type of the name.

Independently from Rausch, another population of *Gymnocalycium* had been found by Borth (BO89) which he found further east near to Zanjitas in Prov. San Luis. This had also been thought to be *G. striglianum*, but most recently, Neuhuber (2007) says that it belongs to his *G. borthii* subsp. *nogalense*. This view agrees with Borth's original opinion at the time he found it, which was that he thought it was a form of his BO55, the type of *Gymnocalycium borthii*, treated here as a subspecies of *G. gibbosum*.

The population growing in the Sierra de las Quijadas was named subsp. *aeneum* by Meregalli & Guglielmone (2006) although it had been known for some time according to Till (1988a).

Commentary

This species is closely related to *G. gibbosum* and is the furthest west of that group of taxa. When I found it near to Ugarteche, Prov. Mendoza, in 1996, I thought it was *G. gibbosum*, but there are certainly sufficient differences to justify its recognition as a separate species.

Cultivation

Surprisingly rare in cultivation, this is a very beautiful plant with its dark brown body and spectacular black spines. It can be tricky to grow, slower and more challenging than *G. gibbosum* forms. The pretty, neat, white flowers are an excellent contrast with the dark colours of the plant. There has recently been a supply of seedlings said to be from Luján, a small town near to Mendoza City. Plants of the subsp. *aeneum* have not yet appeared in commerce but they promise to be an attractive form to grow.

Illustrations

Neuhuber (2007) makes the point that Figs. 10 to 13 in Meregalli & Guglielmone Cactus & Co. X(2): pp.77-78 (2006) are, in fact, his *G. borthii* subsp. *nogalense* not *G. striglianum* subsp. *aeneum*.

New Cactus Lexicon Atlas (2006) Illn. 274.3

Fig.251 *G.striglianum* GC219.03 from near Ugarteche, Prov. Mendoza, Argentina

Gymnocalycium taningaense Piltz

First Description (Piltz)

KuaS 41(2): pp.22-26 (1990)

Gymnocalycium taningaense Piltz spec. nov.

Body solitary (or only offsetting in age) flattened globular, 4.5 to 6cm diameter, 2.5cm high, up to 4cm in cultivation, silky-matt lead-grey to slate-grey, with fibrous roots, crown slightly sunken; ribs 9-11 flat, up to 12mm wide in the middle of the plant body with a weak hump below the areole and over the cross-furrow. Areoles round, somewhat sunken, 2-3mm diameter, with white felt, 5-7mm apart.

Spines 7-11, of which one or two are arranged as centrals, these up to 11mm long pointing upwards, red-brown, grey-brown to grey in the new growth, darker at the base, needle-thin, the radial spines coming from the middle of the areoles spread sideways and upwards at an obtuse angle to the body, straight to slightly curved, 3-8mm long, red-brown to grey, darker at the base, also needle-like.

Flowers:- 4.0-5.5cm long, 3.0-4.5cm wide, thin cup-shaped, pericarpel 1.4-2cm long, 4-7cm wide, olive to dark green, with a bloom, scales up to 4mm wide, up to 3mm long, half round with small tips or spatulate, olive green with a dirty white edge; receptacle up to 11mm long, 6-8mm wide at the bottom, 11-14mm wide at the top, inner wall greenish white, sometimes even reddish above the nectar chamber; transitional petals 7-22mm long and up to 5mm wide, brownish green, edges dirty white, with a silky sheen, outer petals 15-22mm long, 5mm wide, blunted lanceolate, white with a weak brownish central stripe outside and inside, inner petals up to 20mm long, 3mm wide, with a lance-like tip, white, at the base pale yellow; filaments up to 1cm long, greenish yellow, inclined towards the style; anthers yellow, roundish, 0.5-1mm long; style pale green, almost white above, 10-12mm long, usually 8 stigma lobes, up to 4mm long, white to yellow; nectar chamber about 2mm high, tubular, pale pink.

Fig.253 *G.taningaense, Piltz* 212 from Salsacate, Prov. Córdoba, Argentina. The type collection.

Fruit 15-30mm long, egg to club-shaped, blue-green with a bloom, splitting open lengthwise in one to three places, still carrying flower remains.

Seeds c. 1.2-1.4mm long, 1-1.2mm wide; testa matt black, with warts, the loose aril remnants of a light brown appearance, still clinging to the surface here and there; hilum-micropyle area sunken crater-like, oval, sometimes with a point at one end; belonging to the series Baldiana (sensu Buxbaum) or to the subgenus Ovatisemineum (sensu Schütz).

Habitat:- Argentina, Province Córdoba, near Taninga, c. 950m.

Holotype deposited in alcohol, (body, flower, fruit, and seed) in Herbarium des Succulenten der Universitat zu Köln (Cologne) with the number P212/6 (leg. Piltz 1980).

Etymology

Occurs near the village of Taninga, Prov. Córdoba, Argentina.

Synonym

G. lukasikii Halda & Kupcak in *Acta musei Richnoviensis* 7(2): p.72 (2000)

Fig.252 *G.taningaense* GC396.02 in habitat to the west of Taninga, Prov. Cordoba, Argentina

Fig.254 *G.taningaense* (lukasikii), *Papsch* 92 74-98 from the Sierra del Morro, Prov. San Luis, Argentina

Fig.255 *G.taningaense* (lukasikii), *Papsch* 92 74-98 from the Sierra del Morro, Prov. San Luis, Argentina

Fig.256 *G.taningaense,* GC396.02 from the west of Taninga, Prov. Córdoba, Argentina

Fig.258 *G.taningaense,* STO 197 from Taninga, Prov. Córdoba, Argentina

Fig.257 *G.taningaense, Neuhuber* 81 378-1264 from Cienaga, Prov. Córdoba, Argentina

Fig.259 *G.taningaense* GC396.02 from the west of Taninga, Prov. Córdoba, Argentina

First Description of *Gymnocalycium lukasikii* Halda and Kupcak

Species near to *G. bruchii*, but in respect of the flowers, larger up to 40mm diameter whitish-pink in colour. Body solitary, low, flattened, greyish-green, 35-60mm diameter, ribs blunt rounded, more or less crenate, areoles oblong, spines irregular spider-like; flowers slim c.60mm long, seeds small, dark, tuberculate. Taproot thin, c.40mm long, 5mm diameter. Ovatisemineum.

Holotype: PR No. JPR. 100/237 leg. J. Prochazka 24.11.1998. (date changed to 24.11.1992 in the 2002 version)

Habitat: southern Argentina, San Luis, Sierra del Morro, growing on steep stony hills at 900m.

A very unusual taxon with prominent slim whitish-pink flowers, is dedicated to Mr. Emil Lukasik from Ostrava.

Commentary on *G. lukasikii*

Papsch (2003) tells us how, in October 1989, he found this plant in the company of G. Hold, J. Prochazka and A. Wiedrich when only three specimens were found. Later, in 1992, a further visit revealed two further individuals. Since Prochazka did not collect material, the designation of the type under his field number needs to be explained.

From the first visit, Papsch had taken plants that flowered and set fruit in 1991. Some of these seeds went to Prochazka which he distributed under his own field number. It must be assumed that the type material originated from this material.

Papsch goes on to be critical of the description, suggesting that elements from other species growing in the area have been incorporated, so devaluing the description, even though he considers the discovery to be a genuine new taxon.

There are various reports of similar plants being found in the general area and when Neuhuber (2007) published it as a variety of *G. taningaense,* he provided a map suggesting a considerable distribution further north to within a few kilometers of a known locality of *G. taningaense*.

G. taningaense subsp. *fuschilloi* Neuhuber in *Gymnocalycium* 20(1): pp. 705-708 (2007)

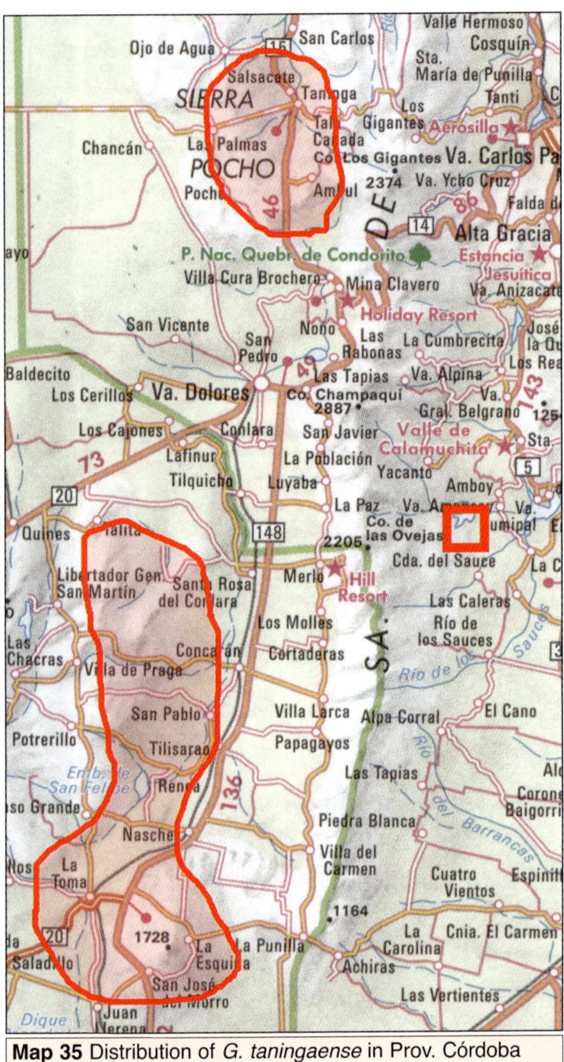

Map 35 Distribution of *G. taningaense* in Prov. Córdoba and San Luis, Argentina

Distribution (Map 35)

The species occurs to the west of the Sierra Grande and Cumbre de Achala in Prov. Córdoba. The northerly limit of the distribution is both east and west of Salsacate. The inclusion of *G. lukasikii* extends the distribution area much further south into San Luis Province. The isolated population called subsp. *fuschilloi* by Neuhuber (2007) occurs on the east side of the Sierra de Comechingones, from Arroyo San Antonio to the Río de la Cruz in Province Córdoba.

Conservation Status

Least Concern. The plant has a large distribution in a stable environment.

History

Piltz (1990) explains that he found this species when he was looking for *G. horridispinum*. It has been accepted as a good species ever since. The combination of *G. lukasikii* as a variety of it by Neuhuber (2007) clarified the relationship of this obscure taxon which remains little known in culture.

Commentary

Neuhuber (2001a) differentiated *G. gaponii* from *G. taningaense* on the basis of its skittle-shaped body, paler epidermis and stiffer spines. The former is here treated as a synonym of *G. erinaceum* but all these could be one variable, widespread species.

Cultivation

G. taningaense is easy to grow and in culture makes a cluster of small stems and flowers easily. Making offsets is a consequence of culture, the plants usually remaining solitary in habitat. In its typical form it has a dark body and fine spines which make it an attractive addition to a collection.

Photo: Massimo Meregalli

Fig.260 *G.taningaense* (lukasikii), *Meregalli* 712 in habitat south of Sierra de la Estanzuela, Prov. San Luis, Argentina

Photo: Massimo Meregalli

Fig.261 *G.taningaense* (lukasikii), *Meregalli* 714 in habitat in the Sierra del Yulto, Prov. San Luis, Argentina

Gymnocalycium uebelmannianum Rausch

Fig.262 The habitat of *G. uebelmannianum* GC979.02 in the Sierra Velasco, Prov La Rioja, Argentina.

First Description (Rausch)

Succulenta 51(4): pp.62-64 (1972)

Simple, offsetting when damaged, flattened-globular; 10mm high 70mm diameter. Epidermis grey-green, sunken in the crown, roots turnip-like. Ribs 8-12, straight, divided into c. 5-8mm long humps by cross-cuts.

Radial spines 5-7, usually one pointing straight down, 5-15mm long, soft and slightly curved, chalky-white, central spines 0.

Flowers 35mm long and wide. Pericarpel skittle-shaped, receptacle funnel-shaped, shiny green with heart shaped scales, brownish with pink points, outside petals wide tapering, brownish with light edges, inner petals smaller, no wider, tapering with a fine point, white, inside yellowish, throat light or dark pink. Filaments white with a pink base, the lower part thickened, loosely arranged around the base of the style, covering the whole of the receptacle; Anthers yellow, style thick, cylindrical, greenish at the base, with 10 stigma lobes, 4mm long, yellow.

Fruit; Broad-globular, 6mm diameter, scales yellowish-brown. Seed pot-shaped, 1mm diameter, hilum basal.

Habitat: Argentina, Sierra de Velasco, 2200-2800m.

Type: *Rausch* 141 in Herbarium W.

Fig.263 *G. uebelmannianum* GC979.02 in the Sierra Velasco, Prov La Rioja, Argentina.

Fig.264 *G. uebelmannianum* GC979.02 in the Sierra Velasco, Prov La Rioja, Argentina.

Fig.265 *G.uebelmannianum, Rausch* 141 from above Pinchas, Sierra de Velasco, Prov. La Rioja, Argentina

Etymology

Named for Werner Uebelmann, Swiss nurseryman and travelling companion of Leopoldo Horst.

Distribution (Map 36)

It grows only at high altitude in the Sierra de Velasco, Prov. La Rioja, Argentina. There are reports of finding the plant at various places in the mountain range once the correct altitude is reached, so it is probably widespread, and not as restricted as Rausch suggested when he described the species.

The climb up the mountain can be difficult because the lower slopes are covered with dense spiny bushes and it is a long steep walk. The higher slopes, where *G. uebelmannianum* is found, support much sparser vegetation with open patches where the plants grow in crumbling granite.

Conservation Status

Least Concern. There is ample suitable habitat in the Sierra de Velasco at the latitude favoured by this taxon and the area is far from human activity and still little grazed. The population I saw was in good condition and contained very many plants of all ages.

Fig.266 *G.uebelmannianum, Rausch* 141 from above Pinchas, Sierra de Velasco, Prov. La Rioja, Argentina

Fig.267 *G. uebelmannianum* GC979.02 in habitat in the Sierra Velasco Prov La Rioja, Argentina.

History

Rausch (1972a) found this plant with Marcus in 1965 and brought back 28 young plants from three small populations together with many seeds. The resulting seedlings flowered in various shades of white, even one with a golden yellow flower. Remarkably, Till (1991) reported that the flower colour was not stable in individual plants and that white-flowered plants could occasionally produce lilac-pink or pale red flowers.

Commentary

A very distinctive species which cannot easily be confused with any other, *G. uebelmannianum* is probably most closely related to *G. andreae* or *G. baldianum*. They are all high altitude species but the others live in a less arid environment

Cultivation

This is a rare species in cultivation, most plants with any data being descendants of the type collection, *Rausch* 141. It is sensitive in culture, the tap root being very susceptible to rot if overwatered or not well enough drained. Perhaps this is because the plant grows in crumbling granite with little organic material in its natural habitat.

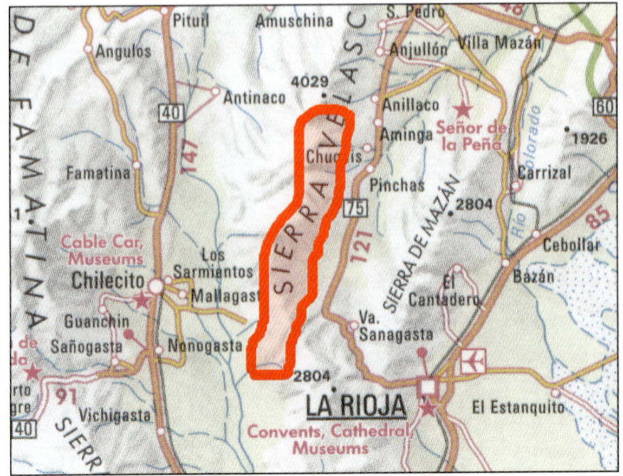

Map 36 Distribution of *G. uebelmannianum* in Prov. La Rioja, Argentina

Subgenus *Microsemineum*

Type Species

G. saglionis (Cels) Britton & Rose

Species and Subspecies (synonyms indented)

Echinocactus Saglionis Cels.

Gymnocalycium albiareolatum Rausch

Fig.268 Habitat of *G. albiareolatum*, GC977.01 near Villa Sanagasta, Prov. La Rioja, Argentina

First Description (Rausch)

Succulenta 64(10): pp.213-214 (1985)

Gymnocalycium alboareolatum Rausch spec. nov.

Simple, flattened-globular, up to 60mm diameter, greenish-grey, often tinted violet, with a turnip-like root. Ribs 9-11, straight, upright, with cross-cuts, that form c. 10mm long, chin-like humps. Areoles, round to oval, 5mm long, strongly white felted in new growth. Radial spines 6-7, awl shaped, curved to the body, brown, rough, up to 10mm long. Central spines absent.

Fig.269 *G. albiareolatum*, GC977.01 near Villa Sanagasta, Prov. La Rioja, Argentina

Flowers appear from close to the centre of the crown, 65mm long and 45mm wide. Ovary long narrow, up to 17mm long, pointed at the bottom: Tube dark green, with wide, whitish edged scales; outer petals whitish-pink with a green central stripe; inner petals silvery-white with a greenish-pink toned middle stripe. Throat, stamens and anthers pink; stamens are set in 2 series, the lower being curved in towards the style. Style is short and thick, with 11 stigma lobes, whitish. Fruit oval to club shaped, thin at the bottom, 25-30mm long, with a blue bloom, scale wide pink, fruit splitting vertically.

Seed round, hat shaped, 1mm diameter, black, rough, with a wide basal hilum.

Habitat: Argentina, La Rioja, near Villa Bustos, 1000m.

Type: *Rausch* 716, deposited in the Städtische Sukkulentensammlung, Zurich.

Etymology

From the Latin 'albus' meaning white, referring to the white areoles.

Distribution (Map 37)

This taxon has a very limited distribution near to Villa Sanagasta (previously Villa Bustos), Prov. La Rioja, Argentina. It is said to be usually buried in the ground making it very difficult to see when not

Fig.270 *G.albiareolatum, Piltz* 221 from Sanagasta, Prov. La Rioja, Argentina

Fig.271 *G.albiareolatum, Piltz* 384 from Cuesta Huaco, Prov. La Rioja, Argentina

in flower or fruit. Rausch later described a smaller, caespitose form as var. *ramosum,* although plants with offsets have been reported from near the type locality.

The type locality is said to be about 2km south of Villa Sanagasta on the way to Dique de Los Sauces. Recent housing development and road construction is thought to have damaged the type locality, although during my recent visit I was able to see hundreds of plants in bud and fruit on a gravelly slope near to the town (Fig.268).

Conservation Status

Vulnerable. Although there are still plenty of plants where there is no obvious immediate threat to them, the total distribution area is believed to be small and near the expanding town of Villa Sanagasta. Recent road development and housing expansion have affected the population near the town, but there is no obvious threat to the remaining plants on the nearby slopes of the Sierra Velasco.

Map 37
Location of *G. albiareolatum* in Prov. La Rioja, Argentina

History

The spelling of this species name was changed from the original by Huxley et al. (1992) supposedly to comply with the ICBN rules. However, although this correction was questioned by H. Till (2005a), the revised spelling is used here.

Although at first it was said to be a member of the subgenus *Gymnocalycium*, it in fact belongs to *Microsemineum*.

Rausch later described var. *ramosum* for plants with smaller heads, but this form occurs in the main population

Commentary

This is one of the best examples of convergence, where two species of *Gymnocalycium* from different seed groups grow near to each other and look so much alike that they can easily be confused. A form of the more widespread *G. kieslingii* (confusingly called forma *alboareolatum)* occurs in a similar habitat nearby. I can find no reference to them actually growing together, rather that there is an altitude gap with *G. albiareolatum* below 1000m and *G. kieslingii* above 1500m.

This taxon appears to be a tap-rooted geophytic relative of the *G. hossei* complex. Like members of this complex, the anthers are red, clearly visible after the pollen has been shed. Meregalli et al. (2000) confirm further similarity in the microrelief of the seeds.

Cultivation

This species is quite rare in cultivation. Most plants derive from *Piltz* 221 originally collected near Villa Sanagasta.

It presents no problems in culture and its similarity to *G. kieslingii,* which is so striking in habitat, is soon lost so that the two are readily distinguishable in cultivation.

Gymnocalycium bayrianum H.Till

First Description (H. Till)

Gymnocalycium bayrianum H.Till nom. inval. KuaS 18(12): pp.222-224 (1967). Validated by H.Till in KuaS 38(8): p.191 (1987)

Root; short taproot. Body flattened, blue-green to green, the examples at hand, c. 6-12, average 90mm diameter. Crown slightly depressed, mostly free from spines.

Ribs; 6-10, later more ribs appearing, straight, very broad at the base (c. 30mm) and very flat, near the apex more raised and weakly edged, above the areoles with weak transverse notch.

Areoles; about 20mm apart, new growth with tufted yellow felt, later greying and disappearing.

Spines; 5 radial spines, c. 25-30mm long mostly round but sometimes with edges, curved over the body, stiff, the 2 uppermost curving upwards, the 2 middle curving downwards, and the lowest pointing straight down. in new growth light brown, sporadically with darker tips, later becoming grey, sometimes 1 central spine, 35-45mm long.

Flowers; in the vicinity of the crown, from the uppermost areoles, funnel-shaped, opening freely, quite long, (average of 4 examples) 62mm long, 40mm wide, cream-white with a silky sheen.

Receptacle; short, 16mm long, 10mm diameter, grey-green, thickly covered with hemispherical, lilac-pink scales, blending into red-brown green, later becoming the same colour as the receptacle but still with lilac-pink margins.

Outer petals; spatulate, c. 22mm long, 8mm broad with a broad metallic central stripe. Inner petals, broad lanceolate, c. 27mm long, 6mm wide, cream-white, with a silky sheen, reddish at the base.

Filaments; very fine, long, white, the lower row lying against the style, so enclosing the nectar gland, the other rows arranged one above the other set in the wall of the perianth tube, reaching over the style. Anthers light yellow.

Fig.273 *G. bayrianum, Neuhuber* 88 69-169 from the Sierra de Medina, Prov. Tucumán, Argentina

Style; greenish-yellow, reaching to half the length of the tube, 11mm long, 16mm long with the stigma, 1.6mm diameter, greenish-yellow, with 11 lobes.

Nectary; deep and narrow, scarcely dilated at the base. Ovary; acorn shaped, 8mm long, 4.5mm diameter, white, full of ramified seed funicles.

Fruit; ovoid, somewhat rectangular, grey-green, with a blue bloom, c. 25mm long, 16mm diameter, with a few broad lilac-pink scales, splitting open lengthwise.

Seeds; small, longish, slightly curved, c. 0.8-1.0mm diameter. Testa matt red-brown, with fine warts, pitted in between. Hilum oval, slightly curved, scarcely depressed, without marginal thickening, micropyle somewhat raised. Seed group *Microsemineae* (Fric).

Fig.272 *G. bayrianum,* GC449.02 in habitat in the Sierra del Nogalito, Prov. Tucumán, Argentina

Fig.274 *G. bayrianum,* GC449.02, a young seedling from the Sierra del Nogalito, Prov. Tucumán, Argentina

Photo: Massimo Meregalli

Fig.275 *G. bayrianum,* CB 213 from La Hoyada, Prov. Salta, Argentina

Origin: Prov. Tucumán near Medina, on top of the mountain range of the same name, 1000-1500m.

The type plants Nos. 446-450, are in the protected collection in Linz Botanical Garden.

Validated by Till (1987) quoting the Holotype as Schickendantz s.n., Argentina, Prov. Tucumán, Cerro Medina, 1000-1500m, cult. HT473 (WU in alcohol)

Etymology

Named for Alfred Bayr, president of the Austrian cactus society Gesellschaft Österreichischer Kakteenfreunde.

Other Names

Gymnocalycium spegazzinii subsp. *bayrianum* (H.Till) Halda in Cactaceae etc, 9(3): p.84 (1999)

Distribution (Map 38)

This taxon was originally known only from the Sierra de Medina and the nearby Sierra del Nogalito. Both are isolated ridges of hills just to the north of Tucumán City and on the eastern extreme of the mountains in this part of Argentina. To the east of here the land is flat. The region has good rainfall and provides rich grazing land for cattle, the cacti being found only in rocky places.

Base Map © Nelles Verlag GmbH

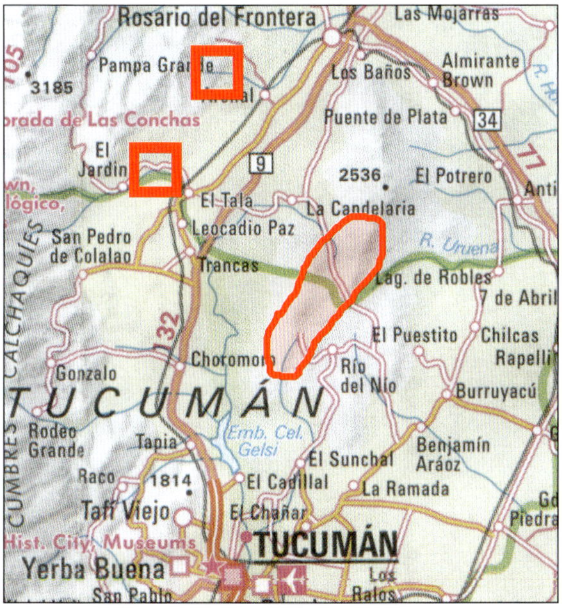

Map 38 Distribution of *G. bayrianum* in Prov. Tucumán & Salta, Argentina

H. Till & Neuhuber (1989) tell us about two new habitats north of the type locality. They explain that Dietrich Herzog found the plants in 1981 near La Hoyada in Prov. Salta at about 1000m growing from cracks in red sandstone similar to where they grow in the Sierra de Medina. Furthermore, they found another similar habitat at El Brete (900m), also in Prov. Salta some 25-30km south of Herzog's locality. Nowhere is this species known to overlap with the distribution of the similar *G. spegazzinii*.

Conservation Status

Least Concern. It is likely that this plant has a much greater distribution than first thought and, although there is evidence that goats eat the plants, there are apparently plenty of healthy individuals.

History

In 1964 Till was visiting Uhlig's Cactus Nursery at Rommelshausen when he noticed a number of Gymnocalyciums amongst a recent importation that he did not recognise. He was told that they had been collected from the Cerro Medina by Schickendantz who lived in nearby Tucumán city. Having purchased 20 individuals, the 5 largest were placed at Linz as nos. 446-450, later to be cited as types in the first description.

This first description of this plant was published in 1967 by Hans Till in KuaS 18(12): pp.222-224. However, it was an invalid publication of the name because of the lack of a preserved type. H.Till later validated the description in 1987 by designating a holotype.

Commentary

This taxon is widely regarded as the closest relative of *G. cardenasianum* from Bolivia and the widespread *G. spegazzinii* which can be found further to the west in Argentina. Although treated as a subspecies of *G. spegazzinii* in the *New Cactus Lexicon* (2006), it is actually more similar to *G. cardenasianum*. Unless all three are regarded as one species, then separate recognition is appropriate.

Cultivation

Of the three species mentioned above, this is the least often seen in cultivation although it is easy to grow and rather quicker than the others. Perhaps because its habitat is away from the most frequently visited *Gymnocalycium* areas, it is not often included on the itinerary of travellers looking for cacti and hence seed is rarely collected.

Gymnocalycium cardenasianum Ritter

Fig.276 The habitat of *G. cardenasianum* GC887.02 near Villa Abecia, Dept. Tarija, Bolivia

First Description (Ritter)

Diagnosen von neuen Kakteen - *Taxon* XIII(4): p.144 (1964) *Gymnocalycium cardenasianum*

Gymnocalycium cardenasianum Ritter, sp. nova.

Body hemispherical, 5-20cm high, 12-23cm diameter, grey-green. Ribs 13-31, 5-10mm high, 2-5cm wide, straight, blunt. Tubercles separated by cross-grooves. Areoles 5-10mm long, 3-7mm wide, 5-15mm apart, grey. Spines almost black, becoming grey, radials 3-6, 3-6cm long, strong, curved inwards, centrals usually absent, later, sometimes 1-2, 5-6cm long, curved.

Flowers from the crown, 5cm long; ovary 11mm diameter, globose; scales semi-circular, reddish, 5mm long, 3mm wide; nectary 3mm long, round, reddish, nearly closed; flower tube 20mm long, throat somewhat narrowed, 16mm diameter, funnel shaped, grey-green, inside purplish, the same as the ovary; petals 20mm long, 6-9mm wide, base narrow, purple, paler at the top, with a central rust-red line; stamens 14mm long, reddish-yellow; stigma lobes 15, pale yellow.

Fig.277 *G. cardenasianum* GC887.02 near Villa Abecia, Dept. Tarija, Bolivia

Fruit ± 20mm long, 17mm diameter, grey-green, narrow at the top; covered with wide short scales. Seed 1mm long, 0.7mm wide, reddish, with fine blackish warts; hilum basal, long, white.

Habitat: Carrizal, Prov. Mendez, Dept. Tarija, Bolivia

Type: *Ritter* 88, 1953, Bolivia, Prov. Mendez, Carrizal (U).

Etymology

Named for Dr. Martín Cárdenas (1899-1973), a Bolivian botanist who described many new cacti.

Other Names

G. spegazzinii subsp. *cardenasianum* (Ritter) Kiesling & Metzing in *Darwiniana* 34: p.404 (1996)

Synonym

G. armatum Ritter in *Kakteen in Südamerika* 2: pp.662-663 (1980)

First description of *Gymnocalycium armatum* Ritter spec. nov.

Body simple, grey-green, with a hard flesh, at flowering age 4-12cm diameter, in habitat only slightly elevated, strong conical turnip-like root. Ribs, 8-15, 15-30mm wide, 4-7mm high with very small narrow cross-cuts; humps missing or up to 4mm high, flat-rounded. Areoles 6-10mm diameter, rounded, on top of the humps, (if any) grey felted, 5-10mm apart. Spines very strong and stiff, away standing, brown to nearly black, radial spines, 3-6, 4-7cm long, stronger at the base, usually all are recurved, sometime a few of the spines are hooked; central spines missing, or 1, rarely 2.

Flowers coming from the crown, 3-4cm long, opening for a few hours a day, only for a few days, then only when the sun is warm, unscented. Ovary semi-globular, 3.5-6cm long, 6-9mm thick, with a

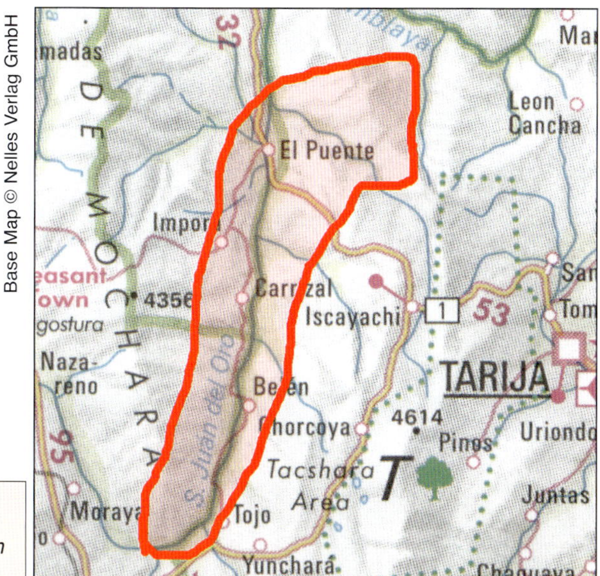

Map 39
Distribution of
G. cardenasianum
in Bolivia

few greenish, white edged, blunt, wider than long scales, 1.5-3cm long. Nectar chamber purple, 1.5-2mm long, with a nectary ring nearly closed at the top with stamens, top part of the tube is beaker shaped, 12-18mm long, at the top nearly as wide, inside purple, outside grey-green with large rounded scales. Filaments light purple in a basal ring, c. 5mm long, leaning against the style, the rest of the filaments are inserted above an intersecting gap of 3mm, up to the seam, curved inwards, the cream-coloured anthers form a hollow globe around the stigma lobes. Style without lobes, 6-7mm long, yellowish, with 8-11, first closed, later, wide spread pale yellow stigma lobes, 3-5mm long. Petals 15-22mm long, 5-7mm wide, very narrow at base, rounded at the top or with a short point, white with a pink edge, central stripe purple towards the bottom, brownish towards the top, inner petals wide, upright, white, pink edged, outer petals spread out. Fruit c. 15-20mm long, 7-10mm diameter, greenish. Seeds 1.2mm long, 0.7mm wide, 0.5mm thick, reddish-black, with very fine humps, dorsal strongly arched, hilum white, long with a large micropyle hole towards the ventral side.

Type: El Paicho, Province Mendez. Bolivia. I discovered it in 1962. Related to *Gymnocalycium cardenasianum*. Number FR1131. Seed only (U).

Gymnocalycium spegazzinii subsp. *armatum* (Ritter) Halda & Sorma comb. & stat. nov in *Acta musei Richnoviensis* 9(1): p. 65 (2002).

Commentary on *G. armatum*

Till & Amerhauser (1999) combined this taxon as a form of *G. cardenasianum* in *Gymnocalycium* 12(4): pp.305-312 claiming that it occurs frequently in typical populations. Then, in the 2002 issue of *Acta musei Richnoviensis* 9(1): p.65, Halda & Sorma published *armatum* as a subspecies of *G. spegazzinii* explaining that plants they saw near Parokia, Río Toraja Puno at 2750m had seeds more like *G. spegazzinii* than *G cardenasianum*.

An interesting article by Beckert (2006), who is better known for his interest in *Parodia*, reports Ritter's writings on this taxon. From these, he was able to visit the type locality and report his findings, concluding that it is probably just a beautiful form of *G. cardenasianum*.

Distribution (Map 39)

This taxon is known from many localities in Bolivia above 2300m, some consisting of numerous individuals. The distribution is mainly in Prov. Tarija, not far from the border with Argentina. The form *armatum* is from the northern part of the range. There is a 200km gap between these populations of *G. cardenasianum* in Bolivia and the related *G. spegazzinii* in Argentina.

Fig.278 *G. cardenasianum, Ramirez* 176 from Carrizal, dept. Tarija, Bolivia

Conservation Status

Least Concern. There is no known threat that could significantly impact the populations of this plant. It is also likely that there are many more populations hidden in the mountains of their remote habitat.

History

This taxon was first discovered by Ritter in 1953 near to Carrizal, Prov Mendez, dept. Tarija, Bolivia. The seeds he sent to his sister Frau Winter were offered in her 1955 catalogue under the provisional name *G. cardenasii* FR88.

Commentary

The appearance of this plant is very variable with regard to the disposition, colour and extent of the spines. Ritter's *G. armatum* is a description of one of these forms with strong, dark outstanding spines, a form which occurs in populations of *G. cardenasianum,* particularly in the north of the range.During dry periods, the plants shrink considerably so that the body is hidden under a dense mass of interlacing spines.

Cultivation

Although slow growing, this plant is of easy cultivation in a more mineral soil. It should be given strong light to encourage the development of its strong spines. The spininess of the parents is reflected in the seedlings, so the most handsome individuals are the progeny of plants from the most densely-spined populations. The first flowering can be expected when the young plants attain a diameter of about 10cm, although the flowers are often prevented from opening fully because of the spines.

Gymnocalycium castellanosii Backeberg subsp. *castellanosii*

Fig.279 The habitat of *G. castellanosii castellanosii* GC953.02 near Olta, Prov. La Rioja, Argentina

First Description (Backeberg)

Kaktus-ABC: pp.287, 416-417 (1936)

Gymnocalycium Castellanosii Backeberg sp. n.

Stem simple, elongated-globular, to 15cm high, 10cm diameter, dull velvety glaucous-green. Crown depressed, with woolly hair. Ribs 10-12, becoming broader and flattener in age, 4cm wide at the base. Areoles 1cm long, c. 2cm apart, at first with dense yellowish wool, more sharp angled. Flowers c. 45mm long, white, with a pink tinge, bell to funnel shaped. North Argentina.

Fig.280 The first illustration of *G. castellanosii* published in Backeberg's *Blätter für Kakteenforschung* 1936

Etymology

Named for Dr. Alberto Castellanos (1897-1968), Argentinian botanist in Prov. Córdoba and assistant of Carlos Spegazzini.

Fig.281 *G.castellanosii castellanosii,* GC953.02 west of Olta, Prov. La Rioja, Argentina

Fig.282 *G.castellanosii castellanosii* (below) GC955.02 growing with *G. saglionis*, south of Malanzán, Prov. La Rioja, Argentina

Fig.283 A young plant of *G.castellanosii castellanosii*, GC953.02 west of Olta, Prov. La Rioja, Argentina

Fig.284 *G.castellanosii castellanosii* (bozsingianum), GC19.02 west of Chepes, Prov. La Rioja, Argentina

Synonym

G. bozsingianum Schütz in *Kaktusy*13(6): pp.124-126 (1977)

First description of *Gymnocalycium bozsingianum* Schütz sp. n.

Body flat-round, simple, not offsetting, 80mm or more diameter, crown sunken, epidermis matt grey-green. Ribs 12, with rounded edges, straight, weakly humped, humps running into each other, not very noticeable, with small chins under the areoles, and indefinite cross-cuts forming between the humps. Areoles round, c. 5mm diameter, with grey-white felt, later becoming bald, 15-20mm apart. Radial spines 5, sometimes only 3, up to 20mm long, awl shaped, brown, later grey, slightly away-standing, central spines mostly missing, rarely 1, similar to the radial spines.

Flowers developing from young areoles in the crown, buds c. 50mm long, flowers funnel shaped, when open 50mm long and 50mm diameter. Flower tube (pericarp and receptacle) c. 20mm long, 8mm diameter, covered with olive-green, pink-edged scales, petals pointed spatulate, outer petals, outside olive-green, inside pink, inner petals pink, flower throat wine red, many stamens developing from the inner wall of the receptacle in the region of the stigma, filaments green, anthers yellow, style c. 20mm long, green, with 10 yellow lobes. Fruit berry-like, up to 20mm long, 15mm diameter, blue-green, with a bloom, splitting open lengthwise. Seeds helmet shaped, 0.8-1.0mm high, 0.6-0.8mm diameter, testa brown, shiny, at c. x10 magnification one can recognise small fine rounded warts, hilum basal, oval, arilus unclear.

Habitat: Argentina, Prov. La Rioja, near the town Chepes Viejo. Holotype: in herbarium SCK.

Gymnocalycium castellanosii subsp. *bozsingianum* Amerhauser & Till

Commentary on *G. bozsingianum*

This plant was found in a consignment of imported plants from Fechser at Uhlig's nursery. Named after Franz Bozsing (1912-1990), the *Gymnocalycium* specialist from Salzberg, Austria. Piltz (1993) combined it as a variety of *G. castellanosii,* then Till (2006) made it a subspecies:

Gymnocalycium castellanosii subsp. *armillatum* (Piltz) Papsch

This combination was initially invalid (Art. 33.4) because the basionym reference was not to the actual pages on which the basionym was published (pages 48-51) but to the whole article (pages 33-52). It was validated in *Cactus & Co.* XII (4): p.200 (2008). This southern form was first made a variety by Piltz (1993) before its elevation by Papsch (2008) to subspecies rank.

Distribution (Map 40 on page 138)

Backeberg did not know where the plant had been found when he described it, and it was not until Piltz (1992) found similar plants growing in the Sierra de Malanzán in 1976, that its habitat was revealed. Since then, its wider habitat range has become clear following observations by other *Gymnocalycium* enthusiasts.

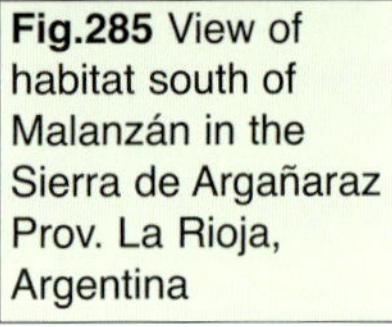

Fig.285 View of habitat south of Malanzán in the Sierra de Argañaraz Prov. La Rioja, Argentina

According to Piltz (1993), the total distribution area includes not only the Sa. Malanzán, but also the Sa. de los Llanos, Sa. del Porongo, Sa. de Abajo, Sa. de Argañarez, and Sa. de Ulapes (Pozo de Piedre), all in Prov. La Rioja, and ranging in altitude from 500m to 1000m. They constitute the mountain chain from Chamical in the north to Ulapes in the south, including Chepes where the form called *G. bozsingianum* grows. The southern and generally more densely-spined form was described by Piltz (1993) as var. *armillatum*.

The plant grows on rocky slopes with little or no vegetation, maybe just a few low bushes. Its habitat is isolated by being surrounded by flat land of an unsuitable nature for the plant to grow.

Conservation Status

Least Concern. The plant is very common and healthy populations occur across its range. Apart from the usual goat damage to its habitat when the rocky slopes are destabilised, there appears to be no threat to its future.

History

Backeberg found this plant in 1933 at Stümer's nursery in Buenos Aires. Three years later he published just a Latin description in *Kaktus-ABC*, naming it for Dr. Castellanos, although he was not the discoverer of the plant.

The original description is not illustrated, but the first picture was published by Backeberg in his *Blätter für Kakteenkunde* in 1936, so it is likely to be a picture of one of the original plants.

The plant remained little known in cultivation until the arrival of plants from Fechser in 1961. Till (2006) gives a detailed account of these Fechser plants, which also included examples from Chepes Viejo which would later be described by Schütz (1977) as *G. bozsingianum*.

Till (2006) also tells us that Backeberg obtained some of Fechser's plants and at the end of the 1960s he showed him which were examples of what he had described as *G. castellanosii*. Till illustrated two of these plants, although the location was still not known since Fechser did not reveal it. It was Piltz (1992) who found that similar-looking plants occur in the north of the distribution area. Till (2006) says the type form can be found between Olta and Ambil and goes on to describe a spiny form as var. *rigidum* from the north-west of the distribution area between Tama and Punta de los Llanos.

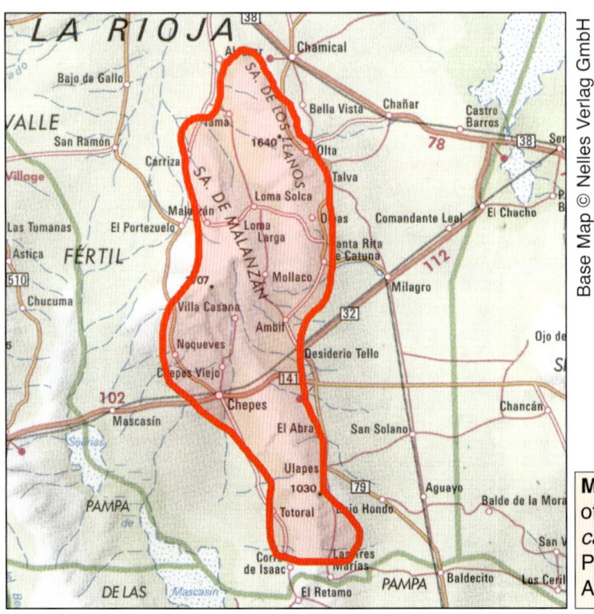

Map 40 Distribution of *G. castellanosii castellanosii* in Prov. La Rioja, Argentina

Fig.286 *G.castellanosii castellanosii, Piltz* 80c from the Sierra Malanzán, Prov. La Rioja, Argentina

Fig.288 *G.castellanosii castellanosii* (armillatum),*Neuhuber* 88 118-288 from Ulapes, Prov. La Rioja, Argentina

Commentary

The species is characterised by its unique seeds which are shiny blackish-brown and have tiny conical warts over their surface. It has been suggested that an origin from hybridisation with a plant from subgenus *Trichomosemineum*, such as *G. bodenbenderianum*, may account for this structure.

The article from Piltz (1992, 1993), published in 'Gymnos' and comprising two parts, gives a good account of the species. It is more variable than the limited number of imported plants available to Backeberg could convey. Thus, the plants from Chepes Vieja, which in fact fit within the overall variation, were described as a separate species, *G. bozsingianum* by Schütz (1977).

The overall shape of the body can be elongated globular or flattened globular while the body colour can be grey-green, dark green or sometimes fresh green. The rib count in the north is from (8-)10 to 12(-17) but in the south, in the Sierra de Ulapes, the count can be up to 19 and with more spines. This form, *Piltz* 217, was described by Piltz (1993) as *G. castellanosii* var. *armillatum* and it certainly retains its different appearance in cultivation. Papsch (2008) later tried to elevate this variety to

Fig.287 *G.castellanosii castellanosii* (bozsingianum), *Piltz* 205 from Chepes Viejo, Prov. La Rioja, Argentina

Fig.289 *G.castellanosii castellanosii, Piltz* 209 from Ambil, Prov. La Rioja, Argentina

subspecies rank but his combination was initially invalid, later corrected in *Cactus & Co.* XII(4).

It is interesting to read the comments by Piltz (1993) about the similarity of its flower to that of *G. bodenbenderianum*. He tells us that he has been able to artificially produce fertile hybrid seed between the two species, which makes him suggest that they are closely related, although traditionally regarded as belonging to different seed groups.

Most recently, a detailed and well illustrated article by Wolfgang Papsch et. al. (2008) has been published in Cactus & Co. It shows the great variability in the appearance of the plants in habitat. There are also a number of seed SEM images which clearly show that *G. acorrugatum* does not belong with *G. castellanosii,* but has the seed of *G. rhodantherum* within whose synonymy it is placed here.

Cultivation

This species requires no special treatment and is easy to grow and flower. Most of the documented plants in cultivation have been grown from seed distributed by Piltz from his various collections across its distribution area.

Illustrations

Blätter für Kakteenkunde 3(7):74/15 (1936)

Gymnocalycium castellanosii subsp. *ferocius* (H.Till & Amerhauser) Charles

First Description (H. Till & Amerhauser)

Gymnocalycium castellanosii subsp. *ferocior* (H. Till & Amerhauser) Charles in CSI20: p.18 (2005)

First description of *Gymnocalycium hybopleurum* var. *ferocior* Backbg nom. inval. (Art. 8.2) in *Kakteenlexikon*: p.168 (1966)

Spines much stronger, ash-grey when dry, chestnut-brown when damp, Radial spines the same as the type, c. 7-9. Central spines (1-)2, standing straight out, or curved upwards, to 3.5cm long, they and radial spines ± compressed and/or ± angular. Collected by Fechser, Uhlig 2167.

Etymology

From the Latin 'ferocia' meaning ferocity and referring to the strong spination. The spelling has been changed because it is a comparative adjective so *ferocius* is the correct neuter ending.

Other Names

Basionym: *G. mostii* subsp. *ferocior* Till & Amerhauser. Type: Argentina, Prov. Córdoba by Agua de Ramón, leg. H. Borth s.n. 1972 cult. in coll. H. Till sub no. 905 (Holotype CORD)

Fig.290 *G. castellanosii ferocius* GC952.02 near Tuclame, Prov. Córdoba, Argentina

Gymnocalycium ferox var. *ferocior* (Backeberg) Slaba nom. inval.

Distribution (Map 41)

Known only from the southern end of the Sierra de Serrezuela in the far north-west of Prov. Córdoba near the border with La Rioja. The type locality is by Agua de Ramón but plants can be found between there and Tuclame. They grow on rocky patches between dense, spiny bushes.

Fig.291 *The habitat of G. castellanosii ferocius* GC952.02 near Tuclame, Prov. Córdoba, Argentina

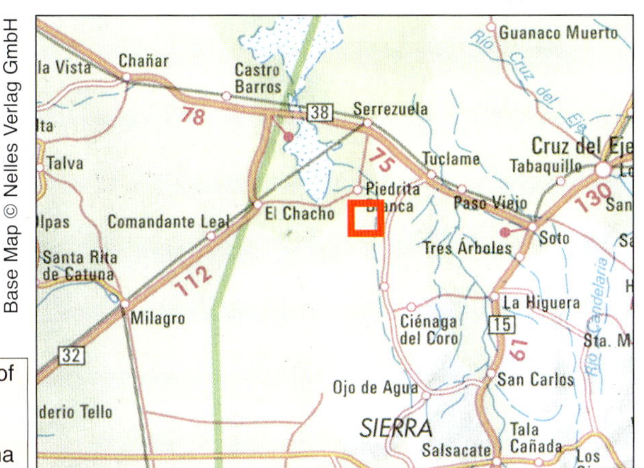

Map 41 Location of *G. castellanosii ferocius* in Prov. Córdoba, Argentina

Conservation Status

Least Concern. The plant is locally abundant and is well protected by dense shrubs, appearing only to be occasionally damaged by grazing.

History

This taxon was first described invalidly by Backeberg as *G. hybopleurum* var. *ferocior* and var. *ferox* in his *Kakteenlexicon* (1966). He described it from imported plants sent by Fechser to Uhlig's nursery which were also seen there by Till, who bought four of the plants with the description 'A.d.R.'. The locality was confirmed as Agua de Ramón when Till and Borth visited the area in 1972.

Slaba (1984) agreed with Backeberg that the plants he thought were varieties of Schumann's *Echinocactus hybopleurus* were a good species, so he endeavoured to describe them as *Gymnocalycium ferox* but he quoted the invalid *G. hybopleurum* var. *ferox* as its basionym, so his publication was invalid as well. Interestingly, Slaba (1984) illustrates the seeds (on page 82) of his proposed species and comments on its different appearance, referring to the prominent tubercles on their surface which are like the seeds of *G. castellanosii*.

The first valid name for this taxon was published by Till & Amerhauser (2002b) in *Gymnocalycium* 15(1): p.435. They described it as *G. mostii* subsp. *ferocior* although they accepted that the seed structure is close to *G. castellanosii*.

The treatment I have followed in this book assumes that *G. mostii* subsp. *ferocius* is a correct name for the plants from Agua de Ramón, but there could be a problem with the application of this name. I think it probable that the plant preserved as the type is not from Agua de Ramón, the type locality quoted in the description, and is a different taxon.

The article where the description appears, Till & Amerhauser (2002b), has a number of errors in the illustration references. For instance, the type designation refers to the type plant as Illn. 25 which does not exist. In fact, as confirmed by Till (pers. com.), the plant in Illn. 24 was preserved as the type.

Fig.292 *G. castellanosii ferocius* GC952.02 near Tuclame, Prov. Córdoba, Argentina

Fig.293 *G. castellanosii ferocius* GC952.02 near Tuclame, Prov. Córdoba, Argentina

Fig.294 *G. castellanosii ferocius* GC952.02 near Tuclame, Prov. Córdoba, Argentina

The article has many illustrations (18-23) said to be as subsp. *ferocius* and all of them look like plants from the area around Agua de Ramón. However, Illn. 24, depicting the type plant, is not the same, and does not have either of the Borth collection numbers 89 nor 91 stated to have been allocated to the Agua de Ramón plants. It has different spines and a different flower. It greatly resembles plants brought into cultivation from the Cerro Pencales (STO 93 829/2), a mountain just south west of

Fig.295 *G. castellanosii ferocius ,Neuhuber* 91 386-1311 from El Espinillo, Prov. Córdoba, Argentina

Fig.296 *G. castellanosii ferocius ,Neuhuber* 91 386-1311 from El Espinillo, Prov. Córdoba, Argentina

Capilla del Monte, about 80km to the east, and well within the distribution of *G. mostii*.

If this error was indeed made, then the name *G. mostii* subsp. *ferocius* must apply to the plants from Cerro Pencales and my combination under *G. castellanosii* would be inappropriate. The plants from Agua de Ramón would then have no separate name and would remain within the scope of *G. castellanosii*. This treatment is favoured by Papsch (2008), who does not recognise this population as a separate subspecies.

Commentary

The prominent tubercles on the surface of the seeds and the shape of the flowers were the main reasons I made this a subspecies of *G. castellanosii* which has a similar seed testa and flowers. My thanks go to Massimo Meregalli for pointing this out and suggesting the combination when we were preparing the *New Cactus Lexicon*. The relationship with *G. castellanosii* has recently been endorsed by Papsch (2008).

Cultivation

This is a very attractive heavily-spined plant which flowers easily in culture. It has no special requirements but a sunny location will help it develop the strong spines that are characteristic of this plant. If you are going to grow just one form of *G. castellanosii*, then this is the one to choose.

Illustrations

Gymnocalycium 15(1): pp. 433-434, Figs. 18 to 23

New Cactus Lexicon Atlas (2006) Illn. 275.6

Fig.297 *G. castellanosii ferocius* GC952.02 near Tuclame, Prov. Cordoba, Argentina

Gymnocalycium chacoense Amerhauser

First Description (Amerhauser)

Gymnocalycium 12(4): pp.301-304 (1999)

Gymnocalycium chacoense Amerhauser spec. nov.

Plant wide-globular, 40-75mm high, 55-80mm diameter, strongly offsetting in age, forming groups of up to 15-25 heads, with copious fibrous, shallow roots. Epidermis light green and slightly shiny. Ribs 8-12, mostly rounded, sometimes weak angled, hardly noticeable tubercles.

Areoles 14-15mm apart, round, 3mm diameter with yellowish-white felt, which remain a long time on young plants, then becoming bald. Radial spines 7-9, the topmost 8-10mm long, the middle spines 16-19mm long and the lowest 13-15mm long, central spines 3, 16-21mm long, standing away from the body straight or slightly curved, in youth light yellow with brown tips, later becoming grey. All spines thin, bristly, stiff yet elastic.

Flowers, many in the area of the crown, thin funnel-form, (25) 35-45mm long, 20-32mm diameter white to pale pink, with a yellowish-brown throat. Ovary flattened-globular, 3mm high, 4mm wide, magenta-pink, with many scales, without hair or spines. Scales half round, pink with a wide paler edge, resolving into the outer petals. Outer petals spatulate, 15-18mm long, light magenta pink with a wide whitish edge, the ends curved outwards, inner petals wide lanceolate, 18mm long, 6mm wide at the widest part, milk white. Filaments thin, pale greenish yellow, inserted over the whole of the tube wall, 9mm long, lying close to the tube wall, the top third curved inwards, overhanging the style. Anthers oval, 1 x 0.5mm diameter, with yellow pollen. Style 7mm long, with stigma 9mm long, 1.1mm thick, light greenish yellow, just overhanging the lower row of anthers. Stigma 2mm long with 7 claw-like, white lobes. Ovary 2mm high, 2.5mm diameter, round, the top cut off, white walled with many white seed funicles.

Fig.298 *G. chacoense, Amerhauser* 990 from the Cerro San Miguel, Dept. Santa Cruz, Bolivia. The type collection.

Fruit small, juicy, 6-8mm high, 4.5mm thick, green, later dry. Seeds stuck together in a jelly-like mass of seed strings. Seeds small 1mm long, 0.7mm thick, red-brown, shiny, with many warts, The micropyle region slightly slanting, with a large elaiosom.

Because of the structure of the seed, *G. chacoense* belongs to the seed group *Microsemineum* Schütz, and because of its similarity seems to be allied to *G. chiquitanum* Cárdenas and *G. paediophilum* Ritter.

Type: Bolivia, dept. Santa Cruz, on steep Mountains Cerro San Miguel, at 790m, 25.09.1995. *H. Amerhauser* HA 990-4 (Holotype: LPB; Isotype: WU).

Etymology

Occurs in the Chaco region of Bolivia.

Photo: Volker Schädlich

Fig.299 Cerro San Miguel, Dept. Santa Cruz, Bolivia, The type locality of *G. chacoense*

Photo: Volker Schädlich

Fig.300 *G.chacoense ,Schädlich* 260 in habitat at the Cerro San Miguel, Dept. Santa Cruz, Bolivia.

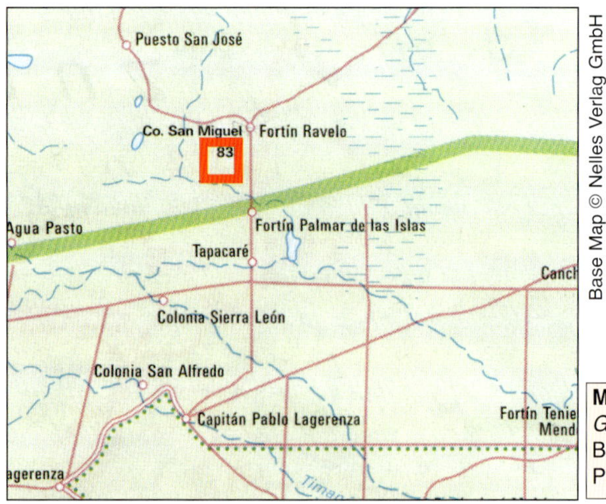

Base Map © Nelles Verlag GmbH

Map 42 Location of *G. chacoense* in Bolivia, near the Paraguayan border

Distribution (Map 42)

Known to grow only on steep rock cliffs at the Cerro San Miguel near to Fortin Ravello, Santa Cruz, Bolivia, near the border with Paraguay.

Conservation Status

Data deficient. Although known only from the type locality, there is no documentation on the extent of the population nor its stability.

History

According to Volker Schädlich (pers. com.) this plant was first discovered by Hans-Jörg Jucker and his wife when travelling in the area during October 1986. They found themselves at the military station Ravello which is near the Cerro San Miguel. Jucker climbed the 800m high hill and discovered the plant growing on rocks.

The plant was then seen in 1995 by Helmut Amerhauser and his companions, who first wondered if they had found a Weingartia species. The plants grow on inaccessible rock faces so he was not able to get photographs until his later visit in 1998.

G. chacoense is clearly related to *G. chiquitanum* and *G. paediophilum* although it has by far the smallest flower. Schädlich (pers. com.) tells me that the seeds of the three species are very similar and that he has been able to produce viable seed by crossing *G. chacoense* with both of the other species.

Commentary

One of a group of related species from the chaco, this species was listed in the synonymy of *G. chiquitanum* in the *New Cactus Lexicon* (2006). I believe it is sufficiently different to justify separate species status because the flowers are very much smaller and the spination much finer and denser. Its rocky locality is isolated by being surrounded by the flat chaco, an unsuitable environment for this plant to grow.

Cultivation

This is one of the best recent introductions to cultivation. It has bright golden spines and a neat habit, offsetting freely to make a cluster of stems. The flowers are small but look in keeping with the body. It does not appear to present any difficulty in culture, although its habitat location suggests that it might be cold sensitive.

Illustrations

Gymnocalycium 12(4): pp.301-304 (1999)

New Cactus Lexicon Atlas (2006) Illn. 276.2

Fig.301 *G. chacoense, Amerhauser* 990 from the Cerro San Miguel, Dept. Santa Cruz, Bolivia. The type collection.

Fig.302 *G. chacoense, Amerhauser* 990 from the Cerro San Miguel, Dept. Santa Cruz, Bolivia. The type collection.

Gymnocalycium chiquitanum Cárdenas

First Description (Cárdenas)

Cactus (Paris) 18(78): pp.95-97 (1963)

Gymnocalycium chiquitanum Card. spec. nov.

Flattened globular, 2-4cm high, 6-9cm diameter, greyish-green in colour, sometimes becoming violet. Ribs 6-7, with tubercles, 5mm high, 2.5cm wide at the base. Areoles 1.2cm apart, circular to elliptical, 5mm diameter, the prominent young areoles are covered with cream felt. The 6 slightly curved and adpressed spines are radiating, 15-23mm long; the single central spine is 18mm long, not always present, all spines are thick, grey, brown at the top, flowers grow from the small depression in the centre of the crown. Flowers pink, 4-6cm long. Ovary spherical, 5mm diameter, with light magenta scales, 3mm diameter, edged white. Tube 2.5cm long, covered with light lilac scales, 5mm wide. Sepals spatulate to pear shaped, pink with a brownish central section, c. 20x6mm. Petals pinkish-lilac, 25mm long, 6mm wide, pointed, violet at the base. Stamens 5mm long; filaments magenta; anthers yellow. Style 22mm long, lilac-magenta, the 8 stigma lobes yellow, each 3mm long, inside of tube is bright magenta. Fruit 2cm long, bluish, later violet, elliptical. Seeds 1mm diameter, light brown, testa granular.

Habitat: Bolivia, Prov. Chiquitos, Dept. Santa Cruz, San José, 600m. December 1961, Father Lorenzo Hammerschmid. No. 5.562. (type) in the Cárdenas Herbarium.

Fig.303 *G.chiquitanum, Metzing* 97 from San José de Chiquitos, Dept. Santa Cruz, Bolivia

Etymology

Occurs in Prov. Chiquitos, Dept Santa Cruz, Bolivia.

Photo: Ludwig Bercht

Fig.304 *G.chiquitanum, Bercht* 3402 in habitat south of San José de Chiquitos, Dept. Santa Cruz, Bolivia

Fig.305 *G.chiquitanum, Bercht* 3402 in habitat south of San José de Chiquitos, Dept. Santa Cruz, Bolivia

Map 43 Locations of *G. chiquitanum* in Dept. Santa Cruz, Bolivia

Distribution (Map 43)

It grows among grass, under bushes or ferns in a warm environment with hot summers in the Bolivian chaco. Its range extends along the Serranía de San José near the town of San José de Chiquitos in Dept. Santa Cruz, Bolivia.

Conservation Status

Data Deficient. There is little information about how widespread and abundant this plant is in habitat.

History

It is thought to have been discovered by Father Hammerschmid, a Franciscan priest at San José de Chiquitos, who sent it to Cárdenas. However, Schütz (1986) tells us that it was in fact another monk who actually found it and gave it to Hammerschmid. According to Schädlich (pers. com.) this other monk was Padre Klinger. Plants were also sent to the Uhlig nursery where Backeberg saw it and described it, invalidly, as *G. hammerschmidii,* also in 1963.

Commentary

The plants are said to very variable in habitat with whitish or nearly black spines which also vary in number between individuals. Slaba (1986) illustrates a number of forms. The maximum diameter of a single head is said to be 15cm. Schädlich (pers. com.) explains that the plants from the west of San José are identical to those at the type locality with regard to flower, fruit and seed, while their bodies are larger and spines straight and shorter. Plants from the lowlands offset more freely and can make clusters up to a square meter.

Fig.306 *G.chiquitanum, Bercht* 3402 in habitat south of San José de Chiquitos, Dept. Santa Cruz, Bolivia

Cultivation

Although it is not amongst the most popular species that people have in their collections, it is not particularly difficult to grow, just preferring a little more warmth in winter and adequate water in summer. It is easily recognised with its glossy body and pretty pink flowers, eventually making a cluster of neat stems.

Illustrations

Slaba, R. *Kaktusy* 21(2): pp.26-28 (1985) pictures of a number of forms

New Cactus Lexicon Atlas (2006) Illn. 276.1

Fig.307 *G.chiquitanum, Metzing* 100 from south of San José de Chiquitos, Dept. Santa Cruz, Bolivia

Gymnocalycium glaucum Ritter subsp. *glaucum*

First Description (Ritter)

Sukkulentenkunde (Jahrbücher der S.K.G.) 7/8: pp.37-38 (1963)

Gymnocalycium glaucum Ritter spec. nov.

Body ash grey-green, the grey tone shows strongly over the green colour; fairly flat, semi-globular in age, flowering plants 5-12cm diameter. Root thick hard conical turnip-like, 10-12cm long.

Ribs 10-16, blunt and wide, becoming thick at the areoles, 7-15mm high, with humps under the areoles and cross-cuts over the areoles.

Areoles with white felt, 7-15mm long, 5-7.5mm wide, on older examples up to twice the length and width, c. 1.5cm apart.

Spines reddish brown when young, later becoming grey, comb-like,in 2-3, rarely 4. sideways spreading pairs, with the odd spine pointing down, all strong, rigid, awl-like, pointing outwards and slightly curved back towards the body, usually 2-4cm sometimes up to 7cm long, the central pair the longest, the lowest pair usually the shortest, sometimes another odd spine appears at the top of the areole.

Flowers from the crown, odourless, when open 3.5-5.5cm long, 2.5-4.5cm diameter, opening before midday. The description was taken from two flowers of two different plants from the type habitat.

Ovary 15-22mm long, 8-10mm thick at the top, 3mm thick at the base. The colour tones are from; reddish, greenish, brownish to bluish, towards the bottom more red, with scales almost the same colour, rounded and white edged, 2-4mm long, 3-5mm wide, with tiny dark tips.

Fig.308 *G.glaucum glaucum,* GC969.01 in habitat in the Sierra de Copacabana, Prov. Catamarca, Argentina

Nectar chamber tubular, 3mm long, 0.5mm wide, purple in the area of the style, the top enclosed by a single ring of stamens, which lean inwards against the style.

Tube funnel shaped, at the bottom 2-3mm, not larger, 7-12mm long, 6-13mm wide at the top, inside purple, outside colour is the same as the top of the ovary, with similar scales that cover up to 1/4 of the tube.

Stamens brownish-purple, the lower ring 3-4mm long, the others 5-8mm long, inserted over the whole of the wall of the tube above the nectar chamber, closer at the top. Anthers leaning inwards, cream-brown to pink, oval, pollen white.

Fig.309 The habitat of *G.glaucum glaucum,* GC968.01 on hot dry hills near San Blas, Prov. La Rioja, Argentina

Fig.310 Habitat of *G. glaucum glaucum,* GC969.01 in the Sierra de Copacabana, Prov. Catamarca, Argentina

Style pale green, reddish at the base, 1.5mm thick, 12-18mm long, of which 2-3mm comprises 9-12 pale yellow stigma lobes, that stand in the middle of the anthers.

Petals 13-21mm wide, the lowest are 3-4mm long, almost straight, 1.5-3mm wide, round at the end, with or without a point, widest at c. 3/4 of their length, narrow at the bottom, widest at the top, the lowest part purple continuing as a thin line towards the top, otherwise white, outer petals shorter and wider, hardly narrower at the bottom, more greenish, pink edged, running into the scales.

Fruit 2.5-3cm long, 1.5-2cm thick, at the base reddish, 0.5cm thick, at the top grey-green to brownish-green, with scales similar to those on the ovary. The flower scar is 12mm to 15mm wide. Fruit splits vertically when ripe.

Fig.311 *G. glaucum glaucum,* GC968.01 on hot dry hills near San Blas, Prov. La Rioja, Argentina

Seeds plentiful, 1mm long, 0.75mm wide, 0.5mm thick. Testa black, finely humped, hilum edge irregularly curved, hilum white, basal, long, slightly downward elongated, somewhat protruding.

Habitat: South East of Tinogasta in the province of Catamarca, near the border of the province La Rioja, Argentina.

Holotype: In the Botanical Museum and Herbarium, Utrecht, Holland. I discovered this species in February 1959 and has my number FR 961.

Etymology

From the Latin 'glaucus' meaning grey-green, referring to the body colour.

Distribution (Map 44)

This taxon comes from a very dry area on the border of the Argentine Provinces of La Rioja and Catamarca. Plants can be found on the Sierra de Copacabana in the west and then further east across the northern edge of the Campo de Velasco, growing on steep slopes of low hills.

Conservation Status

Least Concern. the plants are generally plentiful where they occur and there are vast areas of nearby similar mountains away from roads which probably have further populations. There has been some collecting near to roads over the years and there is recent evidence suggesting the practice is continuing.

Fig.312 *G. glaucum glaucum,* GC210.02 from San Blas, Prov. La Rioja, Argentina

Fig.313 The splitting fruit of *G. glaucum glaucum,* GC210.02 from San Blas, Prov. La Rioja, Argentina

History

In 1959 Ritter discovered the plant he later described as *G. glaucum* in 1963. I can find no indication that he considered *G. mucidum* as the possible identity of his discovery. It was Metzing (1985) who first suggested that these two plants and *G. ferrarii* belong together.

Gymnocalycium mucidum was described in 1937 by Oehme from a single individual without knowledge of its finding place. He got the plant from the nursery of Fortenbacher from an importation of plants from South America. No type was placed in a herbarium and Oehme's collection was destroyed in the war.

Backeberg, in his *Die Cactaceae* (1959) expressed the opinion that *G. mucidum* is related to *G. mazanense*, a name treated here as a synonym of *G. hossei*. He probably formed this opinion because of its yellowish-pink filaments.

G. glaucum is generally regarded as belonging to the species complex around *G. hossei* from which it differs by having a large tuberous root.

Commentary

Some recent treatments accept *G. mucidum* as an earlier name for this taxon. According to Piltz et al. (1994), there is complete agreement between its description and that of *G. glaucum,* and the picture published with the first description is a reasonable likeness for *G. glaucum*.

This opinion was refuted by H. Till (2000) who made a case for *G. mucidum* actually being a relation of *G. capillaense* and so belonging to subgenus *Gymnocalycium*. Oehme, himself, actually suggested this relationship in his first description of *G. mucidum*. Till illustrated some plants found in Córdoba which he claims to be a perfect match for the description of *G. mucidum*.

Even more recently, Papsch (pers. com.) suggested that the recently described *G. prochazkianum* could be the same as *G. mucidum*, another reason to reject the name as indeterminate. (see page 262)

Although I personally have some sympathy with the view that *G. mucidum* is the oldest name for *G. glaucum*, I do not think the evidence is strong enough to justify replacing the well typified and familiar name *G. glaucum*. The added confusion caused by Till's alternative opinion further supports the view taken by The *New Cactus Lexicon* editorial team who decided to reject *G. mucidum* as being of uncertain application

Cultivation

This plant has a reputation of being difficult in cultivation. A clue to why this may be can be found by examining its habitat. It grows in a very dry place on hot slopes which consist of crumbling granite, often with full exposure. Where bushes do exist, the plants can often be found around their bases, but this is probably more to do with stability than protection since the bushes are leafless for much of the year.

It is therefore a good idea to grow these plants in a largely mineral soil with little organic material. They also need as much sun as you can provide and less than average watering. In these conditions, the plants will grow slowly and produce their flowers that are often a lovely pink. The body is brown or greyish and can develop a whitish bloom making this one of the most distinctive species in the genus.

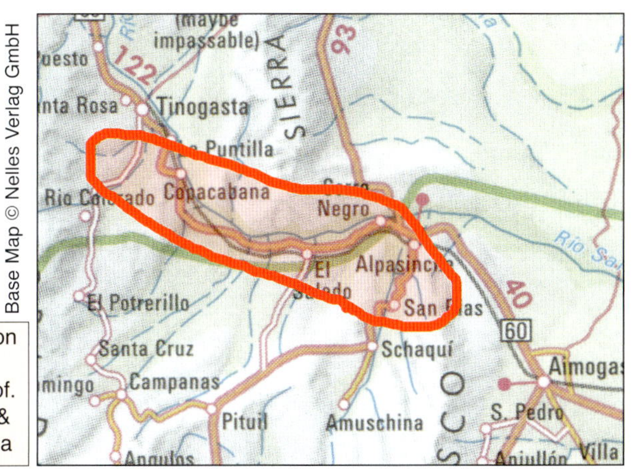

Map 44 Distribution of *G. glaucum glaucum*, border of. Prov. Catamarca & La Rioja, Argentina

Gymnocalycium glaucum subsp. *ferrarii* (Rausch) Charles

Fig.314 The habitat of *G. glaucum ferrarii* GC982.02 near Villa Mazán, Prov. La Rioja, Argentina

First Description (Rausch)

Gymnocalycium ferrarii Rausch

KuaS 32(1): pp.6-7 (1981)

Simple, flattened-globular, 30-40mm high, and 90mm diameter, greenish-grey, with a turnip-like root up to 15cm long. Ribs 10-14, vertical, wavy, humps under the areoles protruding chin-like. Areoles sunken, sitting in the humps, oval, up to 7mm long, grey felted, later becoming bald. Radial spines porrect, in 3 pairs, all slightly curved, up to 30mm long, awl-shaped, thicker at the base, brown to blackish, later becoming grey, the top most pair and the single one at the bottom are a little weaker.

Fig.315 *G. glaucum ferrarii* GC26.04 near Villa Mazán, Prov. La Rioja, Argentina

Flowers 45mm long, 35mm diameter. Ovary (skittle shaped, up to 10mm long), tube yellowish-green, with wide whitish-pink scales, outside petals round, dirty pink with a brownish-green central stripe, inner petals round and ragged, whitish-pink with a pink central stripe. Throat pink. Filaments whitish-pink, anthers yellowish-white. Fruit and seed type, the same as *Gymnocalycium mazanense* Backeberg.

Habitat: Argentina, Catamarca, near Santa Theresa, on the rough stony sandy hillsides.

Type, *Rausch* 718, deposited in the Städtische Sukkulenten-Sammlung, Zürich.

Etymology

Named for Omar Ferrari, Argentinian horticulturist, cactus collector and friend of Roberto Kiesling who often accompanied Rausch on his journeys.

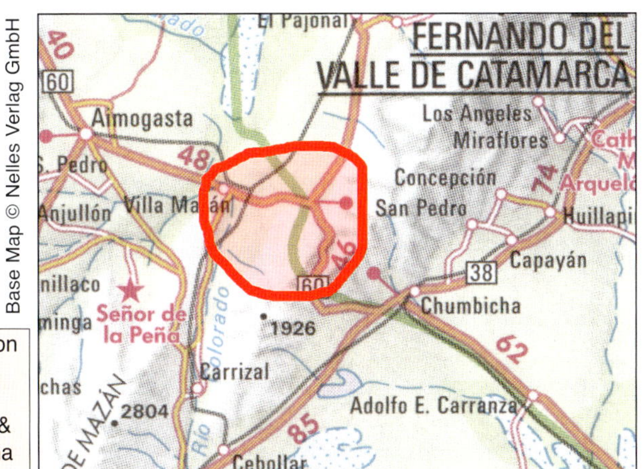

Map 45 Distribution of *G. glaucum ferrarii*, border of. Prov. Catamarca & La Rioja, Argentina

Other Names

Gymnocalycium ferrarii Rausch in KuaS (1981)

Distribution (Map 45)

This plant is found along the southern edge of the Pipanarco Basin on the Sierra de Mazán where it grows on steep exposed hillsides in crumbling granite. It often shares its habitat with the most northerly populations of *Pyrrhocactus bulbocalyx*. The habitat receives low rainfall, mainly in summer storms which probably occur just when the fruits are ripe. Meregalli et. al. (2000) state that the most easterly location is on the Cuesta de Sebile. It looks similar to *G. hossei,* which also grows there, and I suspect that I collected mixed seed under my number GC27.02.

Conservation Status

Least Concern. Plants at the best known locality near to the main road are less plentiful now than years ago, presumably due to collecting, but there are likely to be unaffected populations elsewhere in the mountains.

History

Often considered as just a synonym of *G. glaucum*, its smaller size and disjunct habitat in the Sierra de Mazán persuaded me to make it a subspecies of *G. glaucum* in *Cactaceae Systematics Initiatives* 20: p.18 (2005) in preparation for the *New Cactus Lexicon* (2006).

Fig.316 *G.glaucum ferrarii, Piltz* 136 from Est. Mazán, Prov. La Rioja, Argentina

Fig.317 *G.glaucum ferrarii* GC982.02 near Villa Mazán, Prov. La Rioja, Argentina

Commentary

Similar to the type subspecies but smaller, this taxon grows in the eastern part of the overall distribution. It grows on very steep, hot hillsides composed of crumbling granite.

Cultivation

The same as for subsp. *glaucum* although the plants are smaller and usually have attractive pink flowers. A deep pot is needed to accommodate the tuberous root.

Fig.318 *G.glaucum ferrarii, Piltz* 136 from Est. Mazán, Prov. La Rioja, Argentina

Gymnocalycium horridispinum Frank ex H.Till **subsp. *horridispinum***

First Description (Frank)

Gymnocalycium horridispinum Frank nom. inval. KuaS 14(1): pp.8-10 (1963). Validated by H.Till in KuaS 38(8): p.191 (1987)

Body solitary, short cylindrical, 60-80mm high, becoming about as wide, dark green, 10-13 ribs in fully grown plants, deep and sharply furrowed with prominent chins under the areoles. Areoles large, oval, 8-10mm long, in new growth yellow-grey felted, later becoming brownish. Spines, 10-12 radials, robust awl-shaped, straight, grey, often tipped dark brown, surface rough, when wet entirely reddish-brown, spreading, somewhat curved towards the body, 20-25mm long. Four central spines arranged in a cross, thickened at the base, very robust, strong, standing out from the body, of the same colour as the radials, 30-40mm long.

Flowers in a circle around the unarmoured crown, emerging from just above the youngest spine bearing areoles. Buds, at first deep purple-red, later clothed in red-tipped scales, thick, sharp pointed. Pericarp and lower part of the receptacle, green, covered with fleshy red tipped scales. These scales, On the upper part of the flower tube, merge into the whitish-green, red edged, outermost perianth leaves. Flowers 60mm long, 60mm in diameter. broadly funnel shaped, the inner perianth leaves, pure white within, violet-pink edges and mid-rib, violet-pink outside, from the second to the third day of flowering, the outer perianth leaves, from the edge inwards, become deeper violet-pink when only the throat remains pure white. Flowers last about one week. Filaments white, anthers light yellow, The style ends with 10 pointed stigma lobes in the upper part of the stamen ring, both are white.

Fig.319 *G.horridispinum horridispinum, Strigl* s.n. from S.W.Salsacate, Prov. Córdoba, Argentina

Fig.320 *G.horridispinum horridispinum, Neuhuber* 91 382-1278 from El Potreo, Prov. Córdoba, Argentina

Fruit ovoid, dark green, 15-20mm, with adhering flower remains. When ripe tinted pale red, splitting vertically, containing numerous seeds embedded in white flesh.

Seeds small, 0.5 x 1mm., oval, flattened, with obliquely cut-away hilum, black testa punctured with fine cavities.

Origin: Argentina, Córdoba, south-west of Salsacate, on grassy hillocks at 700-800m.

Etymology

From the Latin 'horridus' meaning prickly, referring to the spines.

Other Names

Gymnocalycium monvillei subsp. *horridispinum* (Frank ex H.Till) H.Till in *Gymnocalycium* 6(3): p.102 (1993)

Distribution (Map 47)

Known from only a small area to the south west of Salsacate and found only by a few people since its description. It is to be hoped that it occurs in more remote habitats which remain undisturbed.

This location is isolated from the much more extensive populations of subsp. *achirasense* which occur in Prov. San Luis to the south.

Conservation Status

Vulnerable. It has a very limited distribution and the increased settlement and farming in the area has seriously affected the plants. Till (1988b) reported that he could find no plants at the type locality. The original collector reported that the plants were very scattered and infrequent. There has been a number of collectors who have found the plant, including Lau who reports it from La Mudana, west of Salsacate, at the higher altitude of 1200m.

Fig.321 *G. horridispinum horridispinum, Neuhuber* 91 382-1278 from El Potreo, Prov. Córdoba, Argentina

History

The first description was of young plants since they can grow much larger, especially in culture. Frank received the first plants from Fechser, just two individuals.

Below: Map 46 The distribution of *G. horridispinum achirasense* in Prov. San Luis, Argentina

Commentary

Although related to *G. monvillei,* and made a subspecies of it by Till (1993), recent observations of *G. horridispinum* subsp. *achirasense* growing sympatrically with *G monvillei* necessitated the recognition of the former as a different species and consequently, this taxon as well.

I am here making *G. achirasense* a subspecies of *G. horridispinum*, rather than the other way around, because Lodé (1995) made it a variety of *G. horridispinum* and created the precedence.

Cultivation

This is an easy plant to grow and has beautiful pink flowers. It tends to grow tall in culture and much bigger than its description suggests. The spination is dense and spectacular in large specimens.

Illustrations

Frank, G. KuaS 14(1): pp.8-10 (1963) 2 B&W pictures with the original description

Above: Map 47 The locality of *G. horridispinum horridispinum* in Prov. Córdoba, Argentina

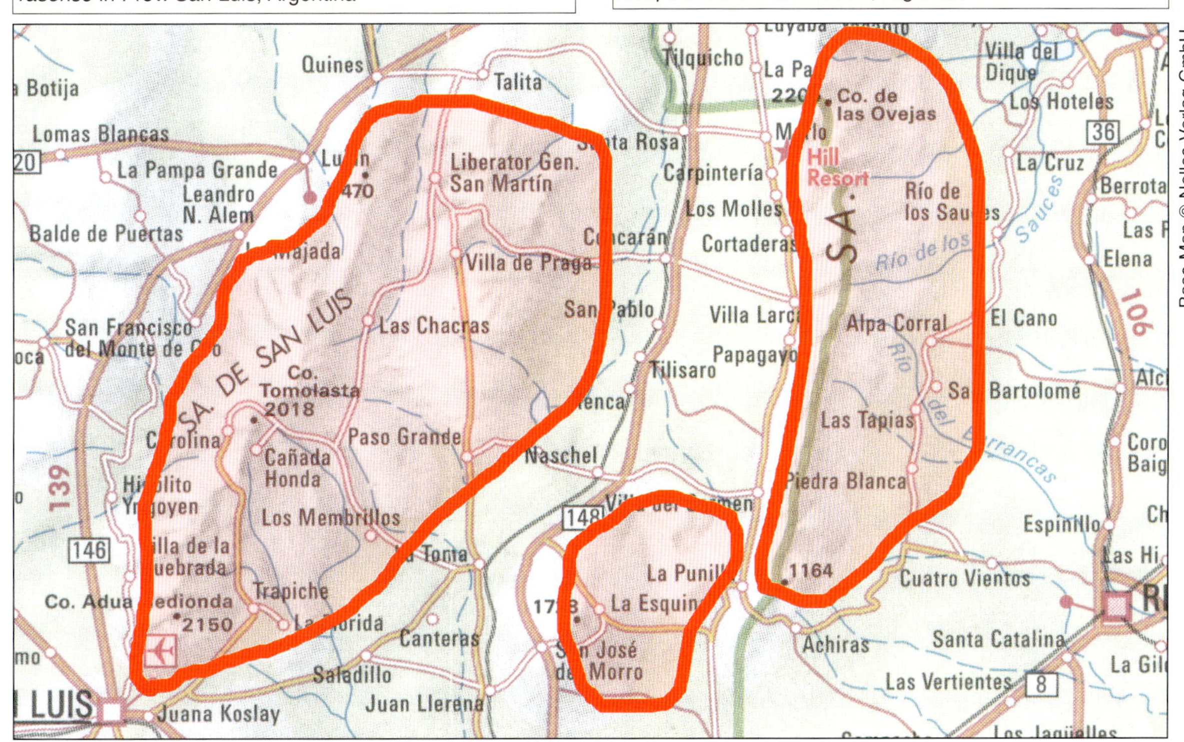

Gymnocalycium horridispinum subsp. *achirasense* (H. Till & Schatzl ex H.Till) Charles

First Description (H. Till & Schatzl)

Gymnocalycium achirasense H. Till & Schatzl nom. inval. KuaS 30(2): pp.25-28 (1979). Validated by H.Till in KuaS 38(8): p.191 (1987).

Shallow rooted, long, relatively thin. Body globular to elongated, in age slightly columnar. The plants in front of me average 5.6cm high, 7cm diameter, dull green to grey-green; crown usually without spines. Ribs 12-15, straight, slightly slanting, tapering into chin-like humps. Areoles oval, 6mm wide, 9mm long, at first slightly felted, later bald, usually elevated but sometimes sunken into the humps.

Radial spines on the areoles in irregular order, comb-like to radiating, strong, slightly flattened with round edges, straight to slightly curved, dark at the base, horn coloured at the tips, sharp, usually lying close to the body, 10 (-12) often uneven pairs in two rows, with one spine pointing upwards and one pointing down, the topmost radial spine takes the position of a secondary central spine, one central spine, in form shape and colour, the same as the radials, straight, sometimes curved.

Flowers funnel-shaped, 58mm long, 71mm diameter. Ovary short, 12mm diameter. and high, light green with a few wide, reddish-brown, pointed scales that eventually taper into the outer petals that are spatulate, 37mm long, 9mm wide, wishy-washy white with a lilac-pink central stripe that is lighter towards the base, darker towards the point, inner petals lanceolate, 34mm long, 7mm wide, same colour as the outer petals. Many stamens, inserted in several rows, yellowish to white, the lower row close to the style. Anthers yellow. Style strong, yellowish-white with 12 radiating lobes, not as high as the topmost petals. Seeds small, c. 1mm (uneven in size, taken from eleven fruits from four plants), roundish, standing diagonally with a slightly lumpy edge. Micropyle somewhat elevated. Testa with fine warts, dark brown.

Fig.323 *G.horridispinum achirasense,* GC930.01 in habitat near the Sierra del Morro, Prov. San Luis, Argentina

Fig.322 *G. horridispinum achirasense* (kainradliae), *Bercht* 3218 in habitat near Carolina, Prov. San Luis, Argentina

Habitat: Argentina, Province San Luis, near Achiras, on the border of the Province of Córdoba, c. 1000m, between grass and stones. Type plant; *Genser* B21 in the protected and type collection of the Botanical Garden Linz. No. BGL. 2275.

In KuaS 38(8): p.191 (1987):

The holotype plant was reported as having been conserved in alcohol and designated by H. Till (WU) together with two isotypes.

New Combination

Gymnocalycium horridispinum subsp. ***achirasense*** (H. Till & Schatzl ex H. Till) G. Charles **comb. et stat. nov. Basionym:** *Gymnocalycium achirasense* H. Till & Schatzl ex H. Till in *Kakteen und andere Sukkulenten* 38(8): 191, 1987. Type: *Genser* B21, Argentina, Prov. San Luis, near Achiras. BGL 2275 (WU).

Fig.324 *G. horridispinum achirasense* (echinatum), *Neuhuber* 91 346 1187 from Est.Grande, Prov. San Luis

Fig.325 *G. horridispinum achirasense* (chacrasensis), *Neuhuber* 88 117-268 from Las Chacras, Prov. San Luis

Fig.327 *G. horridispinum achirasense* (echinatum), *Neuhuber* 91 341187 from Est.Grande, Prov. San Luis

Etymology

Occurs near the town of Achiras, Prov. Córdoba, Argentina.

Other Names

Gymnocalycium monvillei subsp. *achirasense* (H. Till & Schatzl ex H.Till) Till in *Gymnocalycium* 6(3): p.102 (1993)

Distribution (Map 46 on page 153)

It is found in rocky places to the east of the Sierra San Luis, this plant grows in the hills and mountains of northern Prov. San Luis and just into Córdoba at the eastern limit of its distribution. It is found from the Sierra San Luis in the west to the Sierra de Comechingones in the east.

Conservation Status

Least Concern. This taxon has a wide distribution without immediate threats to its survival.

History

This taxon was first described invalidly as a species by H.Till and Schatzl in 1979. They acknowledged its relationship with *G. monvillei* and *G. horridispinum* but still decided to describe it as a separate species. The description was validated by H. Till in 1987 who later made it a subspecies of *G. monvillei*, see Till & Neuhuber (1993).

Fig.326 *G. horridispinum achirasense* (kainradliae), *Bercht* 480 from La Verbena, Prov. San Luis, Argentina

Recent reports of this taxon growing sympatrically with a southern form of *G monvillei* (Papsch & Bercht pers. com.) show that its combination under *G. monvillei* is not correct.

Commentary

This taxon is similar to subsp. *horridispinum* except that it is larger, more applanate, and is distributed over a large area some way south of the isolated habitat of the type subspecies..

I first saw this plant as imported specimens sent by Frau Muhr under the number B21 which had been collected by Genser and was to become the type collection. There have been many varieties and forms of this variable plant described since then by Till and Neuhuber.

It was treated as a subspecies of *G. monvillei* in the *New Cactus Lexicon* (2006) but it grows sympatrically with the southern form of *G. monvillei* (subsp. *gertrudae*) so must be a different species. Bercht (pers. com.) tells me that he has seen *Bercht* 3218 *G. achirasense* (var. *kainradliae*) growing alongside *Bercht* 3217 *G. monvillei* (subsp. *gertrudae*) at a place called La Verbena, North of El Trapiche, Prov. San Luis, Argentina.

I am here making it a subspecies of *G. horridispinum*, rather than the other way around, because Lodé (1995) made it a variety of *G. horridispinum* and created the precedence.

Cultivation

Easy to grow, this plant readily produces its large, usually pink, but sometimes white flowers. It is less densely spined than *G. monvillei* but the spines are usually stronger, some clones being spectacularly armoured. Large plants are very impressive but rarely seen in collections.

Gymnocalycium hossei F. Haage

First Description (F. Haage)

Kakteen Preisverzeichnis: pp.14-15 (1927) (original text in English)

Gymnocalycium hossei sp.n. Haage jun.

Named in honour of Prof. C. Hosseus, Córdoba. Splendid novelty, nearest related to E Schickendantzii, but much easier, already as young plants flowering. Flesh of dark colour. 12 to 30/-

Caption of picture (from German): Echinocactus Hossei, New. Named in honour of Prof. C. Hosseus, Córdoba. Dark fleshed. Habitat Argentina. (Seed No. 137)

The first more detailed description appeared in Berger *Kakteen* (1929). Haage (1971) tells us that this description was made from a plant he gave to Berger that was part of the original importation:

Echinocactus Hossei Friedrich Adolph Haage Jr (1927). – *Gymnocalycium* Berger

Somewhat flattened-globular, dark brown-green, slightly sunken in the crown, unarmed but with longish rounded warts. Ribs 13, cross-cuts forming humps, chin-like under the areoles. Areoles 10-14mm apart, longish, 5-6mm long, with some white grey-white felt. Spines mostly 7, spreading and somewhat curved back, strong awl shaped, flattened, at first brown, later grey, tip darker, fine, the lower spine up to 15mm long, the bottom spine curved upwards, of the same length. Flowers coming from the crown, with a short tube, petals deep pink.

Argentina: Córdoba.

[It is possible that Berger assumed that the plant came from Córdoba because Hosseus lived there. Hosseus (1939) tells us that it is found in La Rioja]

Fig.328 The first illustration of *G.hossei* from Haage nursery catalogue of 1927

Haage (1971), commenting on Berger's description, says that he should have mentioned that the central spine on old plants can grow to 26mm long and radials up to 18mm.

Fig.329 The illustration of *G.mazanense* from Backeberg's seed list of 1933

Fig.330 The illustration of *G.mazanense* from Backeberg's Bildcatalogue of 1934, repeated in 'Blätter für Kakteenforschung' (1935)

Fig.331 The habitat of *G. hossei*, GC980.02, east of Anillaco, Prov. La Rioja, Argentina. This view is dominated by *Tephrocactus alexanderi*

Fig.332 *G. hossei*, GC980.02, east of Anillaco, Prov. La Rioja, Argentina.

Etymology

Named for Prof. Dr. Carl C. Hosseus (1878-1950), German botanist who lived in Argentina and was professor of botany at Córdoba University.

Synonyms

G. mazanense (Backeberg) Backeberg in *Blätter für Kakteenforschung*, Lief.2, Bl.74-6 (1935). Basionym *Echinocactus mazanensis* Backeberg (1932)

First description of *Echinocactus (Gymnocalycium* Br. u. R.) *mazanensis* Bckbg. n. sp. in *Der Kakteen-Freund* (1)12: pp. 133-134 (1932)

Flat globular, single head, matt grey-green, sometimes with a brownish tint, flower and spine colour unusually variable. Ribs 10-11, fairly low, 2.5-3cm wide, chin shaped humps that are bordered below by sharp cross-cuts that form flat triangular areas. Areoles, on the example 2-2.5cm apart, with long spines, covered with white felt, oblong, up to 8mm long, sometimes a little smaller with less felt. Radial spines usually 7, 3 pairs pointing to left and right, one pointing downwards, more or less away-standing, curved, on the type species more or less interlaced horizontally; central spines at first often missing, up to 3.5cm long, in the top areoles away-standing, on older areoles, slightly curved, reddish, when damp a shining fluorescent colour, later grey.

Flowers, short, tube dark green to brownish-green, with light edged scales, outer petals are rounded at the top, not very wide, pale cream colour with a dark central stripe, (according to the colour of the tube) inner petals with a small fine point, slightly curved inwards, light cream colour; style and stamens yellowish-white. Habitat: Mazan, Catamarca, north Argentina.

Commentary on *G. mazanense*

Following his first description of *G. mazanense*, Backeberg explains that he found it in the nursery of Stümer in 1932 and he goes on to say that it probably belongs to *Echinocactus hossei*.

G. nidulans Fric ex Backeberg in Backeberg & Knuth: *Kaktus-ABC* pp.293, 417 (1936)

First description of *Gymnocalycium nidulans* Bckbg. sp. n.

Elongated-globular, up to 10cm high, 11cm diameter, matt brownish-green, with short hair in the crown. Ribs, up to 17, with large areoles on swollen humps, that are separated by sharp cross-cuts, the humps are 6 sided with a small triangular flat on the top. Areoles, c. 11mm apart, elongated-round with pale yellow-grey wool. Spines, c. 6 radials, two pairs pointing sideways, 1 pointing straight up and 1 pointing down, ± curved upwards, dirty yellow, stiff, elastic-like. Flowers bell to funnel shaped, pale pink, darker in the throat. Northern Argentina, Mazan.

Commentary on *G. nidulans*

This is another Backeberg name from the same general area. Recent explorations suggest that plants matching the original description grow near Señor de la Peña, a shrine to the east of Anillaco in Prov. La Rioja, Argentina.

G. weissianum Backeberg in Backeberg & Knuth: *Kaktus-ABC* pp.296-297, 417 (1936)

First description of *Gymnocalycium Weissianum* Bckbg. sp n.

Body flattened-globular, up to 14cm diameter, 9cm high, dull grey-green. Ribs, up to 19, at first 8mm wide, later 2cm wide, with areoles on swollen tubercles, below the areole is a triangular flat on a chin-like protrusion. Areoles 2cm apart, 1cm long, moderately covered with felt. Spines, usually 8, radials, 4 to each side of the areole, slightly curved, nearly 3cm long, only one central, of the same length, upright. Flowers bell to funnel-shaped, pale pinkish-brown, throat darker, tube short.

Habitat: Argentina, near Mazan.

Commentary on *G. weissianum*

Another taxon described by Backeberg from plants sent by Ernst Stümer, Buenos Aires. Recent habitat explorations suggest that the locality is near San Blas in Prov. La Rioja, to the north west of the Sierra Velasco. Bercht (pers. com.) claims that this form has white anthers rather than pink.

Distribution (Map 48)

This is a successful species, often common and occurring on most suitable habitats within its range. It thrives on gravelly slopes as well as the flatter land in the north east of Province La Rioja, Argentina. Its range extends from Mazán in the north to Anillaco in the south and can be found at the northern end of the Sierra Velasco and on the Cuesta de la Cebila.

Conservation Status

Least Concern. This plant is common over its distribution range. It often grows under bushes in rocky places or in crumbling granite. There are no obvious threats to its known habitats, being unsuitable for agriculture and only in very dry periods would goats eat them.

Fig.334 *G. hossei*, GC984.02 at the Cuesta de Cebila, Prov. La Rioja, Argentina

History

The name first appeared in Haage's 1927 catalogue with a picture of a young plant. Haage (1971) explained that only a few plants were sent to their firm by Prof. Hosseus who lived in Córdoba, Argentina. Hosseus regularly sent plants to the Haage nursery at Erfurt.

The illustration Haage published in his catalogue was a young plant, because two small plants from the importation flowered first in 1926. The larger ones that had longer spines were too dehydrated to flower so soon. Haage (1971) comments that the description "dark fleshed" would have been better expressed as "body dull grey-green". He also tells us that the plant had "magnificent blue fruits".

The current view is that this is the oldest name for plants from the Sierra Mazán which have subsequently been given various names such as *G. mazanense* Backeberg. Eggli (2005) also includes

Fig.333 *G. hossei* (polycephalum), *Piltz* 223 from the Sierra Velasco, Prov. La Rioja, Argentina

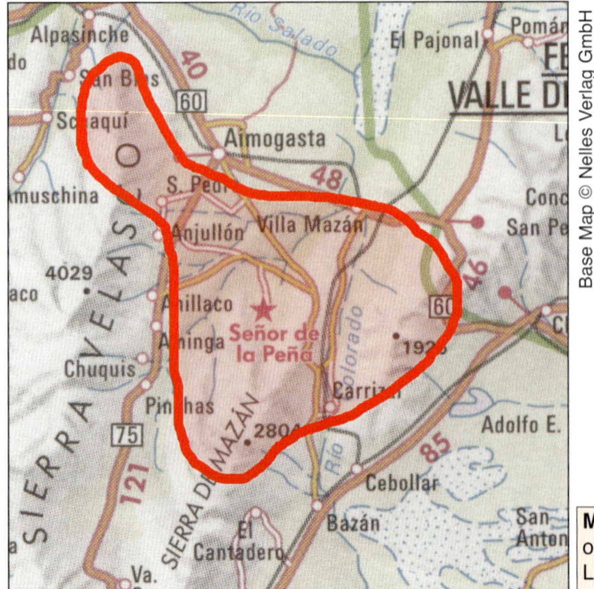

Map 48 Distribution of *G. hossei*, Prov. La Rioja, Argentina

Fig.335 *G. hossei,* STO124-1 from the Sierra Mazán, Prov. La Rioja, Argentina

Fig.336 *G. hossei,* GC27..02 from the Cuesta de Cebila, Prov. La Rioja, Argentina

other species in *G. hossei*, for example *G. rhodantherum* which is here kept separate. It is indeed similar and certainly closely related.

Meregalli et al. (2000) endeavoured to use seed characters to help the understanding of the taxonomy and relationships within this group of related species. Their study showed that seed from the same plant was identical from year to year. They also showed that some characters were extremely constant within a population and could be used to differentiate populations.

Based on their investigations of seeds from 100 documented plants, they were able to identify three main groups which corresponded with the geographic distribution.

1. Sierra de Mazán

The oldest name for these plants is *G. hossei*. Interestingly *G. nidulans,* which is said to come from Señor de la Peña, is the only population in the group to have grey anthers rather than the usual red.

2. South-eastern side of the Pipanaco Basin

G. glaucum subsp. *ferrari* is found here as far east as the Cuesta Sebile where one can also find a form of *G. hossei* and on the east side of the pass, *G. ambatoense*. Cuesta de Sebila is an interesting place where these three taxa come together and can easily be confused because they look so alike. It was noted that no plant having seeds with intermediate characters was found.

[Seed I collected on the summit of the Cuesta Sebila in 1992 (GC27.02) produced a few plants with pink buds and flowers which were different from the majority of the seedlings. I wonder if I inadvertently collected some seeds from *G. glaucum* subsp. *ferrarii*?]

3. Sierra de Famatina (and north to Belen and Andalgalá)

The oldest name for the plants in the south of this area is *G. rhodantherum* (= *G. guanchinense*) and at high altitude *G. ritterianum*. Further north the plants are here treated as forms of *G. pugionacanthum* (*G. catamarcense*).

Commentary

This taxon is another example of an early name which has proved difficult to apply with certainty to plants we see in habitat today. It is tempting to reject the name, but we do have a reasonable description and a contemporary picture, even though it shows a young plant. Backeberg appears to have misunderstood *G. hossei* and perhaps, as a result, described three new taxa from the Mazán area. These represent a single variable species for which most authorities now think *G. hossei* is the earliest name.

If one wants to take a wider concept of a species then it would be sensible to include *G. rhodantherum* as a synonym of *G. hossei* which would extend its distribution further west to include the Sierra de Famatina.

There is a particularly attractive form of *G. hossei* from the northern end of the Sierra Velasco which makes clumps of large heads and has been distributed as *G. mazanense* var. *polycephalum* by Piltz under his number *Piltz* 223 (Fig.333).

Cultivation

This is a species commonly seen in cultivation. It is easy to grow and makes a spiny plant, enjoying a sunny locality and flowering easily. It is a 'classic' *Gymnocalycium* and can easily be confused with other similar-looking species.

Illustrations

Hosseus, C.C. (1939) *Notas Sobre Cactaceas Argentinas* Plate V, Fig.5 according to Haage (1971) is not the right plant.

Backeberg, C. (1959) *Die Cactaceae* III p.1734 Fig 1669, according to Haage (1971) is not the right plant.

Gymnocalycium monvillei (Lemaire) Britton & Rose **subsp. *monvillei***

First Description (Lemaire)

Cactearum Aliquot Novarum pp.14-15 (1838) as *Echinocactus monvillii*

Globular, with many large hump-like tubercles, crown flat, without wool-felt; tubercles not arranged in ribs but in seventeen almost vertical series, forming the whole; intense light green.

Tubercles very closely packed together, particularly near to the crown, hexagonal at the base, very faintly angled where closely pressed together, marked with white dots, very broad (measuring 10 by 8 lines or more) and with a stout hump at the outer end as if abruptly truncated, flattened on the upper side and armed with a long, ovate areole, furnished with short white wool, never deciduous, porrect; spines very long, 12 or 13 in number, bright yellowish, purple and slightly subulate at the base, attaining 18 or more lines long, regularly arranged in two rows, flexible, with faint narrow transverse striations and slightly flattened, presenting a most pleasing appearance, ten longer laterals, the 11th above very short; the 12th below somewhat shorter than the laterals; the 13th central, very long, straight, often two inches long and more, and like the 11th, sometimes missing; grooves between ribs pale green, with a distinctive dark green wavy line.

Sometimes this Echinocactus brings forth lateral offsets, of a shape like that of other species of the genus, without doubt very handsome and not yet flowering, although it now attains 8 and more inches in height and two feet and more in circumference. Without hesitation this plant may be placed among the Echinocactus, simply by looking at it, and named after Mr. Monville, who first cultivated this plant and distributed it at great expense.

Habitat: Paraguai, Cordillera.

Later description of the flower from *Iconographie Descriptive des Cactées*, Part 8, by Charles Lemaire (1841-47):

Flowers very large, almost odourless, coming from around the crown, of a brilliant white, washed pink at the tips of the petals. Opening in the spring and lasting 5-6 days. More then 9cm long, 10-11cm diameter at anthesis.

Etymology

Named for Baron Hippolyte Boissel de Monville (1794-1863), French industrialist and plant collector.

Other Names

Echinocactus monvillii in *Cactearum Aliquot Novarum* (1838)

Fig.337 The original image of *Echinocactus monvillii* from Monville *Cactearum Aliquot Novarum* (1838)

Fig.338 The illustration of Echinocactus monvillii from Lemaire *Iconographie Descriptive des Cactées* (1847). Probably the same image as in Fig.337 with flowers added

Synonyms

G. brachyanthum (Gürke) Britton & Rose in *The Cactaceae* 3: p.159 (1922). Basionym: *Echinocactus brachyanthus* Gürke (1907).

First description of *Echinocactus brachyanthus* Gürke from MfK 17: p.123 (1907):

Body simple, flattened-globular with a fairly deep sunken crown which is completely unarmed, up to 18cm diameter and 7cm high, dull green, becoming slightly grey-green. Ribs 22, tapering into humps separated by cross-cuts, these 5-6 sided shapes are up to 3cm long and 2cm wide with a very strong protruding chin. Areoles small, elliptical, up to 10mm long, 2-3mm wide, with short white-grey wool-felt that stays on for a long time but soon becomes

Fig.339 The illustration of *Echinocactus Monvillei* from Schumann *Blühende Kakteen* (1901)

grey. Radial spines regular, 5 or 7 in 2 or 3 pairs pointing sideways, with a single spine pointing downwards, curved, very strong and sharp, on the new growth, stone yellow, later horn-coloured, slightly grey, the side pairs 15-20mm long, the lower spines 10-15mm long. Central spines, usually missing, as an exception to the rule, only 1.

Buds globular. Flowers in the area of the crown, funnel shaped, a full length of 6mm. Ovary without tube, running into the receptacle, dark green, shiny, no hairs, lighter at the bottom, with 15-18 scales. Scales on the ovary, triangular to semi-circular, light green with whitish transparent edges, 3-6mm long, 4-7mm wide, scales on the lower part of the receptacle short spatulate, dark green, thick, fleshy, with transparent, curved back edges and a brown-red tip, 10-16mm long, 7-9mm wide. Flowers 5.5-6cm in diameter. Outer petals, spatulate, light green with a green central stripe on the outside, brownish towards the tip, 3cm long, 10-12mm wide. central petals also spatulate, with a short point. completely pale pink, the central rib on the outside slightly protruding, a little darker in colour, 3cm long, 12-14mm wide; innermost petals similar to the central petals but 20-25mm long, 9-10mm wide. Many anthers, light chamois yellow, very small, 1.5mm long; stamens 8-10mm long, white. The inner hollow of the ovary is spinning-top shaped, 1cm long and wide. Style (including stigma lobes) 18mm long, very strong, stigma lobes, 12, light yellow, 5mm long.

This new species that originates from Argentina, is cultivated in the Botanical Gardens Dahlem. It belongs to the subgenus *Hybocactus* K. Schumann, (*Hybogoni* Salm-Dyck) and stands next to *E. Monvillei* Lem. The ribs of this species are lost in 5-6 sided humps, also as with *E. Monvillei*, only radial spines are present, on *E. Monvillei* there are usually a greater number of spines (at least on the examples in front of me), spines only up to 25mm long, on *E. Monvillei* up to 40mm long. The flowers on this new species, very short, 6cm long, the flowers will develop in dull weather but only become 36mm long, well developed flowers on imported plants of *E. Monvillei* develop up to 9cm long, the rest of the flower parts on both species show, a relative to size relationship. The style on *E. brachyanthus* is 18mm long, on *E. Monvillei* 40mm long, this difference in the flower form is recognisable in the buds, they are nearly round in the new species, whilst on *E. Monvillei* much longer and pointed.

Commentary on *G. brachyanthum*

There is no contemporary picture to help identify what this plant might have been, but plants thought to be this taxon had been found in the Sierra San

Fig.340 The illustration of *Echinocactus multiflorus* which accompanied the first description in *Curtis's Botanical Magazine* 71: plate 4181 (1845)

Luis resulting in the following combination and designation of a neotype.

Gymnocalycium monvillei subsp. *brachyanthum* (Gürke) Till in *Gymnocalycium* 6(3): p.102 (1993)

Neotype: Argentina, Prov. San Luis, near Suyuque, 1320m leg. G. Neuhuber GN91-77/1183 5.12.1991 (WU). Designated by Till & Neuhuber in *Gymnocalycium* 6(3): p.102 (1993)

Later, Neuhuber (1999a) changed his mind, abandoned his neotype, and applied the name *G. brachyanthum* to plants growing near Carlos Paz in Prov. Córdoba, a more likely location considering the date of Gürke's description. He goes on to suggest that they are the same as *G. grandiflorum*, a Backeberg name of uncertain attribution. Since the San Luis plants then had no separate name, he called them *G. monvillei* subsp. *gertrudae* var. *confusa*. Considering all this confusion, *G. brachyanthum* is best just treated as a synonym of the type subspecies.

G. multiflorum (Hooker) Britton & Rose in *Addisonia* 3: p.5 (1918). Basionym: *Echinocactus multiflorus* Hooker (1845)

First description of *Echinocactus multiflorus* Hooker from *Curtis's Botanical Magazine* 71: plate 4181 (1845)

Our only specimen is of the size here represented, globose, depressed at the top, green, slightly glaucous. Tubercles large, irregularly placed, upper ones only in an imperfect vertical series, and those oblong or oval, very prominent, obscurely angled. Areoles oval, woolly; bearing five nearly equal spines, about an inch long, diverging, but not on all sides, two opposite pairs laterally and the lower one towards the base of the plant; all are so much spread and decurved that they may almost be said to be appressed strong, of a yellowish colour, purple at the base. Flowers large (for the size of the plant), numerous, several opening at one time, so as to cover and conceal the upper surface of the plant. Calyx-scales green, gradually enlarging and becoming petaloid, till at length they pass into the spreading, obovate, almost white petals. Stamens

Fig.341 *G. monvillei*, GC941.02 in habitat on the east side of the Sierra Grande, Prov. San Luis, Argentina

numerous. Anthers small orange. Lobes of the stigma white, or nearly so.

Commentary on *G. multiflorum*

Hooker described this species from a single specimen in the collection of Mr. Palmer from Stockwell, London. There have been recent unsubstantiated suggestions that this plant came from Paraguay and that plants have been found there which match the description. It is suggested that these plants belong to the *Macrosemineum* seed group.

Till & Till (2007) resurrect the old name *Echinocactus ourselianus* and attempt to associate the name with plants from Rausch said to come from Corrientes or Santa Fé. They go on to publish *G. multiflorum* var. *ourselianum* but without any reference to a definite habitat population. It is a regrettable example of unhelpful guesswork based on old plants in cultivation with no reliable provenance.

G. schuetzianum H. Till & Schatzl in KuaS 32(10): pp.234-236 (1981)

First Description of *Gymnocalycium schuetzianum*

Flat rooted. Body flattened-globular, 9-11cm diameter, (on older plants up to 17cm diameter, 15cm high), light bluish green, offsetting in age. Crown somewhat sunken, mostly with entwined spines. Ribs 9-11, on older plants up to 17, running straight down, thickened around the areoles, with chin-like humps. Areoles oval, 8-9mm long, with thick white wool-felt, which is persistent. Spines on new growth light yellow to golden-brown, light at the base, becoming darker towards the tip, later becoming paler, awl shaped, ± slightly recurved to the body, sometimes twisted. Radial spines mostly 5, on older plants up to 7, the lowest and the bottom pair up to 40mm long, the top pair mostly shorter, on older plants the upper pair are also as long as the lower pair. Central spines rare on young plants, on older plants almost always 1.

Flowers from the centre of the crown, variable in form and colour, from narrow to wide funnel shaped, sometimes with several rows of petals, usually the flowers appear half open, pale to deep pink, the bottom always darker, 60mm long, 60-80mm diameter. Ovary 10mm long, 10mm diameter, the same colour as the body, with spatulate, pink scales, becoming longer and paler before reaching the outside petals; outer petals spatulate, curved outwards, dirty pink with a metallic green central stripe, 20mm long, 7-8mm wide; inner petals lanceolate, with ± small points, ragged, from pale to full pink, with a darker middle stripe, c. 30mm

Fig.342 *G. monvillei* GC941.02 in habitat on the east side of the Sierra Grande, Prov. San Luis, Argentina

Fig.343 *G. monvillei* GC941.02 in habitat on the east side of the Sierra Grande, Prov. San Luis, Argentina

Fig.344 *G. monvillei,* GC1000.01 in the Sierra Chica near to Ongamira, Prov. Cordoba, Argentina

long, 7mm wide. Stamens short, in several rows, lying close to the tube wall, the lowest two rows lying onto the style, pink; anthers yellow. Style 25mm long, 1.5mm thick, pink, with 12 pale yellow stigma lobes standing over the upper stamens. Fruit nearly globular, c. 14mm diameter, juicy. Seeds numerous, reddish-brown to blackish-brown similar to *Gymnocalycium monvillei* (Lem.) Br. & R. This species is listed in the same family group, the seeds of both species are very similar, therefore very difficult to separate. Seed group Microsemineum (Fric & Schütz). Habitat; Argentina, Prov. Córdoba, near Cruz del Eje (FR-No.430).

Holotype: In the Herbarium of the Botanical Institute of the Vienna University, (alcohol preserved and colour photographs), isotype (dry prepared in WU). Living isotypes can be found in the Botanical Gardens Vienna and Linz, Austria.

Commentary of *G. schuetzianum*

Seeds of this plant, collected by Ritter in Córdoba, were first offered under the number FR430 in the Winter catalogue (III) of 1956. It was listed in the subsequent catalogues up to No. IX. The plant was said by Ritter to look like *G. valnicekianum* which grows near Cruz del Eje. Schütz (1978) described his experience growing plants from this seed, being surprised that they did not look like *G. valnicekianum*. He published a colour picture of one of these plants with its spectacular pink flower.

It seems likely that a mix-up happened with the seeds, so making it likely that the plants in collections under this name and number probably came from somewhere else, but where?

Fig.346 *G.monvillei,* GC935.01 in habitat in the Sierra San Luis at Suyuqui Nuevo, Prov. San Luis, Argentina

The plant remains mysterious. Many people have looked near Cruz del Eje to try to find the population but without success. The only recent reference I can find is a tantalising statement in Till & Neuhuber (1993) where they say "Near where one finds the isolated population of *G. horridispinum*, one finds *G. schuetzianum* in a relatively small area". This must be west of Salsacate more than 80km from Cruz del Eje, but no more appears to have been written about this sighting.

G. grandiflorum Backeberg in Backeberg & Knuth: *Kaktus-ABC*: pp.289, 417 (1936)

First description of *Gymnocalycium grandiflorum* Backbg. sp. n.

Flattened-globular, green, sunken in the crown. Eight flat ribs with swollen humps formed by strong cross-cuts, c. 15mm wide at the base. Radial spines 5, lying close to the body, 1 pointing downwards. Flowers pure white, 7cm diameter.

Habitat: Northern Argentina.

Commentary on *G. grandiflorum*

Backeberg got this single plant from Stümer without locality data. He though it looked like *G. mostii* (kurtzianum). H. Till (1990b) made it a

Fig.345 *G. monvillei,* GC994.03 in habitat near to Sauce Punco, Prov. Córdoba, Argentina

Fig.347 *G. monvillei,* GC1001.02 in habitat to the east of La Cumbre in the Sierra Chica, Prov. Córdoba, Argentina

Fig.348 *G.monvillei* (coloratum), *Neuhuber* 90 243-775 from La Sierrita, Prov. Córdoba, Argentina

Fig.350 *G. monvillei, Neuhuber* 88 102-257 from Colonel Pringles, Prov. San Luis, Argentina

variety of *G. monvillei* but Backeberg's description suggests it was a form of *G. mostii*.

Neuhuber (1999a) designated a neotype, a form of *G. monvillei*; Argentina, Prov. Córdoba, west of Icho Cruz, 1050m, leg. G. Neuhuber GN90-215, 2.12.1990 (BA). It is therefore listed here as a synonym of *G. monvillei* subsp. *monvillei* in accordance with this neotype.

Distribution (Map 49)

Although the original description gives the locality as the mountains of Paraguay, the name is now usually applied to a taxon found in the mountains of the Argentinian Provinces Córdoba and San Luis for which it is a good match.

Neuhuber (Chileans 59, p. 55) states that from his observations, the habitat range of *G. monvillei* extends from the southern Sierra Ambargasta to Achiras. H. Till (ibid) says that it extends over the whole of the Sierras Grande and Chica. He also quotes Piltz describing its large distribution area as extending from the Río Tercero in the south, north of the Río Cuarto up to the Sierra Tulumba.

Fig.349 *G.monvillei,* RS from Tanti, Prov. Córdoba, Argentina

Conservation Status

Least Concern. The type subspecies is very common and is not threatened by any discernable factor.

History

It is unclear who first sent this plant to Europe. It could have been Gillies who is known to have travelled to Córdoba and is likely to have seen it when crossing the Sierra Grande. He could have sent the plants to Rosario for export.

So why was the original finding place given as Paraguay? It is probably an example of the quoted locality being derived from the departure point of the ship which may have originated in Paraguay before collecting freight in Argentina on the way to Europe. Supportive of this belief is the fact that

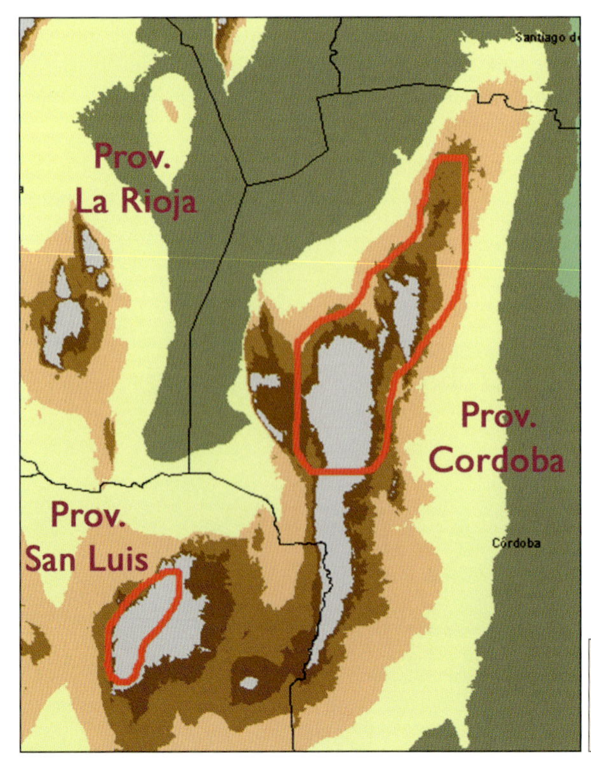

Map 49 Distribution of *G. monvillei* in Prov. Córdoba & San Luis, Argentina

Fig.351 *G. monvillei, Neuhuber* 91 77-1184 from Suyuqui Nuevo, Prov. san Luis, Argentina

Fig.352 *G. monvillei, Bercht* 967 from Los Quebrachos, Prov. Córdoba, Argentina

Paraguay was effectively closed to outsiders by the dictator Francia at the time of the first collection of this species.

The first description (which excluded the flower) was accompanied by a fine, large engraving of a plant without flowers. The same image appeared some years later (1847), reversed, coloured, and with three flowers added, in part 8 of Lemaire's 'Iconographie Descriptive des Cactées' as *Echinocactus monvilleanus*. The illustration [presumably of the added flowers] was drawn from a plant flowering in the collection of Mr. Saglio.

Flowering of the plant was a problem since the huge Monville plant had not flowered by 1840 nor had the specimen in the Salm-Dyck collection by 1850, which was said to be 20cm in diameter.

It was in 1903, when a consignment of various plants from Dr Sprenger of Volmero, Argentina, reached Schumann, proving that Córdoba was the correct locality for this species. So the mystery of its true origin was solved as a result of the first importation for decades.

Commentary

This is an easily recognised taxon if you accept that the Córdoba plants are in fact the correct application of the name. I have not seen a candidate from Paraguay that really looks like the original picture of this nor *G. multiflorum*.

Cultivation

Here is one of the commonest species is cultivation. It is easily and quickly grown into an imposing specimen, especially if a few offsets are formed. Most clones first bloom while they are still relatively small, and the flowers are large, usually pale lilac-pink, but can also be clear white. It is the golden spination which makes this a good plant to have in your collection. Plants from the south of the range tend to have finer spination and smaller flowers.

Fig.353 *G.monvillei, Muhr* B17 from Pampa Poncho, Prov. Córdoba, Argentina

Illustrations

Meyer, R. (1917) MfK 17(12): photograph p.171

Lengdobler (1919) MfK 19(7): photograph p. 81

Britton & Rose (1922) *The Cactaceae* III: p.161 Fig.170, contributed by Spegazzini, is not the right plant. This same illustration appears in Kiesling (1987) labelled as *G. schroederianum* and said to have been collected in Entre Rios.

Werdermann, E. (1936) *Blühende Kakteen*: Plate 110

Gymnocalycium mostii (Gürke) Britton & Rose **subsp. *mostii***

First Description (Gürke)

Echinocactus Mostii Gürke n.sp. - *Monatsschrift für Kakteenkunde* 16(1): p11-12 (1906)

Body globular, dark to blue-green with a flattened, bald, spineless crown, 6-7cm high, 13cm in diameter. Ribs at first 11, at a later stage up to 14, vertical, later running spirally downwards, tapering, through cross-cuts especially above forming humps that protrude chin-like under the areoles; the vertical grooves are sharp and narrow, flattened towards the base.

Areoles 2-3cm apart, elliptical to almost round, 6-8mm in diameter, covered with distinct yellowish-white, short felt, soon becoming bald. Radial spines 7, of which, 2 pointing upwards are 6-7mm long, 4 pointing to the side to 22mm long, and 1 pointing straight down to 15mm long, mostly pointing towards the body but not closely pressed. Central spine 1, vertical, with the tip usually pointing upwards, 18-20mm long. All spines are strong, somewhat angular, light horn coloured, shiny brown at the tip and at the base; the top areoles are free of spines.

Flowers beaker-shaped to rotate with outwards curved petals, 7-8cm in diameter at the widest part. Flower tube short and completely filled with the ovary, 10-11mm long and thick, the outside a shiny blue-green, completely bald, covered with 8-10 scales. Scales, a shiny blue-green, 3-10mm long, 5-10mm wide at the base, with a narrow transparent edge and a reddish spot at the top of the blunt edge. Petals 34-38, the outer 12-14 spatulate, up to 35mm long, 10mm broad, (the lower petals are the shortest, gradually resolving into scales), very thick and fleshy, green, the thin transparent edge is tinted pale red, slightly dirty red at the base; the intermediate 14-16 petals are also spatulate but longer and narrower, up to 40mm long and 8mm

Fig.355 The illustration of *Echinocactus mostii,* Plate 93 from Gürke *Blühende Kakteen* (1907)

Fig.356 The illustration of *Echinocactus kurtzianus,* Plate 97 from Gürke *Blühende Kakteen* (1907)

Fig.354 The illustration of *Echinocactus Kurtzianus* published in MfK 17(8): p.126 (1907) by Gürke

Fig.357 *G.mostii mostii* colonising a rocky slope south of Ascochinga, Prov. Córdoba Argentina

broad, light salmon red, the tips darker, with a delicate pale green central stripe. Many stamens, inserted in the area above the ovary in a 15mm wide ring shaped zone. Filaments up to 15mm long, curved towards the centre, pale yellow. Anthers longish, light yellow. Style comparatively short and very thick, mostly 12mm long (without the lobes), 4mm diameter at the top, light yellow-green, the 12 lobes pale yellow, 6mm long, close together while the flower is in full bloom.

The Royal Botanical Garden, Dahlem, received this species from Mr Carlos Most, Córdoba, Argentina.

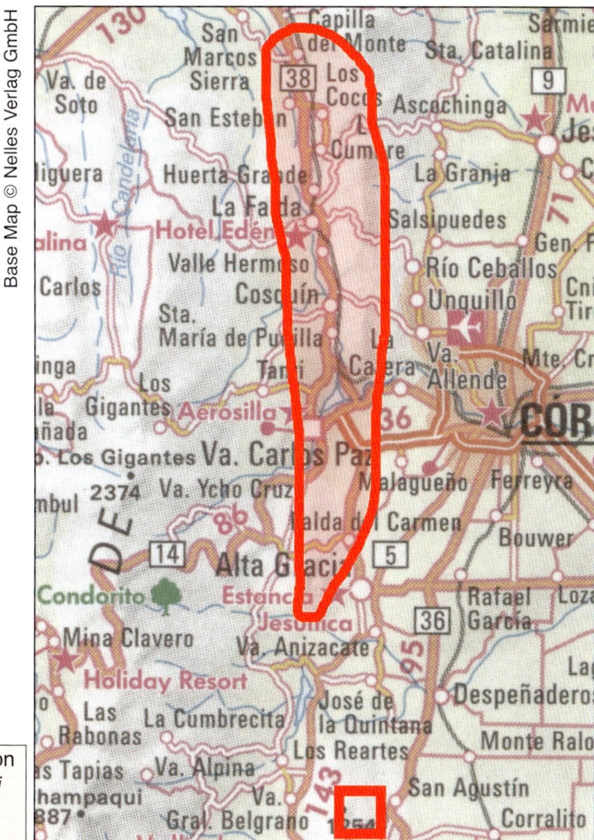

Map 50 Distribution of *G. mostii mostii* in Prov. Córdoba, Argentina

Fig.358 *G.mostii mostii,* GC940.01 in habitat just north of Carlos Paz, Prov. Córdoba, Argentina

Etymology

Named for Mr. Carlos Most, a cactus collector from Córdoba who sent the plants to Europe.

Other Name

Echinocactus Mostii Gürke in *Monatsschrift für Kakteenkunde* (1906)

Synonym

G. kurzianum (Gürke) Britton & Rose in *The Cactaceae* 3: p.163 (1922). Basionym: *Echinocactus Kurtzianus* Gürke (1906)

First description of *Echinocactus Kurtzianus* Gürke n. sp. in MfK 16: p.55 (1906)

Body semi-globose, with a slightly sunken, humped, bald, spineless crown, light green, up to 10cm high, 15cm in diameter. at first with 10-12 ribs, later, through further insertions, up to 18 ribs, mostly vertical, separated into clear humps that are protruding, chin-like from just below the areoles by cross cuts, the cross cuts are sharp and shallow. Areoles 2-3cm apart, elliptical, 10mm long, 5-7mm wide, at first with thick yellowish-white wool-felt, soon becoming bald. Radial spines, 8, of which one points towards the top, 3 pairs pointing sideways and 1 pointing down, 25-40mm long, mostly curved, not lying close to the body, central spines 1, strongly curved upwards, up to 30mm long, all pairs very strong, stiff, horn-grey, the tips mostly brownish.

Flowers, few, in the area of the crown, full length 7-8cm, receptacle wide funnel shaped, tube short and thick, 15mm diameter, shiny light green, bald and smooth, with about 20 scales, the lower scales semi-elliptical, 5-6mm wide, 3mm long, the top scales semi-circular, up to 4mm long, dull green, with a white translucent edge, eventually resolving into petals. Petals long spatulate with rounded tips, sometimes slightly twisted, 35mm long, 10mm broad, inner petals lanceolate, a little shorter, c. 30mm long, only 5-7mm broad, sometimes slightly tapered and more pointed than the outer petals, all petals are thick, nearly fleshy, white with a tint of red, outer petals with a wide green central stripe, inner petals with a reddish central stripe, all are strawberry red at the bottom, as is also the inner wall of the receptacle to the base of the style. Filaments, many, c. half as long as the receptacle, inserted in a circular zone, c. 2cm diameter, on the inside of the receptacle, rising and curving over the stigma lobes, yellow, darker towards the base. Anthers longish, 2mm long, sulphur yellow. Pericarp pear shaped, flat on top, the inside 10-12mm high; 10mm diameter. Style 12-13mm long, greenish-white, very strong; the 14 stigma lobes 5-6mm long light yellow.

Commentary on *G. kurzianum*

Sent to Europe by Mr. Carlos Most in 1905, it was named for Dr. Fritz Kurtz (1854-1920), professor of botany at Córdoba. It was made a variety of *G. mostii* by Backeberg in *Kaktus-ABC* : p.292. (1936)

Fig.359 *G. mostii mostii, Neuhuber* 245-780 from Monslavo, Prov. Córdoba, Argentina

Distribution (Map 50 on page 167)

The extensive overall distribution area of this species extends from the Sierra Chica, just to the west of Córdoba City, to Dean Funes in the north and yet further to the north-east, close to the border with Santiago del Estero. An isolated population near El Mirador south of the main distribution was described as var. *miradorensis* by Till & Amerhauser (2002a).

The typical form has a preference for growing on or amongst rocks. It can be very effective at colonising rock cliffs, even artificial ones, such as road cuts where hundreds of plants of all ages can sometimes be seen.

Conservation Status

Least Concern. The plant is widespread and locally common with no immediate threats.

History

The first description of *Echinocactus mostii* by Dr. Max Gürke appeared in MfK (1906). Later in the same issue, he described a similar plant as *Echinocactus kurtzianus*. Both were illustrated two years later in *Blühende Kakteen* with very competent illustrations, showing the differences, that were drawn by the author's wife Frau Toni Gürke.

Fig.361 *G. mostii mostii,* STO 829 from the Cerro Pencales, Prov. Córdoba, Argentina. The form which was probably designated as the type of *G. mostii* subsp. *ferocior.*

Britton and Rose moved the plants to the genus *Gymnocalycium* in Addisonia 3: p.5 (1918). Rose had seen the species near Carlos Paz in Prov. Córdoba, Argentina, when he was there in 1915. A plant he collected flowered in New York and was used for the fine colour plate XVII(2) in *The Cactaceae* III (1922).

Commentary

To the west, the isolated population of heavily-spined plants which Till (2002b) named *G. mostii* subsp. *ferocior* is here treated as a subspecies of *G. castellanosii*.

Cultivation

A frequently cultivated species, it readily grows into a large, usually solitary, specimen. The plant body is an attractive dark green and the strong grey spines have red bases when young. The petals of the large white flowers have pink bases and tend to be rather untidy.

Illustrations

Britton & Rose Plate XVII (2) in *The Cactaceae* III (1922)

Krainz, H. in *Die Kakteen* Lief. 59 (1974), 2 pictures provided by G. Frank.

Fig.360 *G.mostii mostii* (kurtzianum), *Piltz* 84 from La Falda, Prov. Córdoba, Argentina

Fig.362 *G.mostii mostii,* GC68.02 from west of Alta Gracia, Prov. Córdoba, Argentina

Gymnocalycium mostii subsp. *valnicekianum* (Jajó) Meregalli & Charles

First Description (Jajó)

Kaktusar 5(7): pp.73-74 (1934)

Gymnocalycium valnicekianum Jajó spec. nov.

Not offsetting, flat-round to globular, dull grey-green, up to 18cm diameter. Ribs 10-13, with projecting blunt humps. Areoles large, light felted, round. Radial spines. 7-9, 2.5-3cm long, grey, very brittle, lying close to the body, 1 central spine, up to 3cm long, like a hook, strong. Flowers funnel shape, 3.5-4cm diameter, dirty white, reddish in the throat, sepals 2-2.5cm long, naked. Seeds semi-matt black. Habitat Argentina. (Uruguay?)

Neotype: Argentina, Prov. Córdoba, in the Sierra de Ischilín, near Santa Catalina, 850m, 9th October 1987, leg. H.Till HT 87-2/1666 (CORD) Designated by Till & Amerhauser (2002b)

Etymology

Named for Dr. Jana Valnicka, a Czech cactus hobbyist.

Other Names

G. valnicekianum Jajó in *Kaktusar* (1934)

Synonyms

G. immemoratum Castellanos & Lelong in *Lilloa* 4: pp.195-196 (1939)

First description of *Gymnocalycium immemoratum* Castell. et Lelong nov. sp.

Body flattened-globular, deep green, sunken in the crown, 9cm high, 12cm diameter. Ribs 10-13, 2-2.5cm wide at the base, 10-12mm high, large humps are formed by transverse grooves, the humps are flattened on top with a cone shaped point below. Areoles elliptical, 1cm long, 8mm wide, 12mm apart, set in depressions on the humps. Humps covered with thick felt and provided with ± 15 spines. Spines long, 2.5-3cm long, entwined, bluish-grey, the point, milk white, swollen at the base, chestnut in colour, sometimes 4 centrals are present, the rest are radials.

Flowers, from the edge of the crown, funnel shaped, 5cm long and diameter. Ovary with deep leaf green scales, lower scales ± 4-5mm wide, higher scales 5mm wide, thick, pale edged. Flower tube

Fig.363 *G.mostii valnicekianum, Piltz* 83 from Capilla del Monte, Prov. Córdoba, Argentina

Fig.364 The original illustration of *G. valnicekianum* which was published with the first description in Kaktusar (1934)

± 4mm thick, inside pale purple; outer petals oblanceolate, ± 3cm long, white, outside purplish, inner petals ± 1.75cm long, spatulate, mucronate, pure white. Stamens coming from the flower tube; filaments almost white, anthers straw yellow; style and stigma pale greenish-white. Fruit green.

Type:Castellanos & Lelong 6.1.1939, El Zapato, Capilla del Monte, N. 30526 (BA)

Commentary on *G. immemoratum*

Collected by Castellanos at El Zapato near Capilla del Monte in 1918. The same plant as *G. valnicekianum* which was published five years earlier. Schütz (1985) explains that when Piltz visited the type locality at El Zapato, he found *G. valnicekianum* there and in other nearby localities.

Gymnocalycium prochazkianum Sorma in *Gymnofil* 28(1-2): pp.2-6 (1999)

Fig.365 *G.mostii valnicekianum* (prochazkianum), the type form from near Quilino, Prov Córdoba, Argentina

Fig.366 *G.mostii valnicekianum,* (prochazkianum) *Meregalli* 824 in habitat,Cerro Horqueta, near Quilino, Prov. Córdoba

First description of *Gymnocalycium prochazkianum* Sorma spec. nov.

Body simple, flat to hemispherical, 40-70mm diameter, epidermis dark grey, matt. Crown a little depressed, and covered with spines. Ribs, 7-9 running straight, divided by sharp cross grooves into four-sided tubercles of dimension 10x15mm. Areoles round, at first covered with a dirty whitish felt. Spines, always three, up to 1 cm long and of equal lengths, two directed sideways and one downwards, adpressed, straight or slightly curved, whitish-grey, pungent. (Flowers emerge from near the apex, funnel-form, 60-65mm long. The outer, lanceolate petals are pinkish-white, the centre of the flower is a deep pink colour. The tube is short and carries some scales.

Fruit is elongated, grey, 15mm long, 7mm wide. Seeds are minute, blackish and matt, (in the Latin diagnosis hood-shaped) about 0.8mm long, 0.5mm wide, hilum not prominent, subgenus Microsemineum Schütz.

Type locality Argentina, Province of Córdoba on low ridges south of the market town Quilino, 600m above sea level.

Holotype JPR95-184/562A, 17.11.1995, deposited in WU. Further conserved material in cultivation in collection of V. Sorma under the number VS141/1997

The species occurs sympatrically with *G. quehlianum* (Haage jr. ex Quehl) Berger sensu Till & Till [= *G. rubustum*] and *G. stellatum* var. *obductum* (Piltz) Till & Till [= *G. quehlianum*].

Fig.367 *G.mostii valnicekianum* (prochazkianum), *Meregalli* 891 in habitat east of the the Cerro Horqueta towards San Pedro Norte, Prov. Córdoba, Argentina

Fig.368 *G.mostii valnicekianum,* (prochazkianum) *Meregalli* 824 in habitat at the Cerro Horqueta, near Quilino, Prov. Córdoba, Argentina

The species is named after its discoverer the present president of the Czech association of Gymnophiles Mr Jaroslav Procházka of Brno.

Commentary on *G. prochazkianum*

Sorma explains that they found this plant while looking for *G. robustum*, south of Quilino, Prov. Córdoba, Argentina. It is said the be rare in that area and they only found about ten plants. This taxon may well be a good species not belonging to *G. mostii*, see Fig 365, in culture.

G. valnicekianum subsp. *prochazkianum* (Sorma) H. Till & Amerhauser in *Gymnocalycium* 15(2): p.452.

G. tobuschianum Schick in CSJ(GB) 15(2): p.44, illn. p.36 (1953)

First description of *Gymnocalycium tobuschianum* Schick spec. nov.

Stem at first simple, later shooting at the base, light green, low growing, 4-5cm high, 8-9cm in diameter. Apex covered with whitish-grey, overlapping spines. Ribs 8-11, somewhat wavy formation, flattening out towards the base, 1-1.5cm high. Areoles 2cm remote, egg shaped or round, 6-8mm long, with a whitish-grey woolly felt covering. Spines horn coloured, usually 9 radials 1.5-2cm long, slightly recurved, the other 2 short and lightly wavy. Centrals 1-3, not very distinct from the radials, 1.5-2cm long, directed upwards, all spines curly, flattened at the top with a light groove, prickly. Flowers and fruit unknown.

Fig.369 *G.mostii valnicekianum,* (prochazkianum), *Bercht* 3366 in habitat north of San Pedro Norte. Plants from here look like '*G. bicolor* nom. inval.'

Fig.370 *G. mostii valnicekianum,* GC948a colonising a gravelly slope near to Capilla del Monte, Prov. Córdoba, Argentina

Habitat: Argentina, near Capilla del Monte.

Fig.371 *G. mostii valnicekianum,* GC948a near to Capilla del Monte, Prov. Córdoba, Argentina

Fig.372 *G. mostii valnicekianum,* GC4.01 west of Capilla del Monte at El Zapato, Prov. Córdoba, Argentina

Commentary on *G. tobuschianum*

The third description of *G. mostii* subsp. *valnicekianum,* named by Schütz for his friend Hermann Tobusch, a watchmaker and jeweller in Chicago.

Distribution (Map 51)

G. mostii valnicekianum and its synonyms occupy an area to the north-east of Capilla del Monte. *G. mostii valnicekianum* (prochazkianum) is found in the west of the distribution. Near its type locality by Quilino it is quite rare and shares its habitat with *G. robustum* and *G. quehlianum*. It is not growing on rocks but under the bushes. Even further north and east, plants looking like the invalidly described *G. bicolor* are found.

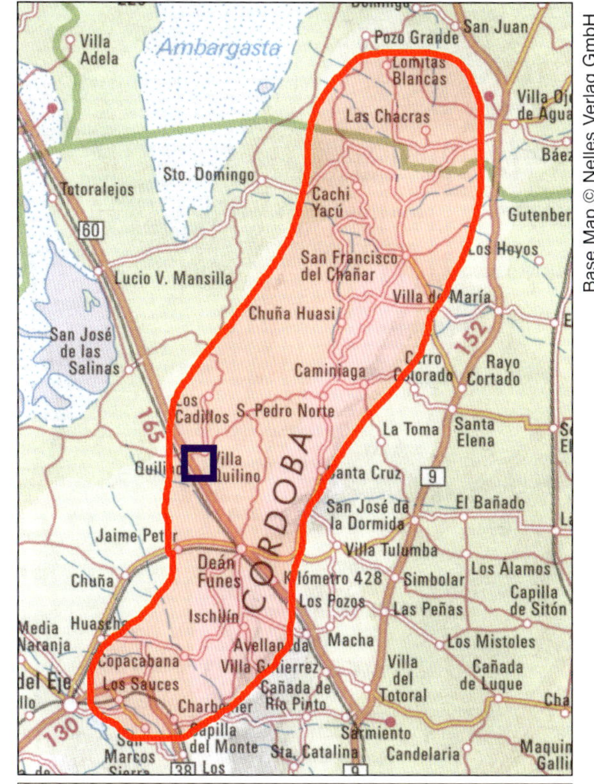

Map 51 Diistribution of *G. mostii valnicekianum* with the type locality of *G. prochazkianum* shown as a blue square

Fig.373 The habitat of *G. mostii valnicekianum.* GC993.02 in the Sierra Ischilín, Prov. Córdoba, Argentina

Conservation Status

Least Concern. This taxon has a large distribution and is frequently abundant in suitable habitats.

History

Before its description, this plant was known by the name *Echinocactus centeterius*, a Lehmann name that was misapplied to *Gymnocalycium* by Hosseus. The name in reality, probably relates to a species of Eriosyce.

Osten (1941) misapplied Schumann's name *Echinocactus froehlichianus* by using it in *Gymnocalycium* for two different species. He did not distinguish *G. valnicekianum* (Plate 67) from *G. monvillei* (Plate 69), both captioned as *G. froehlichianum.*

The publication of *G. bicolor* by Schütz in *Friciana* 1(7): pp.2-3 (1962) is regarded as invalid since no type appears to have been preserved. The name was published again as a variety of *G. valnicekianum* by Till and Amerhauser (2002b). Although published as a new combination, the basionym is invalid so the publication is accepted as a new taxon with the type being the plant the authors designated as the neotype.

In the 1960's Wilhelm Andreae, a nurseryman from Bensheim, Germany, received plants from Frau Dorothea Muhr under the number B44. He intended to describe them as *G. genseri* to honour the collector José Genser who sent the plants to Frau Muhr. However, Andreae died before he could make the publication. Till & Amerhauser (2002b) explain that two different plants were sent with the number B44, one was '*G. bicolor*' and the other, a similar-looking plant came from between Dean

Fig.374 The habitat of *G. mostii valnicekianum,* GC993.02 in the Sierra Ischilín, Prov. Córdoba, Argentina

Fig.375 The habitat of *G. mostii valnicekianum,* GC993.02 in the Sierra Ischilín, Prov. Córdoba, Argentina

Fig.376 *G. mostii valnicekianum, Piltz* 380 from the Sierra Tulumba, Prov. Córdoba, Argentina

Fig.377 *G. mostii valnicekianum* (bicolor), GC5.02 from Capilla del Monte, Prov. Córdoba, Argentina

Funes and the Sierra Ischilín. This latter was the plant Andreae had planned to call *G. genseri*. It fits reasonably well within the variability of *G. valnicekianum,* so another name would have been superfluous.

Till & Amerhauser (2002b) maintain that the original description by Jajó referred to plants from the Sierra Ischilín and so designated their neotype from there. There is no evidence for this view and it is certainly possible that the type locality is near to Cruz del Eje where plants like the original illustration can be found. However, the neotype locality is within the overall range of the taxon as accepted here.

The last taxon to be named was *Gymnocalycium prochazkianum* which first appeared in *Gymnofil* in 1999. It was made a subspecies of *G. valnicekianum* by Till & Amerhauser (2002b). A useful commentary on this taxon is given by Papsch (2002) who adds more details to the original description and publishes useful pictures.

Following its treatment as a synonym of *G. mostii* in the *New Cactus Lexicon* (2006), Meregalli and Charles (2008) gave *G. valnicekianum* a new status as a subspecies of *G. mostii* in CSI 24.

Commentary

G. valnicekianum is usually freely offsetting, not solitary as stated in the description. The plant illustrated with the first description is only young. Schütz (1949) published a var. *pluricentralis* because the original description called for a single central but, as Schütz (1985) later tells us, in fact the population usually has this larger number of centrals. Piltz (1979) is the first to give clear information about this plant in the wild. He confirms that Jajo's type plant was untypical by having only a single spine, a problem you can have when describing a plant without seeing it in habitat.

G. prochazkianum, recently described from near Quilino, has been made a subspecies of *G. valnicekianum* by Till & Amerhauser (2002b). I have not seen this plant in habitat so have accepted their combination, even though pictures of plants growing at the type locality have a grey colour and look rather different from *G. valnicekianum*. It is quite likely that this is, in fact, a separate but related species. Some pictures (22 & 24) in Till & Amerhauser (2002b) claiming to be this new subspecies are surely just *G. mostii* subsp. *valnicekianum,* supporting the view that *G prochazkianum* is restricted to the immediate vicinity of Quilino. It appears that at the type locality it is quite rare and is really distinct from *G. mostii* with fewer ribs; fewer, shorter spines; different flowers and a grey bloom on the body (Fig.366).

Cultivation

This is an easily-grown plant that is widely represented in collections, usually as the form with many central spines giving individuals a densely-spined appearance. It grows quickly into a cluster of stems and can make handsome specimens, large clumps sometimes being exhibited at shows. Many of these originated from imported plants sent by Frau Muhr in the 1970's under her field number B165.

The northerly form from the Sierra Ischilín is only occasionally seen, usually represented by *Piltz* 380 from the nearby Sierra Tulumba.

Fig.378 *G. mostii valnicekianum, Piltz* 83 from Capilla del Monte, Prov. Córdoba, Argentina

Gymnocalycium nigriareolatum Backeberg

First Description (Backeberg)

Blätter für Kakteenforschung, Liefg. 5, Bl. 74-1 (1934)

Solitary, spherical, up to c. 15cm diameter. stumpy, pale green. c. 10 broad ribs divided into sharp-edged, low tubercles. Areoles c. 6mm across, with thick white wool at first, later becoming quite black. c. 7 radial spines, light flesh-coloured, somewhat curved and one central spine up to 3cm long bent upward. Flowers porcelain white, half open, green throat. Fruit round, seed dull black. Occurrence: N. Argentina near Mazan.

Etymology

From the Latin 'niger' meaning black and referring to the areoles that become infected with a black mould in habitat when the plants grow in damp places under trees.

Distribution (Map 52)

Backeberg, in *Die Cactaceae* III (1959), revised the locality to Catamarca on the advice of Stümer. It is now known to be found on both sides of the Sierra de Graciana which lies to the east of Catamarca City.

Conservation Status

Least Concern. The plants are plentiful in habitat and there is no particular threat to their survival.

Fig.379 *G.nigriareolatum,* GC29.02 near the road climbing up to the Cuesta El Portezuela

History

Backeberg first described this species from plants he received from Ernst Stümer. It remained little known in cultivation until re-introduced with plants imported from Fechser. Much later it was collected again by Borth, Knoll, Rausch and Piltz.

Fric offered seeds of his *G. curvispinum* n.n. Portezuelo in his catalogue of 1929. It is thought that this was the longer spined form of *G. nigriare-olatum* that can be seen growing there today.

Till (1998) created a variety *densispinum* for a form which offsets freely and can be found between Pomancillo and Catamarca City. It was said to be a

Fig.380 View northwards from the habitat of *G.nigriareolatum,* GC29.02 near the Cuesta El Portezuela Prov. Catamarca, Argentina

Fig.381 The black areoles of *G. nigriareolatum*, GC29.02 growing under trees with *Parodia microsperma* near the road climbing up to the Cuesta El Portezuela.

validation of the name which was invalidly published by Backeberg in a catalogue of 1937.

He also described a variety *simoi* for a form originally found at Palo Labrado by Borth and more recently discovered between Portezuelo and La Merced, where it sometimes shares its habitat with the red-flowered fa. *carmineum*.

Commentary

The closest relation of this species is perhaps *G. pugionacanthum* which grows further north. There is a red-flowered plant reported from the populations of *G. nigriareolatum* in the north of the range (fa. *carmineum*) that Till (1998) claims is not related to the nearby occurring *G. oenanthemum*, but looks very similar to me. If *G. nigriareolatum* fa. *carmineum* is actually *G. oenanthemum*, then, since fa. *carmineum* is recorded as growing with *G. nigriareolatum* (var. *simoi*), that would support the two being separate species.

Fig.382 *G.nigriareolatum*, GC29.02 from near the road climbing up to the Cuesta El Portezuela, Prov. Catamarca, Argentina. Cultivated plants do not have black areoles.

Cultivation

This species is not common in collections today, most of the plants being derived from the seeds sold by Piltz. It poses no particular difficulty in cultivation and thrives in the usual regime for the genus. The pink-flowered plant distributed as fa. *carmineum* is very pretty but is probably identical to *G. oenanthemum*.

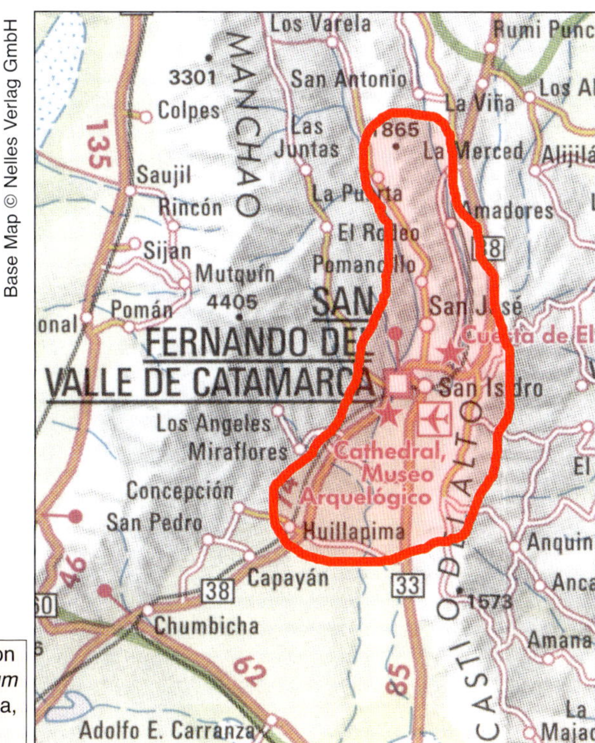

Map 52 Distribution of *G. nigriareolatum* in Prov. Catamarca, Argentina

Fig.383 *G.nigriareolatum*, GC30.02 from east of Catamarca City, Prov. Catamarca, Argentina

Fig.384 *G.nigriareolatum* (carmineum), STO255 from La Puerta, Prov. Catamarca, Argentina

Gymnocalycium oenanthemum Backeberg

First Description (Backeberg)

Blätter für Kakteenforschung Liefg. 9: Bl.74-4 (1934)

Dull pale grey green solitary, c. 11 ribs, up to 2cm wide, sharply angled, protuberances above the areoles with sharp grooves below. Radial spines 5, reddish, slightly curved, up to 1.5cm long, from narrow elongated grey areoles. Wool in areoles yellowish. Central spine wanting. Flowers pale claret-coloured (!) large, up to 5cm long, scales pink edged; fruit pale green, slightly mealy. With the exception of *Gymnocalycium venturianum* the only species so far known with red flowers. Occurrence: Mendoza (Argentina).

Etymology

From the Greek 'oinos' meaning wine and 'anthemon' for flower, referring to the dark red flowers.

Synonyms

G. ambatoense Piltz in KuaS 31(1): pp.10-13 (1980)

First description of *Gymnocalycium ambatoense* Piltz spec. nov.

Simple, flattened-globular to globular, slightly sunken in the crown. Epidermis dark green, matt, ± spotted on the top of the humps, up to 15cm diameter, 5-10cm high. Ribs: 9-17, mostly vertical, (sometimes running into irregular humps that are slightly spiralled), at half way up the plant, c. 2cm wide and 1cm high, at the base 3cm wide and flatter with chin-like humps below the areoles, with ± a horizontal cross-cut below each hump. Areoles 2.5-3.5cm apart, oval, 5-15mm long, 3-7mm wide, in new growth the areole felt is light grey to creamy-white, on older areoles, grey-black to black, at the base, again becoming grey. Radial spines: (5)-11, tough awl-shaped, cross section usually oval, mostly in order of pairs with 1 pointing downwards, curved towards the body or irregularly away standing, pink to pink-grey, in cultivation becoming paler, in the new growth dark red-brown, lighter at the tip, 1.5-3(-4)cm long. Central spines; very rarely missing, usually 1, up to a maximum of 3, also tough awl-shaped, straight or slightly curved, away-standing, same colour as the radial spines, dark pink at the top, (2)-2.5(-5)cm long.

Fig.385 *G. oenanthemum,* GC985.03 from south of Concepción, Prov. Catamarca, Argentina

Fig.386 *G. oenanthemum,* GC31.02 from near El Rodeo, to the north of Catamarca City, Prov. Catamarca, Argentina

Flowers; from the areoles near the crown, 2.5-4.5cm long, 3-4cm diameter, bell-shaped. Pericarp; 0.9-1.5cm long, c, 1.5cm diameter, dark green. Scales; semi-circular, up to 6mm wide and 3mm long, pink, light edged with a dark red spot on the tip. Nectar chamber; 1.5-2.5mm high, lilac-red. Receptacle; inside, dark carmine, outside, dark green to olive-green, 18-23mm diameter, 10-15mm long. Scales; at the bottom, 7mm wide, 5mm high, at the top, 9mm wide, 10mm long, olive coloured with pink edges. Petals; outers blunt spatulate to round, 7-8mm wide, with an olive green centre and a pinkish edge, middle petals blunt spatulate, elongated up to wide lanceolate, a shiny silky white, with a red-brown central stripe, up to 2cm long, 7-8mm wide, inner petals pointed spatulate to lanceolate, silky white with a bright pink to light olive central stripe, red at the base, 17-19mm long, 3-6mm wide. Stamens; primaries in a single ring leaning onto the style, secondaries inserted over the complete inner wall of the receptacle, leaning inwards, the topmost leaning over the lobes of the style, all c. 6mm long, yellow, reddish at the base. Anthers; a dirty lilac. Pollen yellow, Style with the lobes up to 15mm long, 2.5mm thick, just below the lobes a little thicker and yellowish. Fruits wide-round, up to 1.7cm long, and 2.3cm wide, matt dark green, splitting vertically from the base. Seeds c. 1mm long, 0.7-0.8mm wide, testa red-brown, to black-brown, not shiny, with globular warts. Hilum deeply sunken, hilum seam strongly curved. Micropyle in the hilum region. Testa without arillus.

Habitat: North Argentina, Province Catamarca, Sierra Ambato, at c. 900-1100m.

Type: in Succulent Herbarium (PH Rheinland, Abt. Köln), Germany no. P22/5

Commentary on *G. ambatoense*

This is a white-flowered form with finer spines from the southern part of the distribution. It may be intermediate with *G. hossei*.

G. carminanthum Borth & Koop in KuaS 27(4): pp.73-79 (1976) possibly invalidly described because the holotype may not have been preserved.

First description of *Gymnocalycium carminanthum* Borth et Koop spec. nov.

Roots; several, short turnip-shaped, stringy. Stem flattened-globular, crown sunken, epidermis matt blue-green to grey-green, coarse, up to 100mm diameter. 55mm high, 6-11 vertical ribs running straight down, flat, only in the area of the crown are there sharp chin shaped humps below the areoles, cross-cuts reach only over 1/3 of the width of the rib, ribs c. 25mm wide at the base. Areoles a little sunken, oval, with short rough felt that is yellowish on young areoles, 6-7mm high, 5mm wide. Radial spines (5-)-7(-9),

Fig.387 *G. oenanthemum* (type collection of *G. ambatoense*), *Piltz* 22 from Concepción, Prov. Catamarca, Argentina

Fig.389 *G. oenanthemum* (type collection of *G. tillianum*), Rausch 227 from Pomán, Prov. Catamarca, Argentina

tough awl-shaped, cross-cuts oval to round, on the top side covered with sharp scales, sharp edged, ordered in pairs slanting upwards, the longest pointing vertically downwards, all curved to the body, lying close, very rarely porrect, dirty pink to light grey, in new growth, orange at the base, higher up red-brown turning to brown and grey; mostly 15 to 25mm maximum length. Central spines usually missing, very rarely 1-2 appear near the areole edge, strongly curved upwards, 8-12mm long. (sometimes they are longer, c. 32mm, on one example, 80mm!).

Flowers, from the areoles near the crown, developing out of thick wool, wide bell-shaped, 58mm diameter maximum, 45mm high, maximum. Ovary; flattened-globular, sunken on top, outside green up to 7mm diameter and 8mm high. Pericarp; wide, beaker-shaped, the outside a darker colour, shiny (darker than the epidermis of the plant), up to 12mm diameter, to 12mm high. Scales; wide round, fleshy, light green with a membrane-like edge, with red tips. Receptacle; inside carmine, outside blue to olive-green, shiny, 25mm diameter maximum, 16-18mm long, scales as on the pericarp but larger and spatulate, green with a red edge. Petals; the outer petals spatulate, rounded to blunt, with a point, greenish to olive-green, with a transparent middle zone, the central petals are wide lanceolate, dark carmine, silky matt, not transparent, with a very dark carmine-red central stripe, up to 25mm long, 8mm wide, innermost petals small, lanceolate, otherwise the same as the central ones, up to 23mm long, 7mm wide. Stamens; primaries inserted at the base, leaning into the style and curving outwards, secondaries inserted over the entire receptacle, curved inwards, all anthers scarlet-red, 11mm long, the stamens are whitish, anthers 1.3-1.4mm long, 0.4-0.5mm wide. Style; cylindrical, becoming thick at the top, carmine below, more vermilion-red above, papillose, 2mm diameter, 12mm long without the stigma. Stigma; 9 lobes, yellow, to 5mm long. Seed formation, ovary completely full. Fruit; pear-shaped, dark to olive-green, to 10mm diameter, 15mm high, opening lengthwise, from the base.

Fig.388 *G. oenanthemum* (carminanthum), *Piltz* 133 from the Sierra Ambato, Prov. Catamarca, Argentina

Seeds; c. 1mm long, inverted purse-shape, testa matt black, fairly rough with globular warts, hilum sunken, strongly curved, spongy, yellowish-brown, with a wide thick edge.

Habitat: Sierra Ambato, Prov. Catamarca. Argentina. 1300-1800m. growing in mineral rich soil, under small bushes, in grass and in rock fissures.

Holotype; (collection number BO 130), in the Schütz and type collection Botanical Gardens, Linz, Austria.

G. tillianum Rausch in KuaS 21(4): p.66 (1970)

First description of *Gymnocalycium tillianum* Rausch spec. nov.

Body simple, wide globular with a flat root, 15cm diameter, 10cm high, blue-grey-green. Ribs; up to 15, running straight down, 15-20mm wide, humps separated by strong cross-cuts, 15-20mm long. Areoles sitting on the lower half of the hump, c. 8mm long, 5mm wide, felty. Radial spines in 3 pairs pointing downwards, radiating and curved back to the body, central spines 0-1, away standing, curved towards the top, all spines up to 30mm long, awl shaped, thicker at the base, black to brown, later grey. Flowers 30mm long, 25mm diameter; Ovary short. Tube green, with pink scales. Petals round, dark red, throat and stamens carmine-pink. Style and lobes (8-9) orange-yellow. Fruit wide, round, greenish-brown with lighter scales. Seeds scarcely 1mm long hat shaped, with fine humps, testa brown with a large growth-like whitish hilum rim.

Habitat: Argentina, Sierra Ambato, 2600-3500m.

Type: Rausch 227 in the Herbarium of the Natural History Museum, Vienna.

Fig.390 *G. oenanthemum* (carminanthum), *Janeba* 118 from El Rodeo, Prov. Catamarca, Argentina

Fig.391 *G. oenanthemum* (type coll. of *G. carminanthum*), *Borth* 130, Sierra Ambato, Prov. Catamarca, Argentina

Map 53 Distribution of *G. oenanthemum* in Prov. Catamarca, Argentina

Distribution (Map 53)

The Sierra Ambato is given as the location for this taxon. It is a long mountain range, running north-south to the west of Catamarca City. The lower slopes are covered in dense forest which is difficult to penetrate and the higher parts are meadow land with rocky places. *G. tillianum* is said to come from these higher regions, on the west side, among the rocks. *G. carminanthum* somewhat lower on the east side and *G. ambatoense* also on the east side towards the south among the trees.

Conservation Status

Least Concern. The species has a large distribution area and occurs in places where the only threat is grazing. The dense vegetation amongst which it often grows offers some protection and allows the plant to continue to thrive and reproduce successfully. It is often locally abundant.

History

Backeberg's first description appeared in BfK in 1934(9), and also Kaktusar in August 1934. The first locality was stated as Mendoza but is surely wrong and Backeberg later suggests it should be Córdoba. It is not stated where Backeberg obtained his plants but it was probably from an importation into Germany without origin data.

Ritter identified his collection FR 437 with *G. oenanthemum* and his finding place was Catamarca, corresponding to where the later taxa *G. tillianum* and *G. carminanthum* were described from. In fact, Backeberg's original picture of *G. oenanthemum* is a pretty good representation of the plants we know from this region, so Ritter's attribution appears to be reasonable.

Rausch found a plant high in the Sierra Ambato [Slaba (1999) says on the west side] to which he gave the collection number *Rausch* 227. He thought it was *G. oenanthemum* but later (1970) described it as *G. tillianum* in KuaS 21(4): p.66. He claimed that his plant has a different seed structure together with other minor spine and flower differences.

Borth & Koop (1976) described *G. carminanthum* in KuaS 27(4): pp.73-76 with many pictures which show the variability of the species. They made no reference to *G. oenanthemum*.

Found in 1976, *G. ambatoense* was described by its discoverer, Piltz (1980) in KuaS 31(1): pp.10-13. It appears to be intermediate between *G. oenanthemum* and *G. hossei* because it has white flowers and some plants have thinner, longer spination. It grows in the south of the distribution area of *G. oenanthemum*.

The explorations of Slaba (1999) make interesting reading. He described a form of *G. carminanthum* he found near Los Angeles as var. *montanum*, based on his collection *Slaba* 35.

Commentary

This species is most often seen labelled *G. tillianum* or *G. carminanthum*, probably because the seed has been offered under these names and the association with the original name *G. oenanthemum* has not been widely accepted. The other red-flowered species, *G. baldianum*, is smaller growing and can be found in the same general area, although in a different type of habitat. It is interesting that this region also has red-flowered forms of *Parodia microsperma*, *Lobivia aurea* and *Parodia* (*Notocactus*) *mammulosa*.

Cultivation

Surely one of the best species to grow because of its beautiful red flowers, a nice change from the white ones which plants of this type usually have. It is perfectly easy to grow when treated like the majority of Gymnocalyciums. Although it will eventually grow to 15cm in diameter, it starts to flower while still in a 9cm pot.

Gymnocalycium paediophilum Ritter ex Schütz

Fig.392 *G. paediophilum, Ritter* 1177 from the Cerro León, Paraguay

Fig.393 *G. paediophilum, Schädlich* 249 in habitat with *G. stenopleurum* (right) at the Cerro León, Paraguay

First Description (Schütz)

Gymnocalycium paediophylum Schütz sp.n. in *Kaktusy* 13(5): pp.100-101 (1977)

Body wide globular, offsetting at the base, 80mm or more in diameter, green. Ribs 7, straight, rounded with weak humps running into each other.

Areoles rounded, c. 5mm diameter, at first with thick white wool, later without wool, c. 20mm apart. Radial spines 8-12, needle-like, up to 20mm long, straight, radiating, at first reddish-brown, later bluish-grey, central spines 3, up to 30mm long, usually above these are 2 shorter spines, all awl shaped, standing straight out, similar to the radials.

Flowers from the crown, funnel shaped, with strikingly few petals, 60mm long, when open 60mm in diameter, pericarp and receptacle bluish-green with white edged scales. Petals spatulate, curved back, white, outer petals rounded at the tips, inner petals pointed, throat reddish; filaments reddish, anthers, style and stamens white. Fruit almost a round berry, c. 2cm long, blue-green with pink edged scales. Seed helmet shaped, up to 1mm long, 0.8mm diameter. Testa dark reddish-brown, shiny, finely humped; hilum basal, narrow; no aril.

Type: *Ritter* 1179, 1963, Paraguay, Cerro León (PZ) [The field number should have been FR1177]

Etymology

From the Greek 'paidion' meaning small child and 'philos' for friend, referring to the freely offsetting nature of the plant.

Distribution (Map 54 on page 180)

It is known only from the type locality, the Cerro León, a range of hills covering forty square kilometers. The area is covered in dense woodland so making determination of the limits of distribution area difficult (Fig.599).

Photo: Volker Schädlich

Conservation Status

Data deficient. There is no information available to assess the extent of the populations but the range appears to be limited to the vicinity of the type locality.

History

This species was discovered by Ritter in September 1963 at the Cerro León in Paraguay. He found it at the same time as *G. stenopleurum* which was also described as a new species. The seeds that Ritter collected were sold by his sister Hildegard Winter under the number FR1177, from which plants found their way into collections.

The first appearance of the name *G. paediophylum* is said to be in a 1965(?) Winter catalogue but I have not seen the entry. The first description, using the same spelling, was by Schütz in *Kaktusy* 1977. He later (1986) corrected the spelling to *G. paediophilum*. Ritter described the plant again in his *Kakteen in Südamerika* 1: p.269 (1979).

The region of Paraguay where this plant grows is the homeland of a tribe called the Moros who were said, at the time, to be dangerous. Perhaps this is why the plant was not rediscovered until 1996 when Amerhauser and Boss found it again.

Commentary

This species was placed in the synonymy of *G. chiquitanum* in the *New Cactus Lexicon* (2006) but, while related, it has many distinguishing features which justify its separate species status.

Cultivation

It was rarely seen in collections until it was offered in the 1986 ISI distribution under the number ISI1593. Most plants in cultivation came from this source, although habitat collected seeds have been offered recently.

The plant offsets freely, so propagation by cuttings is an easy way to get new plants. Old clusters of stems can get untidy so breaking them up occasionally helps to keep them tidy. Like other species from the Chaco, it needs more warmth in winter to thrive.

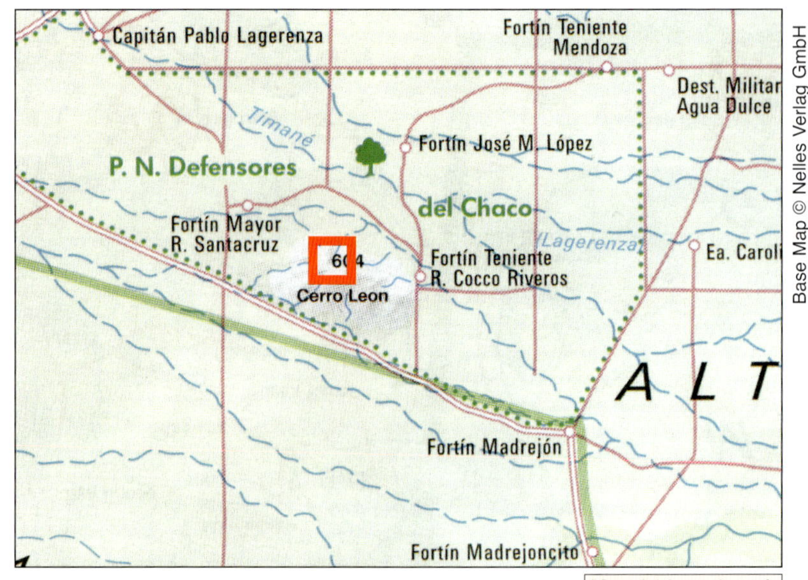

Map 54 Location of *G. paediophilum* in Paraguay

Illustrations

Schütz, B. in *Kaktusy* 13(5): p.100 (1977) Colour picture with the type description

Fig.394 *G. paediophilum, Ritter* 1177 from the Cerro León, Paraguay

Fig.395 *G. paediophilum, Bercht* 3026 in habitat at Cerro León, Paraguay

Photo: Ludwig Bercht

Gymnocalycium pflanzii (Vaupel) Werdermann **subsp. *pflanzii***

First Description (Vaupel)

Zeitschrift für Sukkulentenkunde 1(8): p.83 (1923)

Echinocactus pflanzii Vpl. spec. nov.

Body flattened-globular, up to 30cm in diameter. Crown bare, spines on the youngest areoles. Testa matt green not shiny. Young plants of 9cm diameter have 8 ribs, strong cross furrows forming large wart-like humps, with no chin-like protrusions below the areoles. Areoles elliptical, in youth with short whitish felt that later disappears.

Spines 8-9, in youth soft, red-brown, in age hard, grey, up to 1.5cm long; radial spines spreading, slightly curved, c. 3 on each side, one below the other, with no essential difference, one at the top and one at the bottom; central spines vertical.

Flowers from the younger areoles, 4cm long when closed, 4cm diameter when open; buds longish; ovary and tube with wide oval scales, brownish, without spines or hairs; outer petals blunt, inner petals pointed, curved backwards immediately on opening, salmon colour, lighter at the edges, tinted violet at the base, the many anthers forming a funnel, the topmost are not curved inwards; filaments violet; anthers yellow, style violet, 1cm long, the 11 spreading stigma lobes are in a mass of anthers. Seeds small.

Fig.396 *G.pflanzii pflanzii,* STO 997 from Carapari - Palos Blancos, Dept. Tarija, Bolivia

Habitat: Bolivia, close to the banks of the Río Pilcomayo at Palo Marcado, 50km down river from Villa Montes. (Karl Pflanz)

Neotype: Bolivia, Dept. Tarija, Palo Marcado; leg. H. Amerhauser HA941-4 (LPB). Designated by H. Till, W. Till & H. Amerhauser, *Gymnocalycium* 13(1): p.344 (2000)

Etymology

Named for Dr. Karl Pflanz who sent cacti to Europe when he was the German consul at Villa Montes, Bolivia

Fig.397 Habitat of *G. pflanzii pflanzii,*.GC853.02 near Lagunillas, Prov. Santa Cruz, Bolivia

Fig.398 *G. pflanzii pflanzii* GC858.05 in habitat south of Camiri, Prov. Chuquisaca, Bolivia

Synonyms

G. pflanzii subsp. *dorisiae Amerhauser in Gymnocalycium* 11(4): p267

Commentary on *G. pflanzii* subsp. *dorisiae*

An isolated population on the steep banks of the Río Paichu far to the west of the main distribution area but differing little from subsp. *pflanzii*.

G. chuquisacanum Cárdenas in CSJ(US) 38(4): pp.146-147 (1966)

First description of *Gymnocalycium chuquisacanum* Cárd. sp. nov.

Globose, somewhat flattened, 5-6cm high, 12cm broad, greyish-green. Ribs 13, broken into rounded, 1cm high, 17-28mm broad tubercles. Areoles 1.5cm apart, circular to elliptic, 7mm diameter, grey to blackish felted. Radial spines 7-10, spreading to laterally curved, 2-3cm long. Central spines one, 2.5-3cm long, upwards curved, all spines horn grey coloured, brownish at tips. Flowers few from the top of stem, urceolate, 6.5cm long, 5cm limb laterally flattened. Ovary globose, 12mm diameter, light green with 5mm broad green, whitish margined scales. Tube widening above, pinkish brown. Outer perianth segments lanceolate, 22x7mm, greenish outside, pink inside. Inner segments lanceolate, acute, pinkish salmon, 20x5mm. Stamens from bottom of the tube to the base of petals, 3mm long, filaments magenta, anthers brownish. Style 11mm long, bright magenta. Stigma lobes 22, yellow, 5mm long. Interior of tube bright magenta and glossy.

Bolivia: Province of Azero, Department of Chuquisaca, near Boyuibe, at 700m. M. Cárdenas, May 1958. No. 6226. (type in Herbarium Cardenasianum).

G. izozogsii Cárdenas in CSJ(US) 38(4): pp.145-146 (1966)

First description of *Gymnocalycium izozogsii* Cárd. sp. nov.

Simple, globose, 4-5cm high, 9-12cm broad, grey green turning reddish. Ribs 9-10, broken in round prominent or horizontally compressed tubercles, 12-15mm high, 20-25mm broad. Areoles 1.5-2.5cm apart, circular to elliptic, 6-8mm diameter, grey felted. Radial spines 10-11, spreading or slightly appressed, 10-25mm long, horn grey. Central spine 15mm long. Flowers funnelform to urceolate 4cm long, 3cm limb. Ovary globose 7mm long, brownish shining, with 3mm broad pink scales. Tube short widening above, brownish shining, with brown, pink margined 6mm long scales. Outer perianth segments spatulate 8-16mm long pink-salmon inside, light brown outside. Inner segments spatulate, 20x7mm, light salmon with a magenta flush at base. Stamens from the bottom of tube to the base of the petals, 3-5mm long; filaments thin, magenta, anthers yellow brownish. Style 8mm long, pink, 2-3mm thick. Stigma lobes 18, light magenta to pink outside, yellow inside, 3mm long. Interior of the tube red.

Fig.399 *G. pflanzii pflanzii* GC853.02 in habitat near Lagunillas, Prov. Santa Cruz, Bolivia

Bolivia, Province of Cordillera, Department of Santa Cruz, Izozog Basin, near El Atajado. Growing on the borders of thorny forest on loose sandy soil at 400m. April 1965. M. Cárdenas No. 2625. (Type in Herbarium Cardenasianum).

G. lagunillasense Cárdenas in KuaS 9(2): pp.22-24 (1958)

First description of *Gymnocalycium lagunillasense* Cárdenas, nov. spec.

Simple, globular, crown sunken, 3-4cm high, 9-14cm wide, grey-green to green, sometimes yellowish to purple tinted. Ribs c. 13 on a normal size plant, low, c. 3cm wide at the base, separated into four cornered humps, by horizontal cross-cuts. Areoles c. 2cm apart, wide elliptical, 7mm diameter, grey felted. Radial spines c. 7, radiating away from the body, central spines 1, no more curved than the radials, all spines whitish at the base, pink in the middle, brown at the tip, 1 to 3cm long. Crown areoles unarmed. Flowers c. 5, forming a circle around the crown, funnel-shaped, 5cm long. Ovary 1cm long with round, greenish to yellowish scales, tube very short, outer petals rounded, middle petals egg-shaped to lanceolate, 12-18mm long, green to whitish, inner petals c. 17mm long, lanceolate, pointed cream coloured, white towards the bottom, magenta red at the base. Stamens from the bottom of the tube up to the base of the petals, filaments magenta-red, anthers deep yellow. Style 12mm long, magenta-red, not reaching over the anthers, with c. 14 pointed lobes, magenta-red. Fruit globular, 2cm diameter, red to light magenta-red, bare with a few 3-5mm white edged, wide scales, fruit flesh, dark red, possessing a carmine-red, easily soluble pigment, this colour on cactus fruit is unusual. Seeds very small, 0.5mm, very shiny brown, not spotted, this characteristic is also unusual for this family.

Habitat: Bolivia, Province Cordillera, Department Santa Cruz. Lagunillas, 1000m February. 1949. M. Cárdenas, No. 5027 (type plant) in Herbarium Cardenasianum, Cochabamba. Cotype in US. National Herbarium.

Fig.400 *G. pflanzii pflanzii* GC858.05 in habitat south of Camiri, Prov. Santa Cruz, Bolivia

Photo: Volker Schädlich

Fig.401 *G. pflanzii pflanzii, Schädlich* 246 in habitat south of Nueva Asunción, Boquerón, Paraguay

Commentary on *G. lagunillasense*

This plant was first collected by P. L. Porte in July 1920 and sent to Dr. Rose who published a description without naming it on page 157 of Volume III of *The Cactaceae* as part of the commentary about *G. saglionis*.

G. marquezii Cárdenas in in KuaS 9(2): pp.26-27 (1958)

First description of *Gymnocalycium marquezii* Cárdenas nov. spec.

At first globular, slightly sunken in the crown, 3-4cm high, 8-10cm wide, bluish. Ribs about 8, low, forming 2x3cm humps. Areoles 3cm apart, elliptical, 6-8mm in diameter, young areoles protruding, white felted, older areoles brown felted. Spines 6-7, curved, standing out, thin, awl-shaped, grey at the bottom, pinkish-brown at the top, 1-2.5cm long, a few areoles have a central spine, all spines are thicker at the bottom.

Flowers, few, developing in the area of the centre of the crown, wide beaker to jug shaped, 4cm long. Ovary globular, 8mm diameter, scales 5mm wide, brownish-salmon colour. Flower tube very short. Outer petals 1cm long, rounded, brown-salmon colour. Middle petals elliptical, 17mm long, pink with a brown central stripe. Inner petals white, lanceolate, 15mm long, pink, magenta-red at the base. Stamens 4mm long, magenta-red, anthers yellow. Style 8mm long, 2mm thick, white to magenta-red, stigma, 10-12 lobes, feather-like, 4mm long, pale yellowish pink.

Habitat: Bolivia. Province Entre Rios, Dept. Tarija, near Angusto de Ville Montes. March 1952. M. Cárdenas. No. 5066. (Type plant) in Herbarium Cardenasianum, Cochabamba.

Commentary on *G. marquezii*

Said by Cárdenas to be smaller than *G. pflanzii*, although it grows quite near the type locality of that species.

Distribution (Map 55 on page 184)

This subspecies is found in the south east of Bolivia and just into Paraguay. The northern border of its distribution is between the Río Grande and Río Pilcomayo. North of this line, one can find only subsp. *zegarrae*. An isolated population west of the city of Tarija was described as *G. pflanzii* subsp. *dorisiae* Amerhauser (1998).

Conservation Status

Least Concern. The species is extremely widespread and frequently common where it occurs. It is a very successful plant, being both widespread and abundant.

Fig.402 *G.pflanzii pflanzii* (marquezii), *Ramirez* 016 from Palos Blancos, Dept. Tarija, Argentina

History

Karl Pflanz, the German consul in Villa Montes, was the first collector to send this plant to Europe. He found them on the sandy banks of the Río Pilcomayo near Villa Montes, Bolivia. Two specimens went to the Dahlem Botanical Garden and flowered there.

In 1935, Werdermann published a picture of a flowering plant as Plate 94 in his *Blühende Kakteen*. It was said to be a grafted offset from the original plant that was photographed in June 1934. A picture of similar origin had appeared in *Kakteenkunde* (1933) p.10.

Some time later, the species was found in Paraguay by Friedrich, also near to the Río Pilcomayo on the border with Argentina. Cárdenas, who described many of the synonyms listed here, found the type form at Ibipopo and nearby, a smaller form growing on red sandstone that he called *G. marquezii* in 1958.

Many local populations of this taxon were given names by Cárdenas, so Donald (1971), realising that they did not justify specific status, combined most of them as varieties of *G. pflanzii*. He did not differentiate the northern subspecies *zegarrae* and, without explanation, omitted *G. marquezii* from his lumping.

Fig.403 *G.pflanzii pflanzii* (marquezii), *Ramirez* 016 from Palos Blancos, Dept. Tarija, Argentina

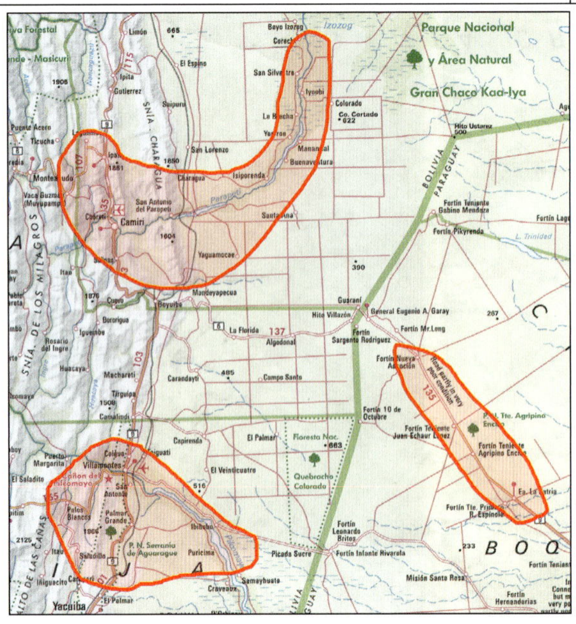

Map 55 Distribution of *G. pflanzii pflanzii* in Bolivia and Paraguay

Commentary

Gymnocalycium pflanzii in all its forms exhibits a unique seed structure which prompted Schütz to place it alone in his subgenus *Pirisemineum*.

The typical subspecies is differentiated from the more northerly occurring *G. pflanzii* subsp. *zegarrae* by its fruits which, unlike most *Gymnocalycium* species, split horizontally rather than vertically (Fig.6). The fruit pulp is also different, usually white in subsp. *zegarrae* and red in this subspecies. Some authors treat *G. zegarrae* as a separate species on account of these differences.

Although usually regarded as a form of *G. pflanzii,* I have omitted *G. eytianum* Cárdenas (1958) from the list of synonyms because the description appears to be mixed with *G. marsoneri,* a species that also grows in the same area near to Eyti. The illustration at the bottom of page 25 of the type description and the flower section drawing look more like *G. marsoneri*.

Cultivation

This is an easy plant to cultivate and flower. It enjoys ample watering when it then grows quickly and occasionally produces offsets. The seed is very small, so the seedlings grow slowly at first and must be kept moist at this time. Some clones produce pretty pink or peach-coloured flowers, often those identified as *G. marquezii*.

It looks a lot like *G. saglionis* to which it is clearly related. They are often confused but can be distinguished with practice.

Illustrations

Werdermann *Blühende Kakteen* Plate 94 (1935)

Werdermann, *Kakteenkunde* (1933) 1(1): p.10

Gymnocalycium pflanzii subsp. *argentinense* H.Till & W.Till

First Description (H. Till & W. Till)

KuaS 39(12): pp.273-277 (1988): Ein neues Gymnocalycium (subgen. *Pirisemineum*) aus Nordargentinien:

Gymnocalycium pflanzii subsp. *argentinense* Till & Till

Simple, body almost globular, in youth somewhat conical towards the crown, the largest plants observed are 7cm tall, 15cm in diameter, in habitat as large as a man's head, epidermis grey-green (somewhat grainy), rough, matt. Ribs, (7-) 8, straight, occasionally somewhat slanting, 20-22mm wide at the base, lost in conical, 12-15mm high humps, formed by deep indentations. Areoles sit on the point of the humps, c. 25mm apart, not sunken, large oval, c. 12mm long, 8mm wide, with thick dirty white, constant wool-felt. Radial spines, mostly 8, radiating away from the body, top spines overhanging the crown, c. 3 pairs radiating sideways and upwards,1 spine pointing downwards, mostly the 2 top pairs are the longest, to 23mm long, central spines (1-)2-3(-4), different lengths, the longest up to 20mm long, mostly shorter than the upper radial spines, all spines stiff, sharp, slightly curved, in youth blackish-brown, in age brownish-grey, granulate and shiny, tips brown, later completely grey.

Flowers from the thick yellowish woolly crown; buds pointed globular, green to brownish-green, with wide, half round, light-edged scales. Flowers urn-shaped, 35-48mm long, 35-40mm wide; ovary 9-12mm wide, 11-12mm high, dark green with lighter almost white, half round scales, that extend to the flower tube, becoming brownish-green immediately

Fig.404 *G.pflanzii argentinense,* GC448.06 from east of Cabra Corral Lake, Prov. Salta, Argentina

above the ovary. Sepals spatulate, whitish-green, iridescent. Outer petals spatulate, at the tip, wide-round with a milled edge, 10-14mm long, 9mm wide, slightly curved back, laterals usually arched, occasionally lightly twisted, blunt at the top end, ± strongly toothed, greenish-white to pale pinkish-white with a wide greenish mid-stripe, inner petals, 15-17 x 5-7mm, blunt lanceolate, often with a short point, white, tinted pale pink. Throat, and the thin short filaments, wine red, the many filaments fill the throat, with the two lower rows leaning against the style, anthers pink, 1.5 x 0.75mm, pollen dark yellow to orange. Stigma in the middle of the anthers, style with stigma, 13-14mm long, 1.5-2mm diameter, stigma with 14-18 lobes, outside light red with pale pink edges, inside pale pink. Ovary in

Fig.405 The habitat of *G.pflanzii argentinense,* GC448.06, east of Cabra Corral Lake, Prov. Salta, Argentina

Fig.406 *G.pflanzii argentinense,* GC448.06 in habitat east of Cabra Corral Lake, Prov. Salta, Argentina

section, U-shaped, with white, (when ripe reddish) developing funiculi.

Fruit round-oval, red, opening crosswise (the same as *G. pflanzii*). Funiculus when fully ripe the pulp is carmine-red. Seeds egg-shaped, c. 0.6 x 0.5mm with a smaller (0.4 x 0.2mm) sloping hilum-micropyle region, this with a pale brown to whitish strophiole-like tissue, testa shiny brown, the border of the hilum-micropyle region, dark brown.

Type: ex cult. HT684, *Anonymous*, Argentina, Province Salta, near Gonzalez, Río Juremento ex coll. K-H Uhlig (WU)

Etymology

Occurs in Argentina.

Distribution (Map 56)

This is the subspecies that grows in Argentina. There is a 300km disjunction from subsp. *pflanzii* where no plants have been found. However, it is possible that there are populations which have not been found in this region with few roads. The plant favours the easterly edge of the mountains which receives significant rainfall resulting in a dense growth of trees, the Tucumán forest.

H. Till, W. Till & Amerhauser (2000a) quote the range of this taxon from Santa Clara south to Lumbreras and Joaquin v. Gonzalez. Neuhuber has found a locality near Tala Yacao in Prov. Santiago del Estero.

The type locality is in the far south east of Prov. Salta near the last mountains before the land drops into the flat chaco.

It looks quite similar to *G. saglionis* but not so much as subsp. *pflanzii* which can grow large like *G. saglionis*. Its distribution is north of that of *G. saglionis* but overlaps at some places south of the city of Salta, Argentina. I have seen it near to the lake of Cobra Corral, a resort area south of Salta city. It makes big clusters of stems under a dense covering of shrubs in an area which is so favourable for cacti that they can even be found growing on tree

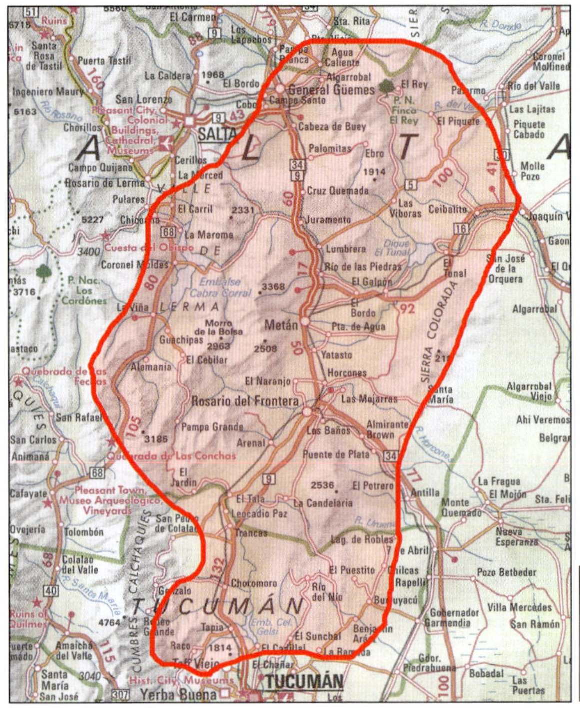

Map 56 Distribution of *G. pflanzii argentinense* in Prov. Salta and Tucumán, Argentina

branches. *G. saglionis* and *G. schickendantzii* subsp. *delaetii* can also be found growing with it there.

Conservation Status

Least Concern. Although its range is limited, the plant is locally abundant and there is no indication of a significant threat to its habitat.

History

Backeberg invalidly described *G. marquezii* var. *argentinense* in his *Kakteenlexicon* (1966) but it was not until 1988 that the plant got its valid name.

Commentary

This is a poorly differentiated subspecies which could probably be included within subsp. *pflanzii*. Its main claim to separate status is its isolated habitat in Argentina, some 300km south of the main distribution area. It differs by its darker green body and much larger areoles.

This taxon sometimes grows with *G. saglionis*, the species which it is most like in general appearance. There is a good B&W illustration on page 276 accompanying the first description by Till & Till (1988) which shows these two plants and *G. pflanzii* subsp. *pflanzii*.

Cultivation

It is just as easy to grow as the other forms of *G. pflanzii* although its habitat is relatively low altitude and consequentially warm. The fruits are blue before they ripen to red and split horizontally to reveal the a large quantity of small seeds.

Illustrations

New Cactus Lexicon Atlas (2006) Illn. 279.1

Gymnocalycium pflanzii subsp. *zegarrae* (Cárdenas) Charles

First Description (Cárdenas)

KuaS 9(2): p.21 (1958)

Die Gattung *Gymnocalycium* in Bolivien - *Gymnocalycium zegarrae* Cárdenas nov. spec.

Simple, globose, crown slightly sunken or flattened, grey-green, 6-10cm high, 11-18cm wide, very seldom larger. Ribs c. 13 on a normal size plant, separated by deep or sharp straight or wavy vertical grooves, which in turn are separated into irregular convex, four, five or six sided 2x3cm humps. Areoles elliptical 1.5-2.5cm apart, 1cm long, grey felted. Radial spines c. 8, somewhat pectinate, 1.5-2.5cm long. Central spines 2.5-3cm long. All spines awl-shaped, strong, curved, whitish with a reddish to blackish point, the crown areoles unarmed. Flowers c. 6-8, in a circle in the crown, zygomorphic, wide bell to beaker-shaped, 3-4.5cm long. Ovary globular, dark green, c. 1cm diameter, with 2-3mm long, wide rounded grey-green scales, edged white. Outer petals wide spatulate, 5-15mm long, green to grey-green, white tinted, inner petals lanceolate, 1.5-2cm long, 4-5mm wide, pure white, magenta-red at the base. Many stamens, from the base of the tube up to the petals, c. 4mm long, undulating, cotton-like, magenta-red; anthers yellowish. Style 10-13mm long, 1.5mm thick, again magenta-red, stigma lobes pointed, 11-14, magenta-red to yellowish, 5mm long.

Fruit round, orange-coloured when ripe, c. 1.5cm long, 1cm wide, with 2mm long membrane-like, wide grey-green, white edged scales, the dried flower remains still on the fruit. Seed very small, 0.5mm diameter, round, shiny brown.

Fig.407 *G.pflanzii zegarrae,* GC919a in habitat south of Mataral, Prov. Santa Cruz, Bolivia

Habitat: Bolivia, Prov. Campero, Dep. Cochabamba, on the road from Perez to Mairana, 1700m. Oct. 1974. M. Cárdenas. 5019. (type plant) in Herbarium Cardenasianum Cochabamba (LIL). Isotype in US. National Herbarium.

Etymology

Named for Ing. German Zegarra Caero, promoter of cactus exploration in Bolivia.

Other Names

Gymnocalycium zegarrae Cárdenas in KuaS (1958)

Fig.408 *G.pflanzii zegarrae,* GC919a in habitat south of Mataral, Prov. Santa Cruz, Bolivia

Fig.409 *G. pflanzii zegarrae* (riograndense), GC913.05 in habitat north of Nuevo Mundo, Prov. Santa Cruz, Bolivia

Synonyms

G. millaresii Cárdenas in CSJ(US) 38(4): pp.144-145 (1966)

G. zegarrae subsp. *millaresii* (Cárdenas) Till & Till in *Gymnocalycium* 13(2): p. 355 (2000)

First description of *Gymnocalycium millaresii* Cárd. sp. nov.

Simple, globose slightly flattened above, grey-green, 6-8cm high, 12-14cm diameter. Ribs 16-19, straight, broken into 6-8mm high, 2-2.5cm diameter, tubercles. Areoles 3cm apart, elliptic, 1cm long, blackish felted. Top areoles unarmed. Spines pectinate; marginals 7-9 curved, laterally somewhat appressed; centrals, 1-2 upwards directed. All spines horn grey, brownish at tips 2-4cm long. Upper spines dark brown, reddish at base.

Flowers from the upper sides of stem, funnelform, zygomorphic, 4cm long, 4cm limb. Tube 2cm wide with broad spatulate light green, lilac margined scales. Outer perianth segments broadly spatulate, 1cm long, pink greenish outside, inner segments lanceolate, 17x7mm, pinkish-salmon, magenta at base. Interior of the tube dark magenta. Stamens from the base of the tube to the base of petals, 5mm long; filaments magenta, anthers yellow. Style 17mm long, magenta. Stigma lobes 13, yellow magenta, 7mm long. Ovary globose, 7mm long, light green lilac with broad round lighter colour margined 2x5mm scales.

Bolivia: Province of Saavedra. Department of Potosi, near Millares, at 2600m. Growing among quartziferous rocks. M.Cárdenas. April 1962, No. 2624 (Type in Herbarium Cardenasianum).

Commentary on *G. millaresii*

This form of *G. pflanzii* subsp. *zegarrae* was described by Cárdenas from a habitat to the south-west of the other populations. He said it is the highest occurring species of *Gymnocalycium* in Bolivia, although the recently described *G. pflanzii* subsp. *dorisiae* was found at 2,300m. The main difference is the small globular red fruit of *G. millaresii* which, like subsp. *zegarrae,* has white pulp and splits vertically.

Till, Till & Amerhauser (2000b) mention two other small populations near Villa Tomina and near Sucre, hence the catalogue names *G. tominensis* n.n. and *G. sucrensis* n.n.

Fig.410 *G.pflanzii zegarrae,* GC919.03 east of Mataral, Prov. Santa Cruz, Bolivia

Fig.411 *G.pflanzii zegarrae,* GC919.03 east of Mataral, Prov. Santa Cruz, Bolivia

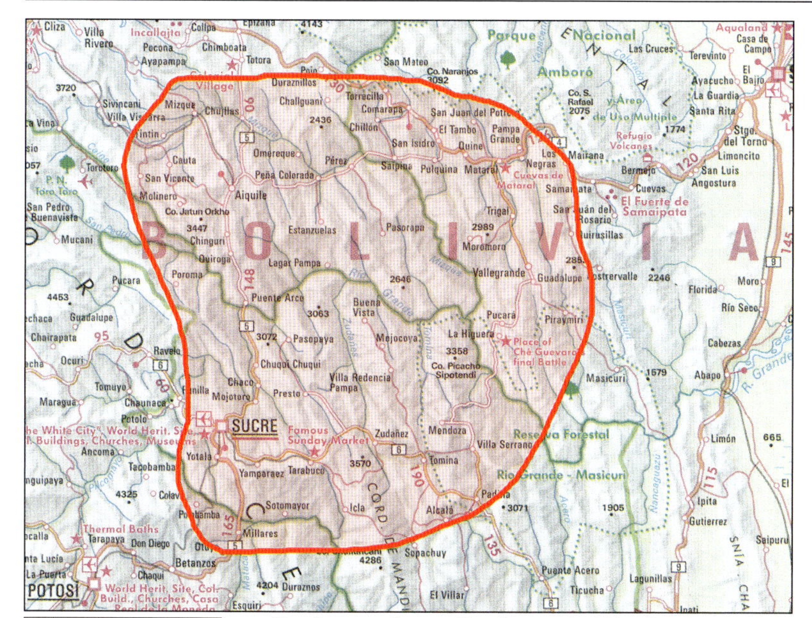

Map 57
Distribution of *G. pflanzii zegarrae* in Bolivia

G. riograndense Cárdenas in KuaS 9(2): pp.24-25 (1958)

First description of *Gymnocalycium riograndense* Cárdenas, nov. spec.

Simple, globular to flattened, 3-6cm high, 7-20cm diameter, shiny dark green. Ribs c.13 on medium size plants, 1cm high, 3cm wide, running into blunt conical humps, 4x2.5cm. Areoles 2.5-3cm apart, upright, elliptical, 8mm long, black felted, young areoles in the crown of the plant with white felt and very short spines, radial spines 8-9, mostly 8, usually comb-like, slightly curved, the shortest 1cm long, inner spines longer, 1.5-2.5cm long, 1 central spine, straight or curved slightly upwards, 1-2cm long, all spines awl-shaped, fairly thin, stiff, grey, young spines in the crown straight, short, brownish at the bottom with a black tip.

Buds develop around the sunken crown, globular, at first magenta-red, later greenish, open flowers zygomorphic, jug-shaped, 4-5cm long. Ovary round, 1cm diameter, with 4-5mm long, green scales, edges white and magenta-red, flower tube wasted above the ovary, becoming wider above. Outer petals long spatulate, 10x7mm green with reddish tips. Middle petals longish, 12x5mm, inside whitish, outside greenish. Inner petals 15x4mm, white, magenta-red at the base. Stamens from the base of the tube, inside of tube, purple-magenta-red, filaments feather like, 5mm long, magenta-red, anthers yellow. Style 8-10mm long, magenta-red, stigma lobes 11-14, radiating, magenta-red, 4mm long.

Habitat: Bolivia, Province Valle Grande, Department Santa Cruz. On the banks of the Río Grande, on the road from Higuera (Santa Cruz) to Ocampo (Chuquisaca), 1400m. March 1955. M. Cárdenas, No. 5064. (Type plant) in Herbarium Cardenasianum, Cochabamba.

Commentary on *G. riograndense*

Found near the banks of the Río Grande, this plant grows in a hot cactus paradise, under the cover of trees. It is characterised by its shiny body although this appears to be an inconsistent feature. It can sometimes be seen under the catalogue name *G. vallegrandensis* n.n.

Distribution (Map 57)

The distribution area of this subspecies lies to the north west of the subsp. *pflanzii* making it the most northerly occurring of all *Gymnocalycium* taxa. It is bordered on the east by the tropical lowland where the dense growth of trees is a natural barrier to further colonisation.

Fig.412 *G. pflanzii zegarrae, Preston-Mafham* 172 from Millares, Dept. Potosi, Bolivia

Conservation Status

Least Concern. This is widespread taxon that is very successful over its range.

History

The plant was first mentioned by Cárdenas when he invalidly published it as *G. saglionis* var. *bolivianum*. The catalogue name *G. bolivianum* appeared in the Winter catalogues from 1956 to 1959, after which Ritter changed the name to *G. pflanzii* var. *albipulpa,* later to be invalidly published in Vol.II of his *Kakteen in Südamerika* (1980).

The valid description as *G. zegarrae* was published by Cárdenas in KuaS (1958). Donald (1971) made it a variety of *G. pflanzii* then I raised it to a subspecies in CSI 20: p.18 (2005), although other authors still regard this as a separate species.

Commentary

This subspecies is differentiated from subsp. *pflanzii* by its white pulp and vertical splitting of its ripe fruits.

Cultivation

An easy plant to grow, this plant rapidly makes a large solitary head and can spectacularly flower in a ring around the crown. It tends to stay flatter than the other subspecies and can make a very large spectacular plant.

Fig.413 *G. pflanzii zegarrae, Ramirez* s.n.

Gymnocalycium pugionacanthum Backeberg ex H.Till

First Description (Backeberg)

Gymnocalycium pugionacanthum Backeberg nom. inval. *Kakteenlexicon* pp.172-173 (1966). Validated by H.Till in KuaS 38(8): p.191 (1987)

Body simple, hemispherical, c.10cm diameter, has been noted, deep bluish-green. Ribs c. 10, with thin transverse furrows, up to 2.5cm wide, 8mm high. Radial spines very stout, somewhat appressed, 4 pairs, pointing directly to each side, one spine pointing downwards, all 10-20mm long, somewhat compressed, at first black later ash-grey or still blackish at the apex. Central spines 0. Areoles c. 11mm long, 16mm apart, dirty white. Flowers 4cm long, 4.5cm diameter, tube only 1.5cm long. Sepals olive-green, with a white border and a red tip. Petals creamy-white tinted light grey, inner petals brownish-pink at the base, pink on the outside; filaments cream; anthers, with or without pollen, pink. In Europe, flowering time is end of April.

Habitat: Northern Argentina (Córdoba?) (Collection of Uhlig and Backeberg U2148, collected by Fechser. No further habitat data is available).

Etymology

From the Latin 'pugio' meaning dagger and 'acantha' for spine.

Synonym

G. catamarcense H. Till & W. Till in *Gymnocalycium* 8(1): pp.142-143 (1995)

First description of *Gymnocalycium catamarcense* Till & Till, spec. nov.

The body form is variable, depending on habitat and soil, discoid to semi-globular, older plants are globular to elongated-globular, occasionally offsetting. Plants in habitat are from 35-60mm high, 75-90mm diameter, very old plants, 120-140mm diameter, and up to 420mm high, with ramified, mostly strong roots. Epidermis matt, not shiny, mostly grey-green, seldom brown-green. Crown somewhat sunken, not covered with spines during the growing period. Ribs (9-)13-26, straight, moderately raised; in the crown mostly sharp edged, flat at the base, 15-23(-27)mm wide, separated in to ± distinct humps, somewhat compressed together with pointed chins. Areoles oval to longish oval, 6-10mm long, 4mm wide, 16-19mm apart, with light yellow to yellowish-brown wool-felt, which soon disappears. Spines (5-)7(-9), flattened, mostly sharp edged, seldom rounded, lightly to strongly recurved, these sometimes lying flat to the body, seldom outwards directed, 15-30(-40)mm long, comb-like; the second or third pair the longest, the bottom spine pointing downwards, mostly shorter than the longest side pair. Young spines in the crown mostly brown, later becoming yellowish horn-coloured to light brown, with darker tips.

Fig.415 *G. pugionacanthum* GC37.02 from north of Andalgalá, Prov. Catamarca, Argentina

Flowers arising from the young areoles near the centre of the crown, funnel shape; 50-55mm long, 40-43mm diameter at full anthesis, washed white to delicate pink, with a carmine-pink throat. Ovary conical, 13-15mm long, 3-4mm diameter at the base, at the top near the nectar chamber 8-9mm diameter, olive green, with half round reddish-green, white-edged scales, which gradually resolve into outer petals. The lowest petals are dark green, with a narrow white edge, the top ones become gradually lighter, with wider white edges, the longest, up to 22mm long, 8mm wide near the tip, spatulate, on the inside white, with a green middle stripe, or, delicate lilac-pink, with a darker pink middle stripe; inner petals, 17-18mm long, 5-6mm wide, pointed spatulate, sometimes lightly dentate, washed white to delicate pink with a weak mid-stripe.

The inner stamens, thin, 4.5mm long, lying onto the style, enclosing the narrow, c. 1.5mm deep nectar chamber, the rest of the stamens are inserted in the tube wall, one above the other, all

Fig.414 *G. pugionacanthum* GC413.03 in habitat north of Belén in the Sierra de Belén, Prov. Catamarca, Argentina

Photo: Geoff Bailey

Fig.416 *G. pugionacanthum, Bailey,* s.n. in habitat near Hualfín, Prov. Catamarca, Argentina

Fig.417 *G. pugionacanthum* GC37.02 in habitat north of Andalgalá, Prov. Catamarca, Argentina

curved towards the style, which they surround and overhang, all stamens are pink. Anthers oval. 1.2mm long, pink. Pollen yellow. Style c. 1.2mm thick, c. 12mm long. Stigma 6mm long, yellow at the base becoming greenish towards the top, with 9-10 stigma lobes. Ovary pear-shaped; c. 8mm long, 4mm wide, white walled, filled with ramified funicles. Fruit pear-shaped, 18mm high, 12mm diameter, dark green, with a bluish bloom, splitting vertically. Seeds round to oval, c. 1mm diameter, red-brown to black, testa surface somewhat variable. Seed group, Microsemineum Schütz.

Occurrence: Argentina, Prov. Catamarca, west of the Sierra de Manchao, and the Nevada del Candado, to a line between Belen and Hualfin. Type Locality: Andalgalá.

Type: Argentina, Province of Catamarca, to the north of Andalgalá, 1340m, collected by H. Till 93-632/2259, 19th February 1993 (WU); Paratype: between Andalgalá and Agua de Paloma, 1050m, collected by H. Till 88-132/1817, 2nd October 1988 about 10km before Andalgalá, 950m; collected by G. Neuhuber 91-475/115 + 116, 20th December 1991 (WU)

Gymnocalycium catamarcense subsp. *schmidianum* Till & Till in *Gymnocalycium* 8(1): p.145 (1995)

Map 58 Distribution of *G. pugionacanthum* in Prov. Catamarca, Argentina

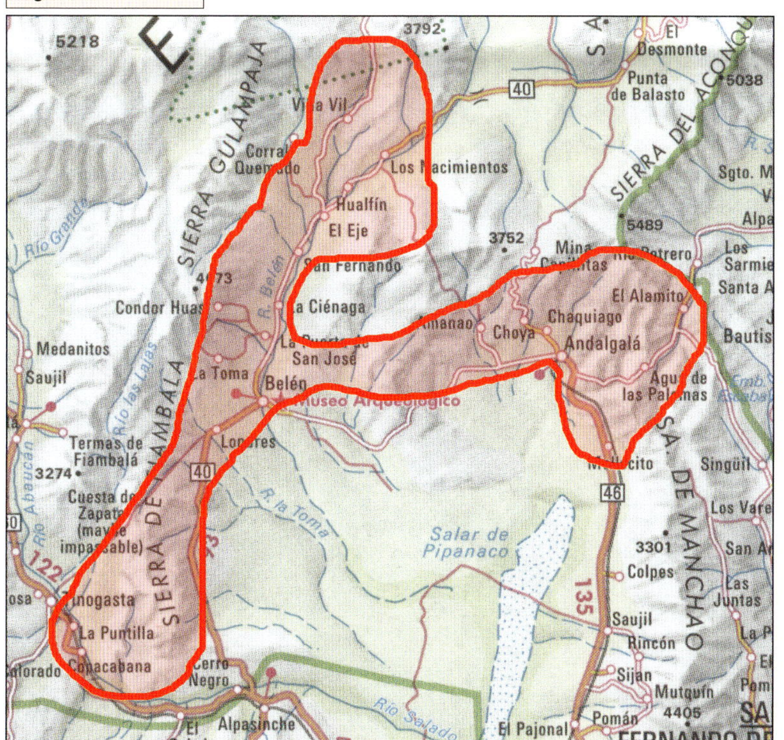

Fig.418 *G. pugionacanthum* (schmidianum) GC213.01 in habitat, Cuesta de Zapata, Prov. Catamarca, Argentina

Commentary on *G. catamarcense* subsp. *schmidianum*

Described from the south west of the distribution area, to the east of Tinogasta and in the Sierra de Zapata.

Gymnocalycium catamarcense subsp. *acinacispinum* Till & Till in *Gymnocalycium* 8(1): p.146 (1995).

Commentary on *G. catamarcense* subsp. *acinacispinum*

Described from the south east of the distribution area, at the base of the west side of the Sierra de Manchao.

Distribution (Map 58)

Plants of this species occupy a large area of Prov. Catamarca around the north and west sides of the Salar de Pipanaco. Their distribution is bounded by the high mountains in the west and the Sierra de Manchao with the Nevada del Candado in the east. In the Sierra de Zapata in the south west it is said to grow with northern examples of *G. rhodantherum*.

Conservation Status

Least Concern. This species is widespread and locally common, with no immediate threats to its continued existence.

History

The plants of the type collection were imported in 1964 by the Uhlig nursery at Rommelshausen, Germany, from Fechser in Argentina. They were seen by Till & Shatzl (1973) who noticed some plants with dark blue-green bodies and hard straight spines that were nearly comb-like, 4 pairs to each areole and one spine pointing downwards, a little flattened, oval in cross section, black at the tip. Fechser had sent the plants as *G. hybopleurum* but two years later, Backeberg described them as a new species *G. pugionacanthum* in Das Kakteenlexicon (1966). Till kept 2 plants and 4 found their way to the Linz Botanical Garden.

Fig.419 *G. pugionacanthum, Piltz* 72 from Andalgalá, Prov. Catamarca, Argentina

Having had the experience of seeing and growing plants from the type collection, Till & Shatzl (1973) make the statement that it is related to *G. spegazzinii*. They further claim that the seeds are from Schütz's section *Loricata*, the group based on *G. spegazzinii*. The only picture published by Backeberg with the first description was Abb 144 in his *Kakteenlexicon*. There is also a picture of a plant from the same collection reproduced in Till & Shatzl (1973) on page 230. Neither of the illustrations particularly resemble *G. spegazzinii*.

It was not until 1987 that Till validated Backeberg's invalidly published name *G. pugionacanthum* by designating a holotype. Till (1987) preserved a plant from the original collection in Vienna University Herbarium.

In his article in the Austrian publication *Gymnocalycium*, Till (1989) changed his mind about the relationship of this taxon. He now maintains it is a very localised form of *G. hossei* of which he claims to have found a small population in northern La Rioja. He gives no location but field number lists suggest it may be near Pituil. He publishes an informative picture (page 19) of one of his original plants from Fechser, now grown on in cultivation

Fig.420 *G. pugionacanthum* (schmidianum) *Piltz* 218 from south of Tinogasta, Prov. La Rioja, Argentina

Fig.421 *G. pugionacanthum, Piltz* 72a from east of Andalgalá, Prov. Catamarca, Argentina

and flowering. He expands on Backeberg's original description and illustrates his supposed locality and plants from there.

Commentary

This is another problematic name, the application of which is disputed. Here I have followed the treatment in the *New Cactus Lexicon*, accepting it as the oldest name for well known plants which grow in Prov. Catamarca, Argentina. This is the best approach to support stability in the naming of taxa. Since Till (1989) believed that it should apply to a form of *G. hossei* from La Rioja, that would leave these plants I am calling *G. pugionacanthum* without a name. So Till & Till (1995) erected the name *G. catamarcense* for these plants.

Having looked at all the evidence, I consider it more likely that this name was originally applied to the plants from Catamarca. They are widespread and must have been found by Fechser, more likely than him chancing upon a small population of an unusual *G. hossei* form in La Rioja. Till's picture of his cultivated original plant (1989) is very much like some plants from the variable Catamarca populations.

For a long time the name *G. hybopleurum* was used for these plants, following an error by Backeberg where he mis-applied Schumann's name. See the commentary on *G. hybopleurum* on page 260.

There have been a number of suggestions that it is related to *G. spegazzinii*, replacing it in the south of its range.

Cultivation

Plants of this variable taxon have no special requirements and make handsome solitary specimens in culture. Some forms can eventually grow into large specimens proving it to be one of the biggest species in the genus. Although slower growing than, for instance, *G. saglionis*, a plant of more than 20cm in diameter is certainly achievable in time and looks handsome indeed.

Gymnocalycium rhodantherum (Bödeker) Backeberg

Fig.422 Habitat of *G. rhodantherum*, GC960.01 west of Cuesta Miranda, Prov. La Rioja, Argentina

First Description (Bödeker)

Kakteenkunde 2(1): pp.13-14 (1934)

Echinocactus (Gymnocalycium) rhodantherus Böd., sp.n.

Body simple, flattened-globular, the typical example at hand is 4-9cm diameter, matt dark leaf green, the bottom of the plant somewhat greyish, crown sunken, not woolly, the spines lean inwards but do not completely cover the top of the plant. Ribs, 10, separated by sharp grooves, shallow at the bottom of the plant, at the centre 8mm high, 15-18mm wide with a sharp ridge, notched by the cross-cuts, so separated into humps that protrude pointed chin-like below the areoles.

Areoles near the middle of the humps, the oldest, c. 2cm apart, rounded to oval, 3-4cm wide, in youth with white wool, balding with age. Spines, all tough, awl-shaped, not thickening at the bottom, matt, ± dark reddish horn-coloured, strongly radiating outwards and upwards. Radial spines 7, (very rarely 9) of which 3-(4) pairs point sideways, and 1 towards the bottom, 2-3cm long, often slightly curved like a hook. Central spines, 1, seldom 2, up to 3.5cm long, the same thickness as the radial spines but more curved outwards, reaching up towards the crown.

Fig.423 *G. rhodantherum,* GC960.01 in habitat west of the Cuesta Miranda, Prov. La Rioja, Argentina

Flowers, few, near the crown, short funnel shaped, when fully open, 3cm long, 4cm wide, flower tube and ovary 15mm long, 4mm thick, olive-green with dark pink scales that are edged pinkish-white, set spirally, 7mm apart, 2mm wide, 2mm high short pointed. Outer petals long and thin, 2cm long, 2mm wide, brown-olive, towards the points the same form and colour as the scales, inner petals a little longer, more lance-like, thin pointed at the bottom, pale red, deep pink in the throat. Filaments curved, many, dark carmine-pink at the base, lighter at the top, white or yellowish. Anthers fire-red to dark violet-red. Style short, yellowish-white with 7 radiating lobes of the same colour or a little lighter, the stigma lobes stand above the anthers that curve in towards it. The fruit and seed are unknown to me.

Habitat: Province of Salta, first dispatched from there by Mr. Fr. Harperath, in the spring of 1931 from the town of Salta.

Neotype: Argentina, Prov. La Rioja, Dept. Famatina, km.525 along Route National south of Pituil, 1250m.s.m., 18.10.1987, leg. H. Till, HT 87-67 [designated by H. Till (2005) in *Gymnocalycium* 18(1): p.602.]

Fig.424 Habitat of *G. rhodantherum* (coloradense), GC957.02 which grows on the granitic rocks in the foreground, Prov. La Rioja, Argentina

Etymology

From Greek 'rhodos' meaning rose-red and 'anthera' for anthers, referring to the red colour of the anthers revealed when the pollen is shed.

Other Names

Gymnocalycium mazanense var. *rhodantherum* (Bödeker) Backeberg

Synonym

G. guanchinense Schütz in *Zpravy Ceskoslovenske Kaktusarske Spolecnosti* 2: p.21 (1947)

First description of *Gymnocalycium guanchinense* Dr. Schütz spec. nov. (Gymn. sp. Guanchin. n.n. Fric Catalogue *Kakteenjäger* c. 1930)

Body flattened-globular, 10cm diameter, brownish-green. Ribs c. 13, low, straight, separated into sharp humps. Areoles elongated, 5mm long, 2mm wide, covered with short woolly greyish-white hair. Radial spines 7-9, 1 curved down, 20mm long, 3 each side of the areole, spreading, c. 25mm long, 2 top spines, short or occasionally missing, all curved, horn-like, porous on the top face; central spine curved upwards, c. 25mm long, sometimes missing.

Fig.425 *G. rhodantherum* (coloradense), GC957.02 in habitat east of Los Colorados, Prov. La Rioja, Argentina

Flowers c. 50mm long and diameter. Ovary 20mm long, greenish-blue, few scales also greenish-blue, edges pink; outer petals oblong, rounded, pink, bluish-green at the bottom with a pinkish edge; inner petals lanceolate, tapering, pink, with no intensely coloured mid-stripe, petals with a silky sheen. Ovary short, with an 8 lobed yellow stigma; many stamens; filaments red to blood-red; anthers pink; pollen yellow.

Fruit thin, conical, blue-green, when ripe splitting longitudinally. Seed; many, small, dull black, elongated, semi-globose, general appearance is flattened, testa somewhat erose.

Habitat: Cordilleras, Argentina, near the Rio de Guanchin

Neotype: Argentina, Prov. La Rioja, Cuesta Miranda near km.516, between El Siziliano and Las Higuertas, leg. H. Till 87-88/639, 20.10.1987, 1800m.s.m. (Neotype BA) Designated by H. Till (2004b)

Gymnocalycium acorrugatum Lambert in *Succulenta* 67(1): pp.4-7 (1988)

First Description of *G. acorrugatum* spec. nov.

Body simple, flattened globular, c. 45mm high, 75mm diameter, dull green to grey-green, crown somewhat sunken, with wool, covered with spines from the neighbouring areoles, with a tap-root.

Ribs 9-10, straight or slightly curved, with distinct vertical grooves, humps rounded, running into each other, without cross-cuts and chins, except on very young plants.

Areoles round, 6-8mm diameter, at first with white wool, later wool becoming black, 17-20mm apart.

Radial spines, 7-9, spreading out in 3-4 pairs, with a single spine pointing down, strong, curved, interlaced with each other, sharp, up to 3cm long, young spines chestnut-brown, yellowish at the base, later greyish-white with a brownish point; Central

Fig.426 Habitat of *G. rhodantherum*, GC959.02 east of Cuesta Miranda, Prov. La Rioja, Argentina

spines, 1-2, strong curved upwards, up to 35mm long, the same colour as the radial spines.

Flowers from young areoles in the crown, funnel-shaped, c. 65mm long, 60mm diameter. Pericarpel c. 15mm long, 10mm diameter, leaf green, sometimes with a purple bloom; scales rounded, 5mm wide, 3-4mm high, white edged with a pink tip, all scales up to the petals are the same, outer petals spatulate, c. 30mm long, 7mm wide, pinkish-white with a brownish-green central stripe, inner petals light pink with a carmine-pink central stripe, spatulate with a point, c. 34mm long, 6.5mm wide, throat a deep carmine colour.

Stamens, primary stamens are inserted around the style, secondary stamens are inserted over the whole of the receptacle wall, curved in towards the style, the topmost stamens overhang the style, white, red at the base; anthers yellow; style greenish-white, 12mm long, 1.7mm diameter, (up to 20mm long with the stigma) stigma light yellow with 9 lobes.

Fruit blunt spindle-shaped, bottle green; scales tinted pink with a fine white edge, 23mm long, 15mm wide. Seeds pot-shaped, dull red-brown, 1.2-1.3mm high, 0.8-1.0mm wide and thick, testa covered with warts, set in a very ordered pattern, warts round to oval, not flattened; hilum elongated oval, curved, slightly sunken, micropyle area somewhat corky, the edge covered with a yellowish-white spongy tissue, not covered with a hilum tissue, aril missing. Seeds belong to the series Microsemineum (Schütz), section Mostiana (Buxbaum).

Habitat:- San Agustín de Valle Fertil, Sierra del Valle Fertil. province San Juan, Argentina. 850m.

Holotype: JL69 (Herbario Universitatis Rheno-Traiecti)

Commentary on *G. acorrugatum*

For some time this taxon was considered to be a westerly subspecies of *G. castellanosii*. The formal combination was made in *Flora de San Juan* II: p.178 (2003) as *Gymnocalycium castellanosii* subsp. *acorrugatum* (Lambert) Kiesling & Metzing. I repeated the combination, by mistake, in CSI 20 (2005) without realising that it had already been made. However, as pointed out by Papsch (2008), it does not have the characteristic seeds of *G. castellanosii* so does not belong in its synonymy.

It was originally described as growing under thorny bushes on stony slopes near the edge of the lake at San Agustín de Valle Fertil where I have seen it. More recent observations suggest that it occurs along the Sierra del Valle Fertil and further south in the Sierra de la Huerta where I saw it growing on the hills behind the village of Las Chacras. These sierras run north-south and lie to the south of the main distribution of *G. rhodantherum*.

Gymnocalycium coloradense Berger in *Gymnocalycium* 19(4): pp.693-695 (2006).

Fig.427 *G. rhodantherum* (guanchinense), *Piltz* 226 from the Cuesta Miranda, Prov. La Rioja, Argentina

Fig.428 *G. rhodantherum* (acorrugatum), GC217.02 from San Agustín de Valle Fertil, Prov. San Juan, Argentina

Fig.430 Habitat of *G. rhodantherum* (acorrugatum), GC217.02 by the lake at San Agustín de Valle Fertil, Prov. San Juan, Argentina

First Description of *Gymnocalycium coloradense*

Body simple, globular, up to 140mm high and to 120mm diameter, with a long stake-like, strongly branched root. Crown not sunken, strongly armed with long projecting spines. Epidermis dull grey-green. Ribs 8-12 (-14), running straight down, separated from each other by vertical furrows, flat at the base, up to 20mm wide, with a weak notch below a small chin-like hump.

Areoles projecting, round, 5mm diameter, 10-15mm apart, the areoles in the area of the crown covered with dirty white wool, older areoles bald. Spines 5-7 (-9), mostly arranged in pairs, (20-25) 35-45mm long, flattened, stiff, comb-like to radiating, mostly standing away from the body, straight to slightly curved, light brown to grey, matt with a shiny brown tip. Central spines 1, (seldom 2), older plants can have 3 centrals, pointing towards the crown, up to 55mm long.

Flowers developing in the crown, white, with a silky lustre, funnel shaped, 35-40mm high, up to 40mm diameter, hardly any scent. Throat pink. Outer petals wide lanceolate, white, 25mm long, 8mm wide, at the base 3mm wide and pale pink, the outer petals are the longest, olive-green with a whitish edge, inner petals lanceolate to pointed lanceolate, white, 18mm long, 6mm wide, pink at the base.

The stamens divided in two basic types, the primaries, 8mm long, lying onto the style, enclosing the narrow nectar chamber, secondary stamens, many rows inserted into the inner wall of the receptacle and curved towards the middle. Filaments white, pink at the base, 8mm long. Anthers oval, 2x1mm, yellow. Pollen yellow. Style thin light pink, c. 18mm long, 2.8mm thick.

Stigma yellow, with 10-12 claw-like lobes, which are overhung by the topmost anthers. Ovary white, 8-10mm high and 5-7mm diameter. Pericarp conical, c. 15mm high, 8-10mm diameter, olive to dark green, shiny, covered with a few, half round, olive-green pink edged scales. Fruit berry-like, dark reddish-brown, smooth, 18mm high, 14mm diameter, with pink edged scales, splitting vertically. Seeds helmet shaped, shiny dark red-brown, 0.7-0.9mm long, 0.4-0.6mm diameter, covered with small humps. hilum-micropyle-region compressed, narrow oval, deep, filled with some elaiosom, hilum edge somewhat raised.

Fig.429 *G. rhodantherum* (coloradense), GC22.02 from east of Los Colorados, Prov. La Rioja, Argentina

Habitat: Argentina, Province La Rioja, Sierra de los Colorados, southern foothills, 700-900m a.s.l.

Type: Argentina, Province La Rioja, Sierra de los Colorados, in parte australi, leg, F. Berger Be97-72/291, 9. 11. 1997. Holotypus; BA; 1.c,.,17.11.2001. Be01.271/1237, Paratype: WU.

Commentary on *G. coloradense*

This taxon is reported only from the southern end of the Sierra de Los Colorados on the north side of Route 74 growing among rocks in crumbling granite.

I saw this plant on my first trip to Argentina in 1992 and at the time considered it to be a form of *G. hossei* which occurs further to the north in similar habitats. I distributed seedlings grown from seed I collected there as *G. hossei* and under my number GC22.02.

Berger (2006) included his species in the relationship of *G. castellanosii* but following the work of Papsch (2008), it is here placed as a synonym of *G. rhodantherum*, which together with *G. acorrugatum*, extends the distribution of *G. rhodantherum* much further to the south.

Distribution (Map 59)

It is found on both sides of the Cuesta de Guanchin, on the eastern foothills of the Sierra Famatina, and in the low hills north to Campanas and Pituil, as well as on the west side of the Cuesta Miranda. At higher altitudes it is replaced by the similar *G. ritterianum*.

Further to the south, its synonyms extend its distribution down the Sierra de los Colorados, Sierra del Valle Fertil and Sierra de la Huerta.

Conservation Status

Least Concern. The species is widespread and locally common. There is no indication of any immediate threat to the populations. The construction of the new road across the Cuesta Miranda has already been very damaging with much destruction

Map 59 Distribution of *G. rhodantherum* in Prov. La Rioja & San Juan, Argentina. Locations of *G. ritterianum* are shown as blue squares

Fig.431 Illustration of *Gymnocalycium rhodantherus* which accompanied the first description in Kakteenkunde (1934)

Fig.432 *G. rhodantherum* (acorrugatum), *Lambert* 63 from San Agustín de Valle Fertil, Prov. San Juan, Argentina

of its habitat, but the distribution area is very large in comparison.

History

When publishing the first description of *Echinocactus rhodantherus*, Bödeker (1934) explained that it was the same taxon as Backeberg (1934) had described as *E. mazanensis* var. *breviflorus*. Backeberg (1959) in *Die Cactaceae* did not accept the species, preferring his *G. mazanense* var. *breviflorum*. The habitat locality remained in doubt since no known plant from Salta has red anthers. The reference to Salta was probably assumed because the collector was said to come from there.

In 1947, Schütz described *Gymnocalycium guanchinense* with a locality near the Río Guanchin. It had first been mentioned in the Fric catalogue of 1930 as "G. sp. Guanchin, sp. n.". Schütz had grown three seedlings from this seed and from these he made the description.

The locality "Río Guanchin" is misleading since the river of that name is near Fiambalá in Prov. Catamarca, from where *Gymnocalycium* have never been found. The town of Guanchin lies just to the west Chilecito. Schütz (1986) in his *Monografie rodu Gymnocalycium* changed the locality to Cuesta Guanchin, near Famatina. In 1972 Rausch collected plants (*Rausch* 568) which look identical to those belonging to Schütz. See illustration 1 in Till (2004b). From this Till deduces that the plants actually come from the west side of the Cuesta Miranda and neotypified the name with a plant from there.

Commentary

In his review of *G. rhodantherum*, Till (2004a) concludes that the plant is the species from the Sierra Famatina and is from a different evolutionary line from *G. mazanense* (= *G. hossei*) which grows further to the east. Although it is not clear as to how he reaches that conclusion, the work of Meregalli (1986-7, 2000) appears to support that view. So, since we need a name for the westerly species, *G. rhodantherum* is the best choice. The minor differences quoted for *G. guanchinense* do not justify a separate species so this name is treated as a synonym here.

Cultivation

This species is not often seen in cultivation although it does not require any special treatment. It can develop attractive dense spination and flowers easily. When seen in cultivation, it is usually labelled *G. guanchinense*.

Gymnocalycium ritterianum Rausch

First Description (Rausch)

KuaS 23(7): pp.180-181 (1972)

Gymnocalycium ritterianum Rausch spec. nov.

Body flattened-globular, 3-4cm high, up to 11cm diameter, with long turnip-like roots, simple or in small groups, the offsets often appear on the top half of the body, light green or tinted violet-brown. Ribs, 10 to 12, vertical, separated into humps by cross-cuts, 15-20mm long.

Areoles sunken, lying on the top half of the humps, oval, 5mm long, white felted, under the areoles the humps form sharp edged chins. Radial spines 7-9 (3-4 pairs and 1 pointing down), radiating and curved towards the body, up to 25mm long. Central spines, sometimes 1, curved towards the top of the crown, up to 30mm long, all spines pinkish-brown.

Flowers appearing near the crown, up to 65mm long, 75mm diameter. Ovary and tube short, dark green with round pink scales, outer petals spatulate, pinkish-white with a cream central stripe, inner petals spatulate, very wide, spread out, ± wavy, irregularly curved, shiny white, throat light violet-pink. Filaments white, anthers brownish-pink. Style thick, 20mm long, lobes 10, yellowish. Fruit pear-shaped, 15mm diameter, tapering towards the top, bluish with whitish-pink scales. Seed pot shaped, 1mm diameter, blackish-brown, globular-warty, hilum twisted, with a distinct spongy white edge.

Habitat: Argentina, La Rioja, near Famatina, 3000-3500m. Type. *Rausch* 126. In Herbarium W. [Not found at Vienna but Eggli (1987) says deposited at ZSS]

Etymology

Named for Friedrich Ritter (1898-1989), German geologist and probably the most successful cactus explorer there will ever be.

Fig.433 *G. ritterianum, Piltz* 219 from Sierra Famatina, Prov. La Rioja, Argentina

Fig.434 *G. ritterianum, Rausch* 126 from Famatina, Prov. La Rioja, Argentina

Synonyms

G. ritterianum subsp. *acentracanthum* Berger in *Gymnocalycium* 18(1): pp.599-602 (2005)

G. jochumii Neuhuber in *Gymnocalycium* 18(1): pp.603-610 (2005)

First description of *Gymnocalycium jochumii* Neuhuber, spec. nov.

Body flat-globular, offsetting at the base, cushion-forming, up to 45mm high and 80mm diameter, lower part, from the middle (below ground) is short and strongly conical. Taproot with very long, strong fibrous roots. Crown lightly sunken strongly armed. Epidermis dark green, matt. Ribs 12, straight, running straight down, separated from each other by wavy furrows 22mm broad at the base, up to 15mm high, separated into strong humps by short straight furrows, humps five-sided with a prominent hatchet-shaped or a pointed projecting chin. Areoles sitting on top, oval, long oval on adult plants, 5-7mm long, 4.5-6mm wide, 14-17mm apart, yellowish-white wool in the crown, not becoming bald, older areoles have white wool. Spines 5-7(-9), mostly arranged in pairs, (14-19)22-31mm long, whereby the topmost are the shortest, round, stiff, straight, more lying onto the body than projecting, light brown to dark red-brown, becoming a little grey, 1 central spine on older plants, pointing towards the crown, 27mm long.

Flowers coming from the crown, light pink, bell-shaped, 47-54mm high, 60-73mm diameter, hardly fragrant. Outer petals wide lanceolate to spatulate, pale pink with a light pink middle stripe, 26mm long, 9mm wide, 2.8mm wide and pink at the base, scales longish with an olive-green to olive-pink middle stripe. Inner petals lanceolate to wide lanceolate, occasionally with a point, white to pale pink with a light pink middle stripe, 20mm long, 6mm wide, pink at the base. Receptacle inside light pink, 12-15mm high, 12-15mm diameter. Nectar chamber light pink, white at the base, 2mm high and up to 4mm diameter. Stamens arranged in 1-2 basal rows, these 4.5mm long, enclosing the nectar chamber, 80% lying onto the style base and style, 5 secondary rows, coming from the base of the receptacle and lying on the wall of the perianth. The two lower rows have small anthers. Filaments white to light pink, 7.5mm long. Anthers pale pink, the connecting point to the filament and the 1.9x0.9mm opening lime yellow, the first 80% inserted over the receptacle base, pollen yellow, style white to light green, 10.5-12mm high, (=100%), 1.5-2.8mm diameter, papillate and somewhat channelled, not free-standing, always straight, white. Stigma yellow, 3.3-4.9mm high, with 12 lobes, the base of the stigma is higher than 1-2 rows of the secondary anthers. Ovary beaker-shaped, white, 5-10mm high, 4-7.5mm diameter. Pericarp pear-shaped, olive to dark green, seldom shiny, often with a weak grey bloom, 8.8-12mm high, 8-12mm diameter, covered with many rectangular, rounded, light edged, light olive-pink to pink scales.

Fruit pear-shaped, dark green with half round, pink scales, without a bloom, 16-21mm high, splitting vertically, also sometimes at the base, barochorie (seeds trickling from the fruit) has been observed.

Photo: Massimo Meregalli

Fig.435 *G. ritterianum, Meregalli* 800 in habitat south of Florentina, Sierra Famatina, Prov. La Rioja, Argentina

Seeds helmet-shaped, somewhat recurved at the Hylum-micropyle-region (HMR). Dark red-brown to black, 0.7-0.9mm long, 0.6-0.8mm diameter, the HMR is inconsistent, (narrow oval or wide oval) and filled with elaiosom (threads of tissue). The dried seeds in the ripe fruit are different from most other seeds in the subsection Microsemineum, fewer, embedded in the sticky pulp. Subgenus Microsemineum Schütz section Saglionis Schütz H. Till subsection Microsemineum series Mostiana Buxb.

Habitat: Argentina, Province La Rioja, southern Sierra de Sañogasta, 1670m.

Type: Argentina, Province of La Rioja, southern Sierra de Sañogasta, 1670m, collected by G. Neuhuber GN96-888/2923, 9.10.1996; Holotype: BA, Isotype: WU.

Commentary on *G. jochumii*

The differences between this plant, its variety *jugum*, *G. ritterianum* and its subsp. *acentracanthum* are well within the expected variation of a species which exhibits so many diverse forms.

Distribution (Map 59 on page 197)

Grows to the west of the town of Famatina, in the foothills of the Sierra Famatina as far south as the Cuesta de Guanchin. Also found further south in the Sierra de Sañogasta (*G. jochumii*). It grows on the mountains above where *G. rhodantherum* is found mainly in the lower lying areas.

However, the quoted altitude of 3,000 -3,500m in the first description cannot be correct. No Gymnocalycium is known to grow at that altitude. In fact, the type locality is at about 1800m in the foothills of the Sierra Famatina, a similar altitude to where Piltz found the very similar *Piltz* 219. Its highest reported altitude is 2300m (*G. jochumii*).

H. Till & W. Till (2004) describe their success at finding this species above the Cuesta de Guanchin, both solitary plants and small groups, some with central spines. They were growing in stony loam soil covered with pebbles and large stones.

Conservation Status

Least Concern. As well as the known populations, there is plenty of similar habitat in the area which could well harbour more populations of this plant. There is no known threat to their continued survival.

Photo: Massimo Meregalli

Fig.436 *G. ritterianum* (jochumii), *Meregalli* 305 from the Sierra de Vilgo, Prov. La Rioja

History

Rausch found the plant during his second trip to Argentina in 1964. In the first description, Rausch tells us that this species is related to *G. guanchinense* but differs because of its grey-green epidermis, almost pink spines and wide petals that are often wavy. H.Till & W. Till (2004) add that it is also different because of its flat-globular body, shorter flowers, tendency to offset, and seeds with a larger more distinct hilum region.

Recent explorations have found plants of this affinity south of the Cuesta Miranda on the Sierra Sañogasta at even higher altitudes and these have been described as *G. ritterianum* subsp *acentracanthum* and *G. jochumii*.

Commentary

This taxon is clearly related to *G. rhodantherum* and some specialists in this genus believe them to be the same species. There are, however, a number of features which differentiate them as described above. *G. ritterianum* had been found in habitat only by a few explorers because its habitats are some way away from roads at higher altitudes than the widespread *G. rhodantherum*. Recent explorations by Berger (2005) and Neuhuber (2005a) have found more habitats further west in the Sierra Sañogasta which appear to be the same taxon.

Bercht (pers. com.) doubts if this taxon is really different from *G. rhodantherum*, stating that the different flowers are found only on plants from the type locality, such as *Piltz* 219.

Cultivation

Most plants in cultivation are seedlings from the *Piltz* 219 collection. The large showy flowers are quite different from those of the related *G. rhodantherum*. It is easy to grow and flower, plants eventually making clusters of heads.

Gymnocalycium saglionis (Cels) Britton & Rose

Fig.437 *G. saglionis* GC966.05 in habitat, east of the Sierra Famatina, north of Chilecito, Prov. La Rioja, Argentina

Fig.438 *G. saglionis,* GC976.04 in habitat north of Huaco, Prov. La Rioja, Argentina

Fig.440 *G. saglionis,* GC978.04 in habitat south of Villa Sanagasta, Prov. La Rioja, Argentina

First Description (Cels)

Portefeuille des Horticulteurs: p.180 (1847)

Echinocactus saglionis Cels, dedicated to Mr. Jos. Saglio, a collector in Strassburg.

Stem hemispherical, apex a little flattened, grey-green, ribs vertical, rounded, very broad, furrows sharp, tuberculate. Tubercles hexagonal. Areoles prominent, oval, well spaced, 3cm apart, provided with whitish, dark grey wool, at first abundant, later short and grey; spines 9, long, recurved, of unequal lengths, at first red-brown, later becoming reddish; 8 radiating, the uppermost spine is usually thinner and shorter, very seldom missing, the lower spine is strong, very recurved, up to 120°, and one central spine, which is awl shaped, stronger at the base, less recurved, and almost erect, the remaining 6 spines arranged symmetrically each side of the areoles. Flowers unknown in Europe.

Fig.439 *G. saglionis* (tilcarense), *Bailey* s..n. in habitat at Purmamarca, Prov. Jujuy, Argentina

Photo: Geoff Bailey

Neotype: Argentina, Prov. Tucumán, Dept. Trancas, about 12km north of Tapia on Route 9 to Vipos. 700m, 24.12.1990, leg. G. Neuhuber, 90-304/1020 (WU). Designated by Till & Till (1997b) in *Gymnocalycium* 10(2): p. 214.

Etymology

Named for Joseph Saglio, a French amateur cactus enthusiast.

Other Name

Echinocactus saglionis Cels in *Portefeuille des Horticulteurs*: p.180 (1847)

Synonyms

Echinocactus hybogonus Salm-Dyck in *Cacteae in Horto Dyckensi cultae anno 1849*: pp.167-168 (1850)

First Description of *Echinocactus hybogonus*

Stem flattened-globular, dark glaucous-green, ribs 13, tuberculate, separated by broad wavy, uneven furrows. Tubercles 5 sided, with an areole, upper tubercles narrow, lower humps wider, flattened, Areoles not close together, almost round, with grey matted hair, radial spines, 8, the lowest usually the shortest, radiating, recurved, brownish-grey, central spines, 1, straight (Nob.)

Adult stems 5-7.5cm high, 10-12.5cm diameter. Flowers average size, 3.75cm high, 2.75-3.1cm wide when open, tube short, broad funnel shaped. Sepals blunt, wide, green, moderately long, lower margins white, pink above, recurved at the tip. Petals blunt, serrated at the tip, erect, recurved, whitish or pale flesh colour, reddish at the base. Many stamens attached over the whole of the inner wall of the large tube, filaments short, purplish, anthers small, yellow. Style thick, short, fistulose, purple, shorter than the stamens; stigma lobes, 14, thick, blunt, spread out from the base, tips curved inwards, cup shaped, erect, yellow.

G. tilcarense (Backeberg) Schütz in *Monographie der Gattung Gymnocalycium*: p.134 (1992). Basionym: *Brachycalycium tilcarense* Backeberg (1942). Nom. inval.

G. saglionis var. *tilcarense* Backeberg nom. inval. in *Kaktus ABC* (1936)

G. saglionis subsp. *tilcarense* Backeberg (Till & Till) *Gymnocalycium* 10(2): p.215 (1997)

Fig.441 *G. saglionis,* GC957.03 in habitat near Los Colorados, Prov. La Rioja, Argentina

Photo: Geoff Bailey

Fig.442 *G. saglionis* (tilcarense), *Bailey* s..n. in habitat at Purmamarca, Prov. Jujuy, Argentina

First description of *Brachycalycium tilcarense* Backbg. in *Cactaceae, Jahrbücher der Deutschen Kakteen-gesellschaft* (1942)

Singular, moderately large, then subcylindric, grooved tubercles, large, spirally arranged, flowers tubeless covered with scales; very short style. Northern Argentina (Tilcara)

Neotype: Argentina, Prov. Jujuy, Dept. Tilcara, Quebrada de Humahuaca, 2250m, 8.2.1993, leg HT93-573 (WU). Designated by Till & Till (1997b) in *Gymnocalycium* 10(2): p. 215.

Commentary on *G. tilcarense*

It is difficult to understand why Backeberg thought this plant was so different from *G. saglionis* that it warranted its own genus. It grows at the most northern locality for *G. saglionis*, in the Quebrada de Humahuaca around the town of Tilcara. This is a 200km disjunction from the main distribution of *G. saglionis*. This form usually has more ribs with more closely-spaced areoles than the main populations in the south.

Distribution (Map 60)

One of the most widely distributed species, it is found in Argentina from Prov. Jujuy in the north to Prov. San Luis in the south. The most northerly populations in the Quebrada de Humahuaca (*G. tilcarense*) have more ribs and more closely spaced areoles giving the plants a denser-spined appearance.

Till & Till (1997a) give a detailed account of the distribution and the variability of the plants within and between the more than 100 known populations. Their illustrations give a good indication of the variability.

Conservation Status

Least Concern. This species is one of the most widespread of the genus and is frequently common. It is a very successful species, often naturally occurring with other Gymnocalyciums. It flowers over a long period and produces huge quantities of seed.

History

The first publication of the name *Echinocactus saglionis* was in 1847 and is attributed to the French nurseryman Jean Cels in *Portefeuille des Horticulteurs*, a rare publication which I have not personally seen.

Gürke (1905) tells us that, following its description, the plant was practically unknown so when Salm-Dyck flowered a plant, he described it as *Echinocactus hybogonus* in ignorance of Cel's publication. A few small specimens arrived in Germany in 1893, possibly from Weber, then in 1901 De Laet received a large consignment which enabled the plant to become common in collections.

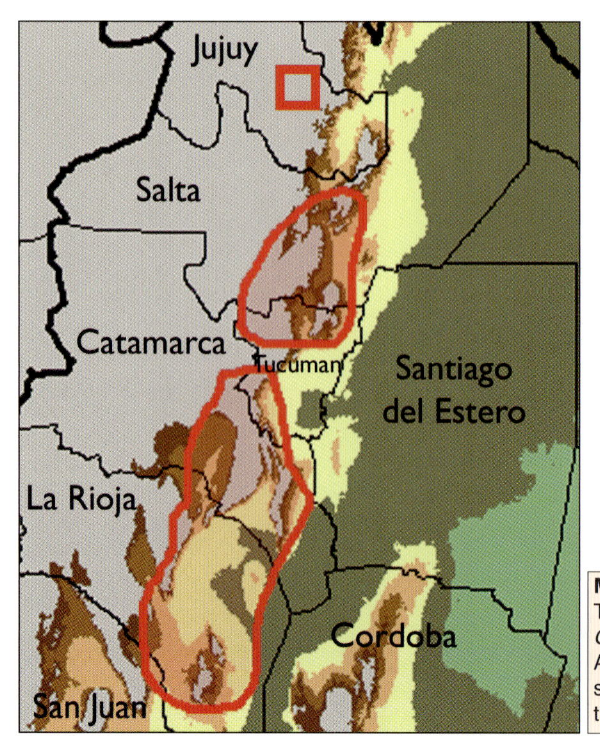

Map 60 The distribution of *G. saglionis* in Argentina. The red square is the location of *G. tilcarense.*

Fig.443 *G. saglionis, Preston-Mafham* 306 from south of Alemania, Prov. Salta, Argentina

Fig.444 *G. saglionis, Till* 90-327 from Sierra Porongo. Prov. La Rioja, Argentina

Fig.445 *G. saglionis,* BLMT 375.04 from south-east of El Arenal, Prov. Catamarca, Argentina

The combination in *Gymnocalycium* was made by Britton & Rose, *The Cactaceae* III: p.157 (1922) but with the incorrect termination 'saglione' which still persists to this day. The plant is very variable with respect to the colour of the body, spination and flower parts so that the attempts to describe varieties based on such differences are pointless and are ignored here.

Commentary

Here we have a plant over which there is no dispute about its identity, even though it is frequently mis-spelt. It is probably the largest species in the genus, capable of growing to 50cm diameter and, remarkably, it has seeds which rank among the smallest, less than 1mm in diameter. Old plants eventually fall over and lie on their side, but they do not produce offsets unless damaged.

It is a very successful plant with a wide distribution in Argentina. It has a long flowering period and plants are often found with ripe fruits containing huge quantities of seeds. They can grow on exposed rocks or under trees and there are usually plenty of seedlings to be found.

The only species with which it can easily be confused is its relative *G. pflanzii* which looks much the same and has similar seeds. The flowers of *G. pflanzii* are bell-shaped in contrast to the urn-shape of *G. saglionis*, which also has longer areoles. See Figs.19-21 in Till& Till (1997a).

In effect, *G. pflanzii* replaces *G. saglionis* in the north of Argentina and its range extends into Bolivia and Paraguay. I know only one place where they grow together and that is south of Salta city, near the lake by Cabra Corral, where the *G.pflanzii* are relatively small and make large groups of heads. In Bolivia, *G. pflanzii* can make very large solitary heads and then looks a lot like *G. saglionis*.

Cultivation

G. saglionis is one of the quickest and easiest species to grow and very popular in collections. In shows, it is the most frequently seen plant in the unlimited pot size class for Gymnocalyciums. The northern form, often called *G. tilcarense*, is especially neat with more closely-packed areoles and makes a particularly attractive plant in culture.

The ultimate size of plants varies from population to population and those which never get really large are the most likely to flower in culture when small. Some clones can first bloom when under 10cm in diameter, such as those from the Dique de Catamarca, see Illn. 279.6 in the *New Cactus Lexicon* Atlas (2006).

The flowers have extremely short tubes and so tend to develop among the spines which often obstruct the opening of the petals. On larger specimens the flowers make a garland around the top of the plant when many can open at the same time.

Illustrations

Schumann, K., MfK (12) 2: p.27 (1902) and repeated in the *Nachträge of Gesamtbeschreibung der Kakteen* p.125 (1903)

Britton & Rose, *The Cactaceae* III: plate XVII, Fig.1 (1922). [Note that Fig.165 on page 157 of the same volume is not correctly identified, perhaps *G. spegazzinii*]

New Cactus Lexicon Atlas (2006) Illn. 279.6, 280.1

Gymnocalycium spegazzinii Britton & Rose

First Description (Britton & Rose)

The Cactaceae Vol. III (1922)

Gymnocalycium spegazzinii nom. nov.

Depressed, globular, 6cm high, 14cm in diameter, greyish-green: ribs 13, broad and low, rounded on the margin; areoles elliptic, spines usually 7, subulate, rigid, appressed to the ribs, sometimes recurved, greyish-brown, 2 to 2.5cm long; flowers 7cm long; inner perianth-segments more or less rose-tinted; filaments and style violaceous; stigma lobes 16, white to rose-coloured; scales on the ovary few, broad.

Type locality: La Viña, province of Salta, Argentina. Distribution: Known only from the type locality.

As this plant requires a new name, it gives me great pleasure to dedicate it to such an enthusiastic cactus student as Dr. Carlos Spegazzini, of La Plata, Argentina.

Etymology

Named for Prof. Dr. Carlos Spegazzini (1858-1926), Italian born pharmacist who worked as a botanist in Buenos Aires.

Distribution (Map 61)

Originally said to come only from the type locality which Spegazzini (1925c) tells us is rocky slopes of the Valle de Lerma, near La Viña, Salta. It is not clear exactly where this location is. There is a town La Viña on Route 68 south of Salta City but, perhaps more likely, is the small resort of 'La Viña' near to Rosario de Lerma, south west of Salta City.

In fact, the species has a large distribution area of almost 500km by 60km covering different types of habitats and many populations which often have recognisably distinct characteristics. It was Piltz (1977) who first defined the extent of its natural range. It is a high altitude species recorded from 1600 to 2800m.

The whole distribution range is within Argentina. The most southerly locality can be found at Capillitas at 2,700m, just north of Andalgalá, while the northern limit is in the Quebrada del Toro, Prov. Salta at about 3000m. The plants from the latter place are particularly spiny and seedlings from this population retain their tendency to grow long spines in cultivation.

After the long dry winter and spring, the plants are drawn down almost level with the ground. This phenomenon is also seen in plants growing in the Quebrada Cafayate, where in drought conditions the spines appear like patterns in the flat sand. (Fig.456). The pictures opposite show the appearance of plants after the arrival of the summer rains.

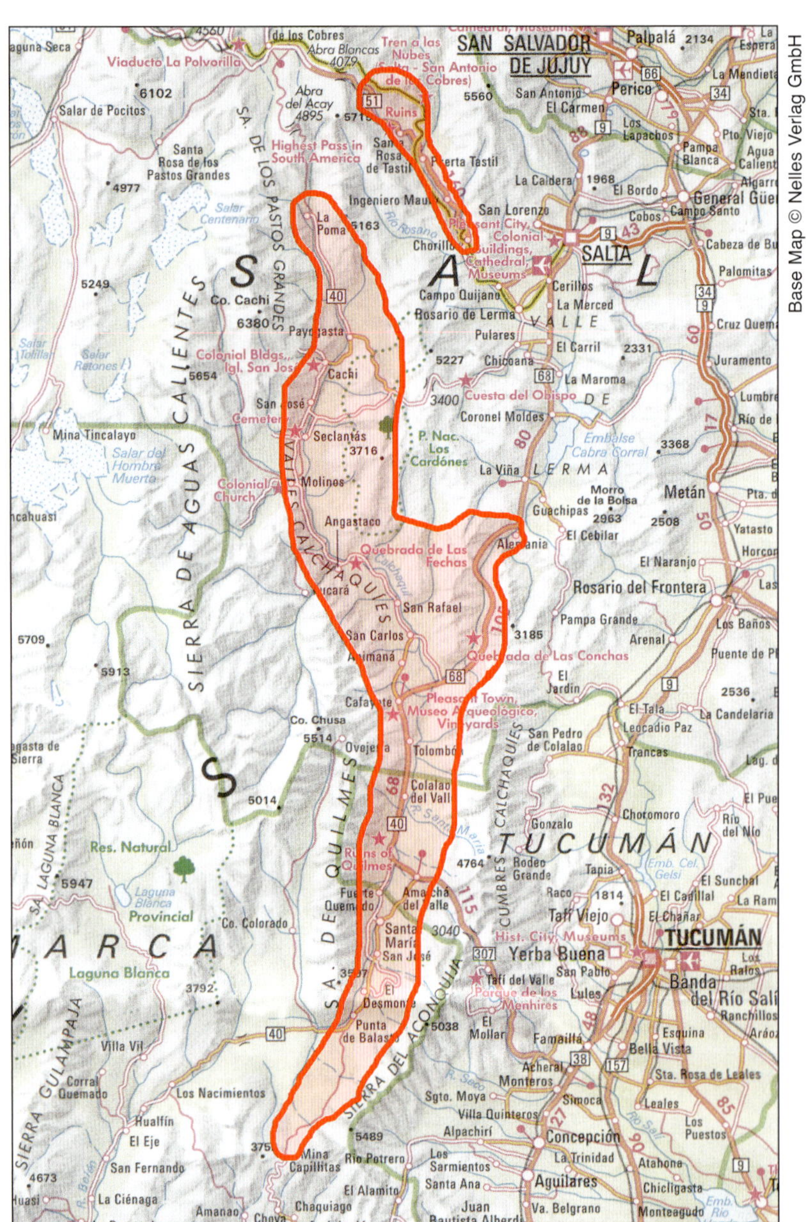

Map 61 The distribution of *G. spegazzinii* in Argentina

Conservation Status

Least Concern. It is regrettable that this species, because of its slow growth, is still collected from the wild by enthusiasts. It is, however, probably not a great threat to its survival because of its widespread distribution, much of it far from roads.

History

The first attempt to name this species was by Spegazzini in 1905 when he described a plant from La Viña, prov. Salta, as *Echinocactus loricatus*. Unfortunately, this name had already been used

Photo: Geoff Bailey

Fig.446 *G. spegazzinii, Bailey* s.n.in habitat at Alfarcito, Quebrada del Toro, Prov. Salta, Argentina

Photo: Geoff Bailey

Fig.447 *G. spegazzinii, Bailey* s.n.in habitat at Amaichá del Valle, Prov. Tucumán, Argentina

Photo: Geoff Bailey

Fig.448 *G. spegazzinii, Bailey* s.n.in habitat at Tolombón, south of Cafayate, in Prov. Salta near the border with Prov. Tucumán, Argentina

Photo: Geoff Bailey

Fig.449 *G. spegazzinii, Bailey* s.n.in habitat at El Obelisco, Quebrada de Cafayate, Prov. Salta, Argentina

Photo: Geoff Bailey

Fig.450 *G. spegazzinii, Bailey* s.n.in habitat at La Angostura, south of Molinos, Prov. Salta, Argentina

Photo: Geoff Bailey

Fig.451 *G. spegazzinii, Bailey* s.n.in habitat at La Cabaña, Prov. Salta, Argentina

Fig.452 *G. spegazzinii,* GC45.04 in habitat south of Alemania, Prov. Salta, Argentina

Fig.453 *G. spegazzinii,* GC56.01, in habitat, Quebrada del Toro, south of Puerta Tastil, Prov Salta, Argentina

Fig.454 *G. spegazzinii,* GC56.01, a dehydrated plant in habitat, south of Puerta Tastil, Prov Salta, Argentina

for another taxon by Poselger in 1853 so rendering the name illegitimate. The valid name *Gymnocalycium spegazzinii* was created for Spegazzini's plant by Britton & Rose in Volume III of *The Cactaceae* (1922) where a clear picture was contributed by the original author.

Fig.455 Habitat of *G. spegazzinii* GC52.02 at 3000m, north of La Poma, Prov. Salta, Argentina

Spegazzini (1925c) tried to restore his original species name by making the combination *Gymnocalycium loricatum* but this was still illegitimate because *G. spegazzinii* was an older name based on the same type. A further superfluous transfer of *E. loricatus* to *Gymnocalycium* was made by Knuth in *Kaktus-ABC* (1936).

Werdermann (1930) gave a more detailed description under the name *Echinocactus loricatus* based on imported plants he received in 1928 from Hosseus.

It soon became clear that the plant had a wider distribution than first thought, and many collectors sent plants to Europe in the 1960's and subsequently Frank (1964) illustrated the great variability among the imported plants.

Commentary

Here we have a species over which there has been little discussion concerning its identity. Even though it is very variable, there is no basis for describing varieties of this widespread species. The two published varieties are var. *major* Backeberg from the edge of the Cachipampa and var. *punillense* Till & Till described from the Río Calchaqui

Fig.456 *G. spegazzinii* (left), GC44.05 in habitat in the Quebrada de Cafayate, Prov. Salta, Argentina. On the right is *Echinopsis* (*Acanthocalycium*) *thionanthum.* The plants are dehydrated in spring before the first rains.

Fig.457 *G. spegazzinii, Kiesling* s.n. from the Quebrada de Cafayate, Prov. Salta, Argentina.
Photographed in the collection of Elisabeth Charles

and Río de las Conchas valleys. It is capable of growing much larger than the first description suggests and splendid specimens up to 25cm in diameter can be found in favourable places. Its habitats often endure long dry periods to which this plant is well adapted by shrinking down even to the point where it can be temporarily covered by blown sand.

Cultivation

The symmetrical appearance of *G. spegazzinii* makes it one of the most popular species in cultivation. Because of its slow growth, it does particularly well at competitive shows, gaining more credit for maturity at a small size than faster growing species. Although slow, it is not difficult to grow, appreciating good light to develop the best spines. Seedlings from habitat collected seed retain some individual characteristics of the population from which they were harvested, for instance the long-spined forms originating from the Quebrada del Toro in Salta Province, Argentina (Fig.454).

Illustrations

Frank, G. KuaS 15(6): pp.116-118 (1964)

New Cactus Lexicon Atlas (2006) Illn. 281.1-3

Fig.458 *G. spegazzinii,* GC419.02 from near Punta de Balasto, Prov. Catamarca, Argentina

Fig.459 *G. spegazzinii, Muhr* B34 from Molinos, Prov. Salta, Argentina

Fig.460 *G. spegazzinii, Vareb* 17 from Cafayate, Prov. Salta, Argentina.
Photographed in the collection of Elisabeth Charles

Subgenus *Trichomosemineum*

Type Species
G. quehlianum (F.Haage ex Quehl) Vaupel ex Hosseus

Species and Subspecies (synonyms indented) — **Page**

Echinocactus Quehlianus Ferd. Haage.
Tafel 105.

Fig.461 Habitat of *G. bodenbenderianum* (paucispinum), GC987.02, south of San Martin, Prov. La Rioja, Argentina.

First Description (Berger)

After the name appeared in the W. Haage catalogue of 1928, the valid description appeared in Berger's Kakteen: pp.221-222 (1929)

Echinocactus Bodenbenderianus Hosseus (1928). – *Gymnocalycium* Berger

Quite flat, almost disc-like, c. 8cm diameter, brownish-grey-green, the slightly sunken crown without spines, with many small humps. Ribs 11-14, low, wide, rounded, 6mm wide, cross-cuts divide the ribs into nose-like pointed humps. Areoles with dirty grey felt when young.

Spines 3(-4-5), robust, at first blackish-brown, later grey-brown, all recurved, the lateral spines sickle-like, 10mm long, central spines weak, the laterals lying onto the humps. Flowers in a circle from the highest areoles on the crown, medium size, very pale pink with brownish central stripes, ovary and tube, bluish, longer than the petals.

Argentina, Córdoba. Introduced by Fr. Ad. Haage jr. Flowers described by Fr. Ad. Haage jr. Stands very close to *E. quehlianus*.

Etymology

Named for Dr. Bodenbender, a German mineralogist who worked in Argentina.

Synonyms

G. piltziorum Schütz in KuaS 33(7): pp.144-145 ((1982). Possibly invalidly described (no known preserved type)

First description of *Gymnocalycium piltziorum* Schütz

Body simple, flattened-globular, crown sunken, spineless, up to 10cm high, 16cm in diameter. Epidermis bluish-green, matt, rough. Ribs, c. 12(10-17), wide, rounded, straight, with humps that protrude chin-like below the areoles, cross-cuts and straight grooves are deep and narrow. Areoles oval, 3-5mm long, 3-4mm wide, at first with short grey felt, later balding. Radial spines (3)-5, at first upright, later radiating, 2 sideways and one pointing down. strong awl-shaped, very stiff, often with length-ways grooves, 25mm long, pink to brownish with a dark point, central spines missing.

Flowers from the areoles near the centre of the crown, funnel-shaped, c. 70(-80)mm long, 60mm in diameter, with sparse grey-green, rounded scales. Petals lilac-pink, outers spatulate, inner petals long lanceolate. Many stamens coming from the bottom of the receptacle, filaments pink, anthers yellow. Style green with c. 12 stigma lobes, yellowish, the same height as the anthers.

Fruit c. 30mm long, 10mm in diameter, narrower at the top and bottom, grey-green with a few rounded scales that are lilac-pink at the tips. Seed c. 1mm high,1.2mm wide, hat shaped, testa smooth, brown and shiny, hilum basal, fairly small with a thick light edged seam. (Subgenus *Trichomosemineum*).

Fig.462 *G. bodenbenderianum,* GC949.03 near road from Tuclame to Agua de Ramón, Prov. Córdoba, Argentina

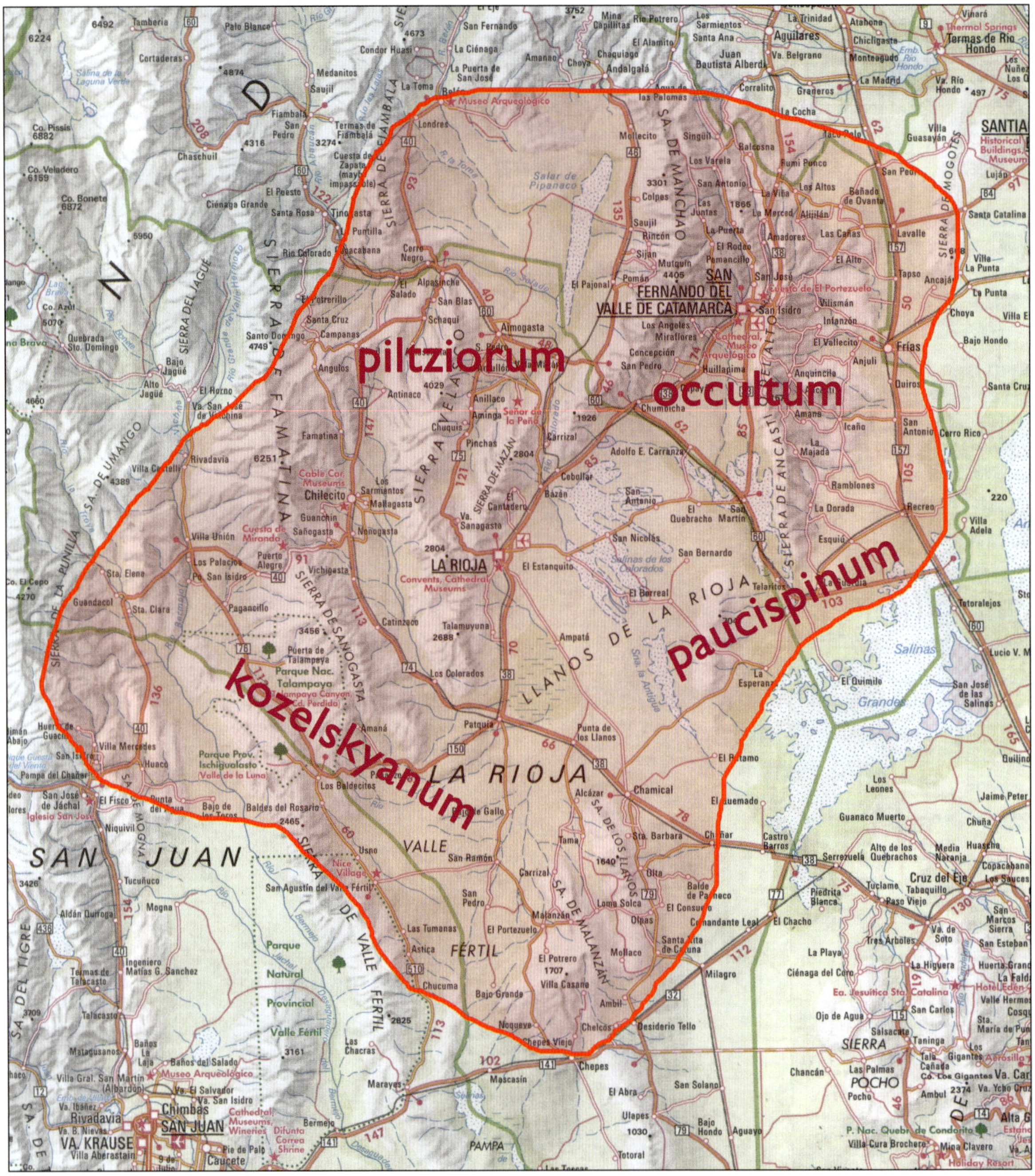

Base Map © Nelles Verlag GmbH

Map 62 Distribution of *G. bodenbenderianum* mainly in Prov. Catamarca & La Rioja, Argentina

Habitat: Argentina. La Rioja, Sierra Velasco. Herbarium material in Karls University, Pilsen, Czech. (PZ)

H.Till & W. Till validly published this taxon as a subspecies of *G. riojense* with a new type HT88-174 from Dept. Castro Barros in La Rioja.

G. riojense Fric ex H. Till & W. Till in *Gymnocalycium* 4(2): pp.48-49 (1991)

First description of *Gymnocalycium riojense* Fric ex Till & Till, spec. nov.

Body semi-globular, in age discoid, 40-50mm high, 75-90(>200)mm diameter, brownish-green to grey-brown-green, with yellowish to light brown wool-felt coming from the areoles in a lightly depressed, unarmed crown. Ribs 10-13(-28), mostly straight, separated by sharp (somewhat wavy by the appearance of new ribs) long furrows, 22-28mm wide at the base, narrower towards the top of the plant, round or nearly sharp, divided into small, distinct, blunt chin-like humps by ± deep cross-cuts. Areoles small, round to wide oval, 3x3-4x3mm diameter, in youth with a little light brown or yellowish wool-felt, later disappearing, older areoles naked and often sunken. Spines mostly 5, rarely only 3, all radials, centrals always missing, hard, 9-15mm long, the top pair spreading horizontally, the lower pair (when present) sloping downwards, the bottom spine pointing towards the base, ± rounded with a slight edge, mostly straight, only a slight curve towards the body, young spines dark brown to black, soon becoming grey, only the tips dark and shiny, very old plants in habitat often lose their spines.

Flowers from the areoles in the crown, bell-shaped, c. 48mm long, 39mm in diameter (average of 10 measurements), pale pink with a darker throat. Pericarp conical, 19mm long, 9mm diameter at the top, outside brownish-green with half round, pointed, often light edged pink scales. Outer petals spatulate, 15-21mm long, 4-6mm wide, pale pink with ± a greenish-brown back stripe, inner petals wide lanceolate 17-19mm long, 4-6mm wide, 2mm wide at the base, pale to whitish-pink, sometimes with a narrow pink middle stripe. Stamens overhanging the style and stigma, the lowest (inner) 1-2 rows lean against the style, therefore enclosing the pink nectar chamber. Filaments thin, 6mm long, light green. Anthers oval, 2x1mm, yellow. Style 12mm long with stigma, 17mm long, 2mm thick, yellow to pink at the base, pale green above, with 9-12 radiating, somewhat claw-like, light yellow stigma lobes. Nectar chamber very narrow, 4mm long, 2.2mm wide, light pink. Ovary longish heart-shaped, 12x5mm, inside white walled.

Fig.463 *G. bodenbenderianum,* GC957.01 in habitat near Los Colorados, Prov. La Rioja, Argentina

Fig.464 *G. bodenbenderianum* (paucispinum), GC989.01 in habitat near Recreo, Prov. Catamarca, Argentina

Fruit barrel-shaped to longish oval, still carrying the flower remains, brown-green to olive-green, bloom somewhat glaucous, opening lengthwise, 24-26mm long, 13-18mm in diameter, with white to greenish-white seed strings. Seeds hat-shaped c. 1.2mm high and wide, 1mm thick with a wide hilum-micropyle region, the edged somewhat horseshoe shaped, raised and curved outwards, elaiosome of the edge only slightly raised, on the long side somewhat thicker. Testa shiny brown with small papillae.

Distribution: In the centre of the Province La Rioja, the region between the Sierra los Colorados, the Sierra Velasco south of the Cuesta Huaco, the southern part of the Sierra Mazán and in the south, the Sierra de Arganaraz, 300-500m.

Type: Argentina, Province La Rioja, near the village of Bazan towards the east of La Rioja city, collected by H. Till No. HT88-122, 29th September 1988 (Holotype: WU)

Fig.467 *G. bodenbenderianum* (paucispinum), GC989.01 in habitat near Recreo, Prov. Catamarca, Argentina

Fig.465 *G. bodenbenderianum* (piltziorum), GC967.02 in habitat west of Sierra. Velasco, Prov La Rioja, Argentina

Fig.466 *G. bodenbenderianum* (paucispinum), GC986.03 in habitat north of San Martin, Prov. Catamarca, Argentina

Gymnocalycium riojense subsp. *kozelskyanum* H. Till & W. Till in *Gymnocalycium* 4(3): p.52 (1991)

Gymnocalycium riojense subsp. *paucispinum* H. Till & W. Till in *Gymnocalycium* 4(4): p.55 (1991)

Gymnocalycium riojense subsp. *piltziorum* H. Till & W. Till in *Gymnocalycium* 4(3): p.51 (1991)

Gymnocalycium stellatum subsp. *occultum* Fric ex H. Till & W. Till in *Gymnocalycium* 9(2): p.176 (1966)

Fig.468 *G. bodenbenderianum* (occultum), GC28.02 in habitat east of Catamarca City, Prov. Catamarca, Argentina

Fig.469 *G. bodenbenderianum,* GC201.03 from between Jáchal and Huaco, Prov. San Juan, Argentina

Distribution (Map 62 on page 210)

Hosseus gave its place of discovery as La Rioja but then Berger said it was Córdoba, now considered to be an error.

The wider concept of *G. bodenbenderianum,* followed here, is that it is distributed over south and central La Rioja, as well as parts of adjoining Catamarca and a small part of San Juan. It often grows under bushes (such as *Larrea*) on gently sloping sandy soil.

The most north-easterly form, which grows near the city of Catamarca, was named, probably invalidly, as *G. occultum* by Schütz (1962). Later validly published as a variety of *G. stellatum* by H. & W. Till (1996). It is of similar appearance but a smaller plant.

Conservation Status

Least Concern. Although impacted by grazing, the plant is widespread and often abundant. Animals do eat the plants and grazing can reduce the numbers of bushes under which it often grows and establishes from seed, but young plants are frequently seen, suggesting that populations are healthy.

History

The name *Gymnocalycium bodenbenderianum* first appeared in a trade catalogue of W. Haage in 1928 where it was attributed to Hosseus. Berger (1929) then published a description and validated the name in his book *Kakteen*. The combination in *Gymnocalycium* was made by Sir Arthur Hill in *Index Kewensis*, Supplement 8: p.105 (1933).

Commentary

G. bodenbenderianum is here taken to include the plants which are considered to be *G. riojense* by Till and others. There is no satisfactory way to distinguish plants to which these two names are applied. The name *G. riojense* was first mentioned by Fric in his 1926 catalogue but not validly published until 1991 when H. Till & W. Till resurrected it for forms of *G. bodenbenderianum* from north of La Rioja and just into Catamarca. They also described a number of poorly differentiated subspecies of *G. riojense* based more on locations than physical characters. Among these was subsp. *piltziorum* which they based on a new type because they considered the original Schütz naming of the species to be invalid since no type was deposited.

The name *G. stellatum* has also been applied, at least in part, to these plants as explained by Papsch (1989) and it would be an earlier name than *G. bodenbenderianum*. The application is certainly not clear from Spegazzini's original description, and here it is taken as a later synonym of *G. quehlianum*.

Till and Neuhuber (1992) claim that plants resembling the illustration of *G. bodenbenderianum* in Hosseus (1939) *Notas Sobre Cactaceas Argentinas* can be found in sandy loam between the Sierras of Abajo and Ulapes in the south of La Rioja. They maintain this as a different species from what they call *G. riojense* from further north.

Fig.470 *G. bodenbenderianum,* GC404.02 from south of Los Colorados, Prov. la Rioja, Argentina

Fig.471 *G. bodenbenderianum* (occultum), *Piltz* 131 from Tres Puentes, Prov. Catamarca, argentina

Fig.472 *G. bodenbenderianum,* GC402.02 from west of Patquia, Prov. la Rioja, Argentina

Fig.473 *G. bodenbenderianum* (piltziorum), *Piltz* 38 from the Sierra. de Velasco, Prov. La Rioja, Argentina

Cultivation

This is a very popular species with enthusiasts, particularly for showing, because of its slow growth and unusual appearance. Plants remain quite flat and usually have a brown epidermis, making them quite easy to identify. Plants take many years to reach a diameter of 10cm diameter, making it among the slowest species in the genus.

Illustrations

New Cactus Lexicon Atlas (2006) Illn. 281.5-6

Fig.474 *G. bodenbenderianum* (paucispinum), STO 120.1 from Telaritos, Prov. Catamarca, Argentina

Fig.475 *G. bodenbenderianum*, *Piltz* 31 from Sierra de Velasco, Prov. La Rioja, Argentina. Photographed in the collection of Elisabeth Charles

Fig.476 *G. bodenbenderianum,* (triacanthum) *Piltz* 124 from the Sierra Ancasti, Prov Catamarca, Argentina

Gymnocalycium ochoterenae Backeberg

Fig.477 Habitat of *G. ochoterenae,* GC937.02 near Quines, Prov. San Luis, Argentina

First Description (Backeberg)

Kaktus-ABC, Backeberg & Knuth: pp. 293, 417 (1936)

Gymnocalycium Ochoterenai Backbg. sp.n.

Flattened, sunken in the crown, naked or with a little opaque wool-felt, olive green to grey (tubercles brownish). Ribs, c. 16, at first c. 5mm wide, later becoming wider and flatter, tubercles above the areoles are short tooth-like, below the areoles, blunted keel shape.

Areoles almost round, covered with short, protruding white or yellowish wool-felt. Radial spines, 3-5, grey or whitish-yellow, with dark tips, 1 spine porrect, 2 obliquely recurved. Flowers c. 35mm long, white, a little reddish in the throat, tube short.

Note: The spelling has been corrected to *'ochoterenae'* in accordance with Art. 60.11 of the Botanical Code.

Etymology

Perhaps named for Prof. Isaac Ochoterena (1885-1950), a Mexican botanist after whom other cacti were named.

Synonyms

G. intertextum Backeberg ex H. Till in KuaS 38(8): p.191 (1987)

First description of *Gymnocalycium intertextum* Backeberg. nom. inval. (Art. 8.2) in the *Kakteenlexicon*: p.168 (1966)

Body simple, semi-globular, plants up to 11cm diameter, have been observed, grey-green. Ribs c. 13-15, c. 1.5-2cm wide, with transverse furrows with chin-like tubercles above the furrows. Spines irregularly bent, entwined ± claw-like, partially compressed, when dry a pinkish-grey, dirty grey or horn colour, rather strong, not thickened at the base. Radial spines 5-7, up to 2.5cm long, central spines 0, sometimes 1, curved upwards, not stronger than the radial spines, occasionally the spines are up to c. 2mm thick at the base. Flower unknown. when wet the spines are reddish-brown.

Habitat: Northern Argentina. In the collection of Uhlig and Backeberg U2176, from a Fechser collection.

Fig.478 The illustration of *Gymnocalycium ochoterenai* published with the first description in *Kaktus-ABC* (1936)

Fig.479 *G. ochoterenae,* GC937.02 in habitat near Quines, Prov. San Luis, Argentina

Fig.480 *G. ochoterenae,* GC937.02 in habitat near Quines, Prov. San Luis, Argentina

Gymnocalycium ochoterenae subsp *herbsthoferianum* H. Till & Neuhuber

Gymnocalycium ochoterenae subsp *vatteri* (Buining) W. Papsch

Gymnocalycium bodenbenderianum subsp. *intertextum* (Backeberg ex H. Till) H. Till

G. vatteri Buining in *Succulenta* 29(5): pp.65-67 (1950)

Fig.481 *G. ochoterenae* (moserianum), GC947.02 in habitat north of Salsacate, Prov. Córdoba, Argentina

Map 63 Distribution of *G. ochoterenae* in Prov. Córdoba & San Luis, Argentina

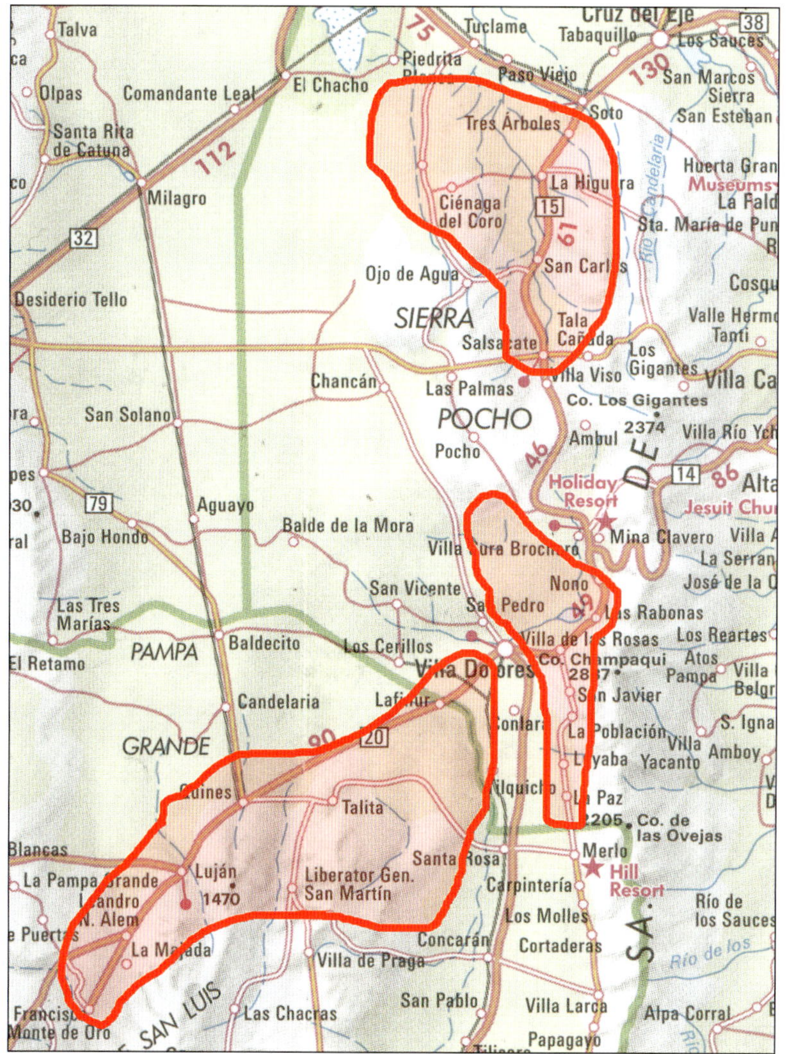

Fig.482 *G. ochoterenae,* GC950.03 in habitat at Agua de Ramón, Prov. Córdoba, Argentina

First description of *Gymnocalycium vatteri* Buining spec. nov.

Plant flattened-globular, dull olive green, 4cm high, 9cm wide. Ribs 11, 2-2.5cm wide at the base and 3cm apart, 10-12mm high, divided into c. 5 pronounced tubercles by cross-grooves, almost conical. Sunken areoles sit on top of the tubercles, 2cm apart, 5mm wide, covered with short dirty grey wool. Spines, only 1 strong spine (occasionally 2) coming from each areole which curves over the tubercle like a goat's horn, flat, thickened at the base, c. 2cm long, 2mm thick. Flowers 5cm long, 4cm diameter, flower tube 3.5cm long, sepals spatulate, white with a green central stripe, inner petals white, 2cm long, 0.8mm wide, lanceolate, tapering.

Fig.483 *G. ochoterenae* (vatteri), GC944.01 in habitat north of Altautina, Prov. Córdoba, Argentina

Fig.484 *G. ochoterenae* (vatteri), GC10.01 in habitat at Las Rojas, Prov. Córdoba, Argentina

Fruit 3cm long, 1cm wide, bluish, splitting longitudinally. Seed cap-shaped.

Habitat: This remarkable plant was found by Mr. Vatter, who discovered this plant in the neighbourhood of the small village of Nono, in the Sierra Grande, Córdoba, 800-1000m. Argentina.

Fig.485 *G. ochoterenae* (vatteri), GC944.01 in habitat north of Altautina, Prov. Córdoba, Argentina

Distribution (Map 63 on page 215)

The location was not stated precisely in the original description, just 'northern Argentina', but Backeberg (1959) later speculated that it might be La Rioja, now considered not to be correct.

It is now accepted that this is the most southerly species of the subgenus *Trichomosemineum*, with the typical form growing in the north of the Sierra San Luis where plants resembling all of Backeberg's varieties can be found.

It is widespread in the west of Prov. Córdoba and in the Sierra San Luis in Prov. San Luis.

Conservation Status

Least concern. It is a widespread taxon, preferring rocky localities where is reasonably protected from the detrimental effects of grazing.

History

First described as *Gymnocalycium Ochoterenai* in *Kaktus-ABC* (1936), the spelling later corrected to '*ochoterenae*' in accordance with Art. 60.11 of the Botanical Code. The description lacked detail, but was accompanied by a picture of a flowering plant

Fig.486 *G. ochoterenae* (vatteri), GC10.01 from Las Rojas, Prov. Córdoba, Argentina

Fig.487 *G. ochoterenae* (vatteri), *Piltz* 108 from Las Rabones, Prov. Córdoba, Argentina

Fig.488 *G. ochoterenae,* GC11.02 from north of San Martin, Prov. San Luis, Argentina

Fig.489 *G. ochoterenae* (scoparium), *Neuhuber* 88 98-244 from Lujan, Prov. San Luis, Argentina

which is the best indication of what the name should be applied to.

Backeberg (1959) suggested that his var. *cinereum*, described at the same time, may be the same as *G. bodenbenderianum*. He differentiated this variety by its more ash-grey body, less-wavy, broader ribs and more, almost entwined black spines. He explains that he found the plants in a consignment of freshly collected plants at Stümer's nursery in Buenos Aires without knowledge of their origin.

Backeberg (1966) subsequently invalidly named a number of varieties of very different appearance, saying that it showed how variable the species could be. Again, there was no indication of the provenance of these plants so we do not know if they all came from a single population. Backeberg's statement that his descriptions were based on single specimens clearly does not take into account the variability of the species.

H. Till & Neuhuber (1992a) provide a more detailed description of *G. ochoterenae* including the variability they have seen in wild populations. They observe that the form Backeberg called var. *cinereum* can be found towards the west near to the town of San Francisco del Monte de Oro. A new variety, var. *scoparium* and the new subsp. *hersthoferianum* were described for minor variants from nearby localities near Luján.

Commentary

This name has often been mis-applied to forms of *G. bodenbenderianum,* for instance *Piltz* 31 from the Sierra Velasco, a form that has been widely grown from commercial seed.

G. ochoterenae is here accepted as including *G. vatteri* which occurs further to the north-east and *G. intertextum* from even further north. All these plants look alike and have similar flowers. The extreme form of *G. vatteri* with a single central spine is a selection from the population where plants with more spines are usually in the majority, particularly on young plants. Note that Backeberg's illustration Abb. 1659 in *Die Cactaceae* III (1959) is

Fig.490 *G. ochoterenae* (intertextum), *Neuhuber* 91 375-1253 from Los Mogotes, Prov. Córdoba, Argentina

not *G. vatteri* as captioned but probably a form of *G. bodenbenderianum*.

The populations of *G . intertextum* include plants with short, close lying spines known as *G. moserianum*, a name invalidly published by Schütz in 1966 (no type preserved) and the subsp. *intermedium* found in the south of the central distribution area.

Cultivation

It is the *G. vatteri* form of this species that is most frequently seen in culture, particularly the selection with a single central spine that has a very distinctive and pleasing appearance. Many plants in collections labelled as *G. ochoterenae* are actually forms of *G. bodenbenderianum*. The majority of provenanced plants in collections probably originate from Piltz seed.

It is very easy to grow although not particularly quick. The white flowers are freely produced after about three years from seed.

Illustrations

New Cactus Lexicon Atlas (2006) Illn. 282.2-4

Gymnocalycium quehlianum (F. Haage ex Quehl) Vaupel ex Hosseus

First Description (Quehl)

Monatsschrift für Kakteenkunde 9: pp. 43-44 (1899)

Echinocactus quehlianus Hge. jun., a novelty from the Argentinian Andes which permits me to discuss more below.

This plant belongs to X. subgenus *Hybocactus* K. Schumann.

A) Ribs with chin-like humps. Ovary bare, with pointed, red-brown scales.

B) Ribs very distinct, resolved into humps; body reddish-grey with humps set close together.

Referred to under number 97a. [in fact 98b] in *Gesamtbeschreibung der Kakteen* by Herrn Professor Schumann.

Body flattened-globular, simple, c. 7cm. diameter, 3.5cm high, reddish-grey, (the colour of ripe grapes of the so-called Rhine wine). Crown sunken, almost naked, with a few close lying spines. Roots thick, turnip-like, similar to Ariocarpus. Ribs 11, straight, divided into humps.

Humps 12-15mm high, 10-12mm diameter at the base, younger humps thinner, because of the narrow form, more or less sharp angled, later wider and flatter, and with the areole sunken on top of the humps, chin-like below the areoles, without cross-cuts. (In contrast to *Ects. Odieri* Lem.)

Areoles round, with white hair until the flower develops, later the hair is missing.

Radial spines 5, spreading out, lying close to the body, the pairs in the middle of the areoles pointing sideways are the longest, 5cm long, the bottom spine, (the shortest) pointing straight down, (appearing in young plants, disappearing with age), between these 3 spines a further spine each side, therefore the upper part of the areole is without spines, all spines stiff, wine red at the base, horn coloured towards the tip, translucent. Central spines missing.

Many flowers around the crown, the plants at hand are in bud.

The following year Quehl (1900) published a photograph with the following text:

Echinocactus quehlianus Ferd Haage jr.

In the continuing tropical heat of July, the plant described on page 43, of *Monatsschrift für Kakteenkunde* Vol. IX flowered, which I illustrate and describe below.

Flowers, many, near the crown, the whole length 6-7cm long. Ovary slender, grey-green, smooth, covered with semi-circular somewhat pointed,

Fig.491 Illustration of *Echinocactus Quehlianus* published in MfK by L. Quehl in 1900

Fig.492 Illustration of *Echinocactus Quehlianus* published in a Haage nursery catalogue from about 1912

Fig.493 Illustration of *Echinocactus Quehlianus* published in the Haage nursery catalogue of 1930

reddish-white scales. Tube differs from ovary only by its larger diameter. Flowers funnel-shaped, covered with reddish-white scales, set like tiles on a roof, that are curved inwards at the tip.

Outer petals lanceolate, pointed, white with a grey central stripe on the outside, inner petals set in three rows, lanceolate, frayed, the innermost relatively smaller and snow white, in the throat the petals and filaments are deep wine red, the rest of the stamens are white and of uneven length, the outer stamens longer than the inner ones. Anthers ochre yellow, style white, stigma lobes ochre yellow, not spreading over the anthers. Anthers surround the deep lying stigma, hardly standing out above the fully opened flower. Flowers open from 1-2 hours each day for a few days in the strongest afternoon sun, (c. 3-5 pm.)

Neotype: R. Kiesling and O.Ferrari 8711 (SI) from Argentina, Córdoba, Dept. Ischilín, between Deán Funes and Cruz del Eje on the secondary road, XI-1995. Designated by Metzing, D.; Kiesling, R. & Meregalli, M. in Hickenia 3(3): pp.9-11.

Etymology

Named in honour of Postmaster Leopold Quehl, cactus collector and founder member of the D.K.G. (1849-1922).

Other Names

Echinocactus platensis var. *quehlianus* (F. Haage ex Quehl) Spegazzini

Synonym

Gymnocalycium stellatum Speg. in *Nuevas Notas Cactologicas - Anales de la Sociedad Cientifica de Argentina*, 99: pp.142-143 (1925)

Fig.494 *G. quehlianum*, GC941.04 on the east side of the Sierra Grande, Prov. Córdoba, Argentina

Previously illegitimately published as *Echinocactus stellatus* Spegazzini in *Cactacearum Platensium Tentamen* (1905)

First description of *Gymnocalycium stellatum* Speg.

This is a good species, mistakenly related to *Gymnocalycium platense* (Speg.), which Britton and Rose claim to be a synonym. In habitat the part of the body below ground, has the form of a cone, the upper, exposed part of the body appears slightly concave or flat. The rows of tubercles show the form of a true star in 7-11 slightly raised ribs. In the more humid atmosphere in cultivation the ribs are more rounded, the plant becoming more or less convex, reminding one, at times of *Gymnocalycium damsii* (K.Sch.). Flowers are 60-65mm high with a club-shaped ovary, 20mm long, 9mm diameter, the exterior of the perianth tube becoming moderately thin, perianth tube is widest at its centre, (15mm long). Inner petals tinted violet, external petals shorter and narrower, 26mm long, 7mm wide; innermost petals 28mm long, 8mm wide, all spatulate with a sharp point, white. Stamens are clearly arranged in 2 series, the upper series being separated from the lower series by a gap, 7-8mm wide, filaments white, anthers yellow; style white, with 12 stigma lobes, also white, reaching to the centre of the upper circle of anthers.

Habitat: Usually on the stony slopes in the Provinces of Córdoba, La Rioja and Catamarca, Argentina

Commentary on *G. stellatum*

This name has also been applied to two different taxa. It is treated here as a younger synonym of *G. quehlianum* but quite recently it was applied to populations of *G. bodenbenderianum* from Prov. La Rioja. The description and Spegazzini's illustration do not permit definitive identification.

G. obductum Piltz in *Succulenta* 69(4): pp.74-78 (1990)

First description of *Gymnocalycium obductum* Piltz sp. nov.

Body simple, matt, reddish to greyish-brown, flattened-globular with a distinct turnip-like root, 5.5-7cm wide, 6mm high, 2cm above the surface of the soil, crown sunken with yellowish to whitish felt. Ribs 13-15, very weakly elevated, separated by shallow grooves. Areoles round, at first white felted, later bald, sunken, 5-7mm apart, on the upper side of the chin-like humps, below small cross-cuts. Spines, 3-5, lying close to the body or slightly porrect, 2-5mm long, light brown, dark brown to black at the foot, all radials.

Flowers 3-4cm long, 3.5cm wide, coming from the crown funnel shaped, a dirty pinkish-white. Ovary dark olive, bloomed, 11mm long, 6mm wide, scales a delicate pink, up to 3mm long and wide. nectar chamber 3mm long, lilac. Flower tube 9mm long, 8mm wide, scales 10mm long, 4mm wide, dark olive with pink edges. Stamens up to 7mm long, yellowish-white, primaries up to 4mm long, leaning onto the style. Anthers yellow. Style deep pink at the base,

Photo: Geoff Bailey

Fig.495 Habitat of *G. quehlianum, Bailey* s.n.near Villa Carlos Paz, Prov. Córdoba, Argentina

yellowish above, 8mm long. Stigma lobes 5mm. Outer petals 1.6mm long, 3mm wide, recurved, a dirty white with wide olive-green central stripes, lanceolate, inner petals up to 1.3cm long, 1.5-2.5mm wide, with a point, dull white with drab pink central stripes.

Fruit up to 1.8cm long, with persistent flower remains, 9-13mm diameter, blue-green, bloomed, grey-green to darker, unbloomed, almost globular, slightly elongated. Seed smooth, shiny brown, 1.3mm long, hilum micropyle area sunken, with a spongy border margin. Belonging to the series Quehliana Buxbaum.

Habitat: Argentina, Province Córdoba, east of the Salinas Grandes c. 300m. The exact location and type placed in the Herbarium of the University of Utrecht under the No. P 121/1.

Commentary on *G. obductum*

Schweitzer (1994) reports finding a population of similar, but larger plants further south, to the east of the Salinas Grandes at 210m.

Distribution (Map 64)

The main distribution area is in Prov. Córdoba, Argentina, centred on the Sierra Chica. Many reported localities, for instance those in La Rioja, are probably not this taxon, but in fact *G. bodenbenderianum*.

Fig.496 *G. quehlianum* (obductum), *Piltz* 121 from Salinas Grandes, Prov. Córdoba, Argentina

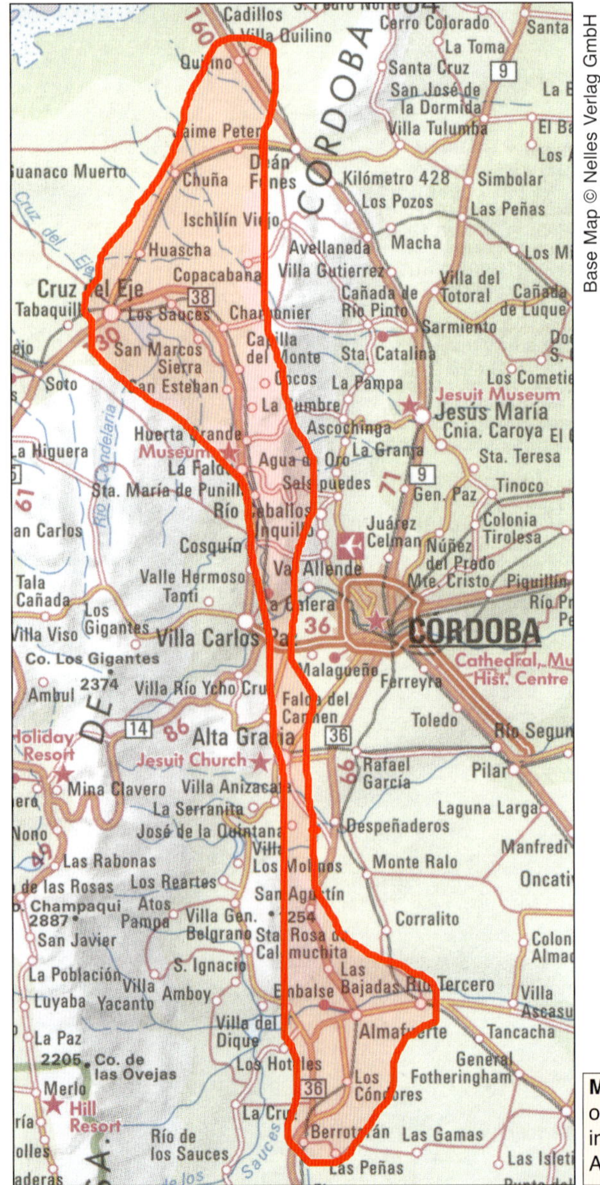

Map 64 Distribution of *G. quehlianum* in Prov. Córdoba, Argentina

Fig.497 *G. quehlianum,* GC995.03 in habitat south of Quilino, Prov. Córdoba, Argentina

Fig.498 *G. quehlianum,* GC940.02 in habitat north of Carlos Paz, Prov. Córdoba, Argentina

Conservation Status

Least Concern. This taxon is widespread and usually abundant. It often grows under bushes and small trees where colonies of all ages can be found. Sometimes it can be seen on exposed rocky places and can colonise road cuttings in great numbers where there is little competition from other plants. The northern form, from near Quilino, can become larger, up to 10cm in diameter in its gravelly habitat, where it often grows under bushes (Fig.499).

History

Echinocactus quehlianus first appeared as a name in the Haage catalogue of 1899 without an illustration, but with the text "New, rare, beautiful South American species". Later the same year, a description was published by Quehl in MfK but without information about the flower or an illustration. The following year, Quehl added to his description by publishing details of the flowers and a photograph of a plant said to represent the type specimen. However, under the rules of the ICBN, these additions cannot be considered as part of the protologue, and furthermore, various authors have pointed out contradictions between the picture and the first description.

Schumann, in his *Nachträge* (1903), gives us a detailed description that better matches Quehl's picture which he reprints. However, as Till suggests

Fig.499 Habitat of *G. quehlianum,* GC995.03 south of Quilino, Prov. Córdoba, Argentina

Fig.500 *G. quehlianum, Neuhuber* 90 221-644 from Villa Serranita, Prov. Córdoba, Argentina

Fig.501 *G. quehlianum, Piltz* 160 from Cruz del Eje - La Falda, Prov. Córdoba, Argentina

Fig.502 *G. quehlianum, Neuhuber* 91 331-1129 from Steinbruch, Prov. Córdoba, Argentina

Fig.503 *G. quehlianum, Papsch* 89 94-125B from Capilla del Monte, Prov. Córdoba, Argentina

(1993), this description may contain elements from other taxa. Schumann goes on to explain:

> "For the knowledge of this plant I must thank Herrn Cand. phil. DAMS, who observed and showed this plant in and around Berlin, and many other places. Herr QUEHL was kind enough to send me his original example for comparison, perhaps it originated from the area of the Sierra de Córdoba, from where, until now, we have received very few cacti".

The 1900 picture bears a striking resemblance to plants from the north of Dept. Córdoba which belong to the subgenus *Trichomosemineum* whilst other old illustrations match the first description rather better and are reminiscent of plants from subgenus *Gymnocalycium*.

The text by Gürke, accompanying plate 105 in *Blühende Kakteen* (1908) states that its habitat includes Paraguay from where Prof. Anisits sent it. Dölz (1957) tells us that this was a mistake by Schumann, who confused it with *E. damsii* which he described in 1903.

Suggestions that it is identical to *Echinocactus platense* Spegazzinii were refuted by Quehl (MfK 1911). Combination of the name in *Gymnocalycium* has been widely attributed to Berger (1929) but is now believed to have been done earlier by Hosseus (1926).

In 1935, Schütz published his system for classifying *Gymnocalycium* on the based on their seed structure. He designated *G. quehlianum* as the type of his subgenus *Trichomosemineum*. Similarly, Buxbaum (1968) designated *G. quehlianum* as the type of his series *Quehliana*.

Till (1993) wrote a comprehensive review of the history of *G. quehlianum* where he explained his opinion that the application of the name to a *Trichomosemineum* plant has been a long standing mis-application of the name. He maintains that it really is the oldest name for a plant he illustrates from the vicinity of Quilino in northern Córdoba. His illustrations 2 to 5 are probably *G. robustum*, a species belonging to the subgenus *Gymnocalycium* and not described until 2002. I find his arguments unconvincing and certainly not strong enough to overturn the long-standing usage of the name.

Metzing et al. (1999) endeavoured to fix the application of the name in the way it is usually applied by designating a neotype for *Gymnocalycium quehlianum* with a plant from the *Trichomosemineum* subgenus. In response, Till (2002) rejected this move, again trying to justify the opinion he expressed in his 1993 article. To further support his view, he published a photograph on page 441 of *G. reductum* in cultivation (identified by its provisional name *G. cordobensis*) which does indeed resemble the 1900 Quehl picture.

Fig.504 *G. quehlianum, Neuhuber* 91 336-1140 from Quilino, Prov. Córdoba, Argentina

However, on page 444, he publishes two habitat pictures of plants near Quilino which he identifies as *G. quehlianum*. My opinion is that these are most likely to be *Trichomosemineum* plants which are common in the area, and not the *Gymnocalycium* seed group plant which I believe is much less widespread. I have seen the two species growing side by side near Quilino, when the differences between them were more obvious (Fig.228).

Commentary

G. quehlianum has had a confused history. Following Kreuzinger (1935), and for most of the 20th century, the name was applied by many authors to plants from Córdoba, Argentina, which belong to the subgenus *Trichomosemineum* Schütz (1969) for which it is the type species. Some authors considered it to be a different species from the more recently named *G. stellatum,* although the quoted differences were trivial. As a result, very similar looking plants in cultivation were seen with both these names.

The situation was greatly complicated in 1993, when Till and others suggested that the original description referred to a similar-looking plant of the subgenus *Gymnocalycium,* for which it would then be the oldest name. Since the original description does not tell us the seed type, it is a question which can never be absolutely settled.

It is possible that the importation of plants from Córdoba to the Haage nursery prior to 1899 included plants which today we would consider belong to more than one species. The description of *E. quehlianus* could contain elements from more than one species and the 1900 picture may well be a *Trichomosemineum* plant, while other illustrations, such as the engraving used in various later Haage catalogues, could belong to subgenus *Gymnocalycium* (Fig.492).

The recently described *Gymnocalycium robustum* Kiesling, Ferrari & Metzing (2002) belongs to subgenus *Gymnocalycium* and Till (2002) has suggested that it is a redescription of the original *G. quehlianum*. It certainly has the

Fig.505 *G. quehlianum, Neuhuber* 91 324-1100 from Santa Rosa C, Prov. Córdoba, Argentina

prominent humps suggested in the first description of *E. quehlianus*. However, it has a limited distribution near to the town of Quilino in the north of Dept. Córdoba, whereas plants belonging to subgenus *Trichomosemineum* are widespread and common from there to much further south, near to the city of Córdoba and beyond. Early explorers would therefore be much more likely to have collected this latter plant. A much more widespread candidate from the *Gymnocalycium* seed group to have been collected at the time is the plant we now know as *G. capillaense,* since this also has a wide distribution area in northern Córdoba.

Because of the confused history of the name *G. quehlianum* and the lack of a type, the editors of the *New Cactus Lexicon* (2006) decided to abandon the name and use *G. stellatum* as the name for the *Trichomosemineum* plant, and the similar-looking *G. robustum* for the species belonging to the subgenus *Gymnocalycium* (with an incorrect illustration 273.3 which is actually *G. quehlianum).*

Although I was a party to that decision, my recent fieldwork in Córdoba has persuaded me to accept *G. quehlianum* as the oldest name for the widespread plants with *Trichomosemineum* seeds. This view is in accordance with that of Metzing et al. (1999), who endeavoured to fix this application of the name by designating a neotype for *G. quehlianum* with a plant from the subgenus *Trichomosemineum*.

Cultivation

Although the plant label often reads *G. stellatum*, this species is widespread in culture. It is easy to grow and flowers easily when small. It is important to give the plant adequate water and even a little in the resting period to prevent excessive shrivelling.

Gymnocalycium ragonesei Castellanos

First Description (Castellanos)

Dos especies nuevas de Cactaceas Argentinas - *Lilloa* 23: pp.5-13 (1950)

Gymnocalycium Ragonesei Castell. sp. nov.

Plant small, flattened-globular, disc-like, slightly sunken in the centre of the crown, ±3-5cm diameter, 1.5-2.5cm high, brick red to greenish-grey. Ribs 10 extremely flat, ±1.5mm high, radiating, ±3mm wide at the crown, 12mm at the periphery, fine cross-cuts form scarcely distinguishable humps. Areoles round, ±1mm diameter, sitting over the small humps, humps ±3mm apart, with wool-felt. Spines 6, spider-like, whitish, all radials, 3mm long, without centrals.

Flowers 2-3 from the crown, ±3.5-4cm high, somewhat greyish; ovary, inverted cone-shape, ±21mm long, 6mm diameter, at the upper part, with half moon shaped scales, that have paler edges, ovary interior ±11mm long, 3mm diameter; perianth tube slightly funnel shape, interior ±12mm long, 7mm diameter, with a tint of wine colour at the base; perianth calyx with 2 rows of petals, the outer row thick, the back having the same colour as the outside of the flower, edges translucent, the inner row longer and more delicate; perianth corolla white, with 2 rows of petals, elongated tongue shaped, tapering to a long point, the outer rows ±16mm long, 4mm wide, inner petals smaller. Stamens, many, filaments greenish, anthers yellow, in 2 series, the lower series inserted at the base of the perianth tube around the style, reaching to the middle of the style, the upper series inserted in the wall of the tube up to the throat of the flower. Style, ±10mm long, 1.5mm diameter, greenish-yellow, with 8 radiating stigma lobes, 4mm long. Fruit and seeds unknown.

Fig.506 *G. ragonesei, Neuhuber* 91 422-1367, Salinas Grandes, Córdoba

Fig.507 *G. ragonesei*, photographed in the collection of Elisabeth Charles

Type: *Ragonese* collected 13 Dec.1949, Argentina, Prov. Catamarca, Salinas Grandes, between Km969 and Totoralejos (LIL 16.120).

Etymology

Named for Arturo Ragonese, an Argentinian botanist.

Fig.508 Habitat of *G. ragonesei* in the Salinas Grandes near the border of Prov. Catamarca with Córdoba, Argentina

Distribution (Map 65)

It appears that the distribution area of this species is very limited, being confined to a place at the edge of a salt lake, the Salinas Grandes, in the south of Province Catamarca near to the border with Province Córdoba. Only a few people have found the plant there since its description, some of those searching unsuccessfully have concluded that it no longer exists in habitat. However, the images on this page were taken by Franz Kühhas in January 2009, so they are still there, it is just a matter of finding them!

Conservation Status

Data Deficient. Although not suffering from any obvious threats, its limited distribution makes it vulnerable to the change in land use which is starting to happen.

History

This species was discovered in 1949 by Arturo Ragonese who then gave plants to Dr. Castellanos, the publisher of its first description the year after.

Commentary

Van Vliet (1971) described his visit to the habitat in January when it was very hot and had been dry. It was not until a sudden thunder storm washed away the surface layer of soil that the plants were revealed. Normally, the crowns of these small plants are level with the surface, even shrinking further down in dry periods, so leaving a small depression which fills with dust and hides the plants. Perhaps this, and the colour of the body, are the reasons that so many enthusiasts have failed to find this elusive plant.

Cultivation

G. ragonesei is very popular in cultivation perhaps because it is one of the smallest Gymnocaly-ciums and presents some challenges to grow well. It easily dehydrates causing the body to become mis-shapen. However, it is very free flowering and grows easily, if slowly, from seed.

A well grown specimen should do well in a show class for *Gymnocalycium* with a small pot size restriction. Like some other miniatures, it appears to be short-lived in culture, and may need replacing with a young seedling after it has flourished for several years.

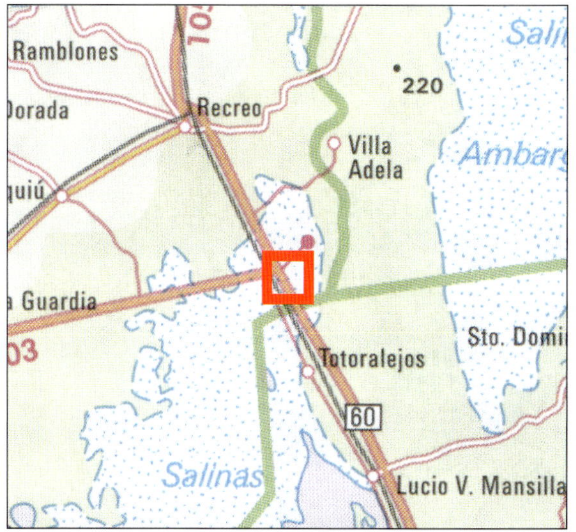

Base Map © Nelles Verlag GmbH

Map 65 Distribution of *G. ragonesei* near the border of Prov. Catamarca and Prov. Córdoba, Argentina

Photo: Franz Kühhas

Fig.509 *G. ragonesei* in habitat at the Salinas Grandes, Prov. Catamarca, Argentina

Photo: Detlev Metzing

Fig.510 *G. ragonesei, Metzing s.n.* in habitat after rain at the Salinas Grandes, Prov. Catamarca, Argentina

Photo: Franz Kühhas

Fig.511 *G. ragonesei* in habitat during the dry season at the Salinas Grandes, Prov. Catamarca, Argentina

Subgenus *Muscosemineum*

Type Species

G. mihanovichii (Fric & Gürke) Britton & Rose

Species and Subspecies **Page**

(synonyms indented)

Echinocactus Mihanovichii Fric̆ et Gürke.

Tafel 101.

Gymnocalycium anisitsii (Schumann) Britton & Rose **subsp. *anisitsii***

Photo: Ludwig Bercht

Fig.512 *G. anisitsii anisitsii, Bercht* 3085 in habitat near Concepción airport, Paraguay

Fig.513 The illustration with the first description. Plate 4 of *Echinocactus Anisitsii* from *Blühende Kakteen* (1900)

First Description (Schumann)

Echinocactus Anisitsii K. Sch. n.sp. - *Blühende Kakteen - Iconographia Cactacearum* 1: pl.4 (1900)

Echinocactus Anisitsii K. Sch. n. sp.

For the past few years, we have had a number of cacti from Paraguay that are near to the two known species, *E. denudatus* Lk.& Otto, and *E. multiflorus* Hook. (*E. Ourselianus* Monv.). They differ from the true *E. denudatus* Lk. & Otto by way of the greater number of ribs, the deeper notches in the body and the stiff outstanding spines. They have been named and described as varieties by Ferdinand Haage.

The 3 plants in front of me are so very different that I see them as a separate species. I thank Herr Prof. Dr. Anisits of Asuncion University, who brought them back from a tour in the northern part of Paraguay and sent them to me. Since this species has not yet been published, therefore I give the following description.

The young plants are globular, tapering to the root, with 8 slightly protruding ribs. The adult plants are short cylindrical, 5.5-8cm high, 7.5-10cm diameter, leaf green, rounded on top, crown slightly sunken, weakly tubercled at the apex but with no wool nor spines. Ribs 11, 1.5-2cm high, tapering to the base where they are widely spreading, separated into narrow humps by horizontal notches, the humps have a chin-like appearance, usually angled at the base. Areoles 1-2cm apart, elliptical, 4-5mm long, somewhat elongated above the spine clusters. Spines 5-7, radial and central spines are not clearly separable, the uppermost spine is usually the longest, 2.5cm long, spines somewhat angular, hardly thickened at the base, always curved, dull white.

Flowers 4cm long. Ovary narrow cylindrical, 1.5cm long, hardly 5mm diameter, scales broadly elliptic, obtuse, rounded, flower tube also has scales. Outer petals pointed egg-shape, inner petals narrowly spatulate, not dentate. Stamens included by half the length of the inner petals. Stigma lobes 7. The red berry is spindle-shaped, 2.5cm long, 1cm diameter, also with scales. Seeds numerous, very small, hardly 1mm long, almost globular, (resembling those of the Bladder Nut, *Staphylea pinnata*). Light brown, with rather dense minute tubercles.

Photo: Volker Schädlich

Fig.514 *Gymnocalycium anisitsii anisitsii, Schädlich* 28 from Concepcion, Paraguay

Photo: Volker Schädlich

Fig.515 *G. anisitsii, Schädlich* 290 from Concepción, Paraguay

Photo: Volker Schädlich

Fig.516 *G. anisitsii, Schädlich* 283 from east of Puerto Murtinho, Mato Grosso, Brazil

Paraguay; By the river Tagatiya-mi. (Anisits cact. n. 21. flowering on January 25th 1898).

E. Anisitsii is at once conspicuous for its long, white, variously curved, angled spines; and the fresh light green colour of the plant. The relationship to *E. denudatus* Lk. & Otto is quite evident in the flowers. The plant does well in cultivation and blooms profusely in the Royal Botanical Gardens and sets seed easily. This beautiful species will no doubt prove to be popular.

Base Map © Nelles Verlag GmbH

Map 66 The original location of *G. anisitsii* in Paraguay is shown with a red square. Plants of a similar affinity have been found by Schädlich in Brazil (blue square)

Etymology

Named for Prof. Dániel Anisits (1856-1911), a Hungarian pharmacist who worked in Asunción, Paraguay and sent cactus plants to Schumann.

Distribution (Map 66)

The limited distribution of the type subspecies is in the valley of the River Paraguay near to Concepción in Paraguay, the type locality being about 50km north of the town. More recent reports (Till & Amerhauser 2003a) state that the habitat extends into Brazil on the north side of the River Apa with its most northerly limit at Porto Murtinho (Till & Amerhauser 2003b) where its distinction from subsp. *damsii* is unclear.

The two varieties, var *griseopallidum* and var. *pseudomalacocarpus* are excluded since here they are treated as synonyms of *G. anisitsii* subsp. *damsii* and *G. marsoneri* subsp. *megatae* respectively

Conservation Status

Vulnerable. The limited distribution area is increasingly under threat from human activities and the plant is becoming difficult to find near the River Paraguay.

Photo: Volker Schädlich

Fig.517 *G. aff. anisitsii, Schädlich* 23 in habitat south-east of Puerto Murtinho, Mato Grosso, Brazil

Fig.518 *G. anisitsii,* from near Concepción, Paraguay

Fig.520 The plant described by Till and Amerhauser as *G. anisitsii* var. *griseopallidum* from near Cerro Miguel, Dept. Santa Cruz, Bolivia *Schädlich* 258

History

Plants of this species were sent by Prof. Anisits of Asunción University to Schumann, who realised that they were different from other Paraguayan taxa known at the time. He published the description to accompany Plate 4 of his *Blühende Kakteen - Iconographia Cactacearum* published in 1900.

There followed much debate about whether *Echinocactus damsii* was different from *E. anisitsii* culminating in a very long explanation by Bödeker in MfK 31 (1921). He tells us that Anisits had travelled northwards up the Río Paraguay, passing the Río Apa on his way to Puerto Olympo. He found the plants on both sides of the river as Hassler had stated.

Fric went even further north and, in about 1905 when he was 23 years old, he sent a consignment to De Laet of more than 1000 plants including *E. mihanovichii* and *E. anisitsii* which were stolen on route. Also in the consignment were forms of *E. damsii* with white and pink flowers, the latter identified by Bödeker with his *E. joossensianum*. Confusion over these plants continued and plants circulated as *G. damsii*, probably originating from Fric, did not match the Schumann description.

Two other populations of plants of this relationship from Prov. Santa Cruz, Bolivia were named invalidly by Backeberg. The first, *G. griseopallidum,* he described in the *Kakteenlexikon* (1966) from plants imported by Uhlig in 1964. The name was validated by Till & Amerhauser (2003b) as a variety of *G. anisitsii* (Fig.520), but it is doubtful if the plant they chose is the same as Backeberg's.

Similarly, the second from further east, Backeberg called *G. pseudo-malacocarpus* also in the *Kakteenlexikon*. These plants had been discovered by Father Hammerschmid and also sent to Uhlig in Germany. This invalid name has also been validated as a new variety of *G. anisitsii* by Till and Amerhauser (2003b). Donald (1971), suggested that it belonged to *G. marsoneri* but no combination under that species has been published. Here it is treated as a synonym of *G. marsoneri* subsp. *megatae* (see page 244) It is striking how similar Till's picture of this (Abb. 22, p.531. Till & Amerhauser 2003b) is to Schumann's original picture of *Echinocactus damsii* in the *Nachträge* (1903)!

Fig.519 *Schädlich* 15 in habitat at the Cerro Chovoreca in Paraguay, near the Bolivian border. A good match for the original *G. griseopallidum*

Fig.521 *G.* aff. *damsii, Schädlich* 286 in habitat north of Río Apa, Mato Grosso, Brazil

Fig.522 *G. anisitsii*, no data

Photo: Volker Schädlich

Fig.523 *G. anisitsii*, *Schädlich* 28 from Concepción, Paraguay

Commentary

Discussion about whether *Echinocactus anisitsii* and *E. damsii* were really different species was happening soon after they were described. The question was discussed at meetings of the DKG in the early 1900s (Till & Amerhauser 2003b) as a result of which it appears that they considered them to be just forms of the same species.

Following recent visits to the habitat, Till & Amerhauser (2003b) maintain that they are separate species and even place them in different series of their classification. They point to differences in the flowers to support their view, but the illustration of this difference makes no reference to the origins of the plants pictured. The suspicion has to be that the plants Schumann described were two different-looking individuals from populations of one variable species as believed by Moser (1969a). It is also possible that the freely offsetting, small-bodied plants seen in cultivation as *G. damsii* may be different from *G. anisitsii* but not the plant that Schumann described as *G. damsii*.

East of Porto Murtinho, in Brazil, a more caespitose form of *G. anisitsii* grows close to *G. marsoneri* subsp. *matoense* which is differentiated by having many more ribs.

Cultivation

Although it has been known for over 100 years, this plant is rare in cultivation. It probably needs more warmth than most to thrive, since it comes from a hot place near the border of Paraguay and Brazil. Plants with location data are particularly difficult to come by, but recent trips to the area may result in more seed becoming available soon.

Illustrations

New Cactus Lexicon Atlas (2006) Illn. 284.2

Photo: Volker Schädlich

Fig.524 *G. anisitsii*, *Schädlich* 290 from Concepción, Paraguay

Photo: Volker Schädlich

Fig.525 *G.* aff. *anisitsii*, *Schädlich* 23 from south-east of Poerto Murtinho, Mato Grosso, Brazil

Gymnocalycium anisitsii subsp. *damsii* (Schumann) Charles

Fig.526 The illustration of *Echinocactus Damsii* published in the Schumann *Gesamtbeschreibung der Kakteen - Nächträge* (1902) from a picture by Mr. Dams

Fig.528 *G. anisitsii damsii,* STO985, Santiago de Chiquitos near the road to Santo Corazón, Prov. Santa Cruz, Bolivia

First Description (Schumann)

Echinocactus Damsii Gesamtbeschreibung der Kakteen, Nachträge 1898-1902: pp.119-120 (1903)

Echinocactus Damsii K. Sch. n. spec.

Body simple, flattened globular, rounded or slightly sunken in the crown, with small humps. Spines coming from the youngest areoles. Ribs 10, later, a few more, straight, separated by sharp grooves, humps protruding chin-like below, in new growth light green, later dark green, sometimes with light green grooves below the areoles.

Areoles 1.5cm apart, elliptical, round when flowering, up to 3cm diameter. Spines, up to 8, not clearly recognisable as centrals or radials, awl shaped, round, up to 1.2cm long, in new growth white with a dark chestnut brown point, later becoming grey.

Flowers numerous, appearing near the crown behind the bundles of spines, the whole length being 6-6.5cm long. Ovary, in comparison, very slender and thin, light yellow-green, with egg shaped, pointed, light green, white edged scales, positioned like the tiles on a roof, without wool.

Flowers funnel shaped, wine red at the base, 5cm at the largest diameter, outer petals green, reddish on the outer edge, lanceolate, pointed; inner petals clear white, a little larger, fine, serrated at the top edge. Stamens in 2 groups, those coming from the ovary shorter than those coming from the wall of the receptacle, filaments white, anthers brownish; style standing above the stamens with 7 white radiating stigma lobes.

Berries small cylindrical, red, covered with scales, 2.5cm long, 6mm diameter. The seeds were sterile.

Type: Anisits s.n., N. Paraguay (B).

Etymology

Named for Erich Dams, secretary of the Deutsche Kakteen Geschelshaft about 1903.

Photo: Ludwig Bercht

Fig.527 *G. anisitsii damsii, Bercht* 3413 in habitat west of San José de Chiquitos, Dept. Santa Cruz, Bolivia

Photo: Ludwig Bercht

Fig.529 *G. anisitsii damsii, Bercht* 3405 in habitat east of Chochis, Dept. Santa Cruz, Bolivia

Synonyms

G. joossensianum (Bödeker) Britton & Rose in *The Cactaceae* 3: p.166 (1922). Basionym: *Echinocactus Joossensianus* Bodeker (1918)

First description of *Echinocactus Joossensianus* Böd. spec. nov. in MfK 28: pp.40-44 (1918)

Body simple, flattened-globular, slightly sunken in the crown, and humped, covered with spines. Ribs, 9-11, straight, sharp grooved, strongly enlarged around the areoles, above which are cross-cuts separating the humps that protrude chin-like under the areoles. Plants in youth, dark matt green on the top edges of the ribs, around the areoles more of a bronze colour, fully grown examples in habitat are a light leaf green in the grooves, matt dark leaf green on the humps. Areoles up to 2cm apart, elliptical, 5mm long, covered with white wool. Spines radiating, 6-7 on young examples, short, regularly positioned, on older plants up to 5cm high, up to 9 spines, irregularly positioned, 2-2.5cm long, the lowest nearly always the shortest, usually there is a sometimes shorter, sometimes longer central spine, all are awl-shaped, in cross section flattened, round, rough, mainly curved downwards, in new growth, yellowish-brown, later a dirty white with red-brown points, not thicker at the base.

Many flowers come from among the spines near the area of the crown, quite long, 4-5cm. The buds are a beautiful carmine-pink, up to 2.5cm long, less than 6cm diameter, conical, thicker at the top, shiny light green, with scales standing 1cm apart, set spirally, rounded with a point, whitish at the tip, the edges light wine red, without wool. Scales gradually resolving into outer petals. Outer petals are spatulate with a point, 1cm long, more than 4mm wide, somewhat fleshy, the lower half green, a light wine red towards the top, with whitish edges, petals further up become darker, the light wine red becomes dark wine red and the white edges become light wine red. Inner petals are longer, up to 2cm, narrow, straight lanceolate, the ends irregularly toothed or frayed, a beautiful lilac-pink. Many stamens, long, curved and lying over the stigma lobes, yellowish-white with very dark grey-black-brown anthers. Style thick, short, yellowish-white, with 6 fairly long, grooved lobes of the same colour. Fruit spindle-shaped, up to 2.5cm long, light green to reddish, with whitish, red-tipped scales. Seeds, according to De Laet light brown-yellow, helmet shaped.

Habitat: The same as the related species complex in Paraguay and Northern Argentina.

[Named for Herr Joossens, head gardener of De Laet.]

Commentary on *G. joossensianum*

Although usually synonymised with *G. anisitsii* subsp. *damsii*, the illustration which accompanies the first description is more like *G. schickendantzii* subsp. *delaetii,* particularly with regard to the position of the flowers and the shape of the fruits. It was included under that taxon in the *New Cactus Lexicon* (2006).

I have included it here because this is its usually accepted relationship, supported by the report that Bödeker is said to have identified pink flowered plants of *G. damsii* sent by Fric as this plant. Since no precise location was ever given, it is impossible to be certain what this name refers to, so it is best abandoned.

Gymnocalycium anisitsii subsp. *multiproliferum* (P. J. Braun) P. J. Braun & A. Hofacker in *Succulenta* 74(3): p.131 (1995) [First described as a variety of *G. damsii*]

Gymnocalycium damsii subsp. *evae* Halda, Horacek & Milt subsp. nov in *Acta musei Richnoviensis* 9(1): p. 58 (2002)

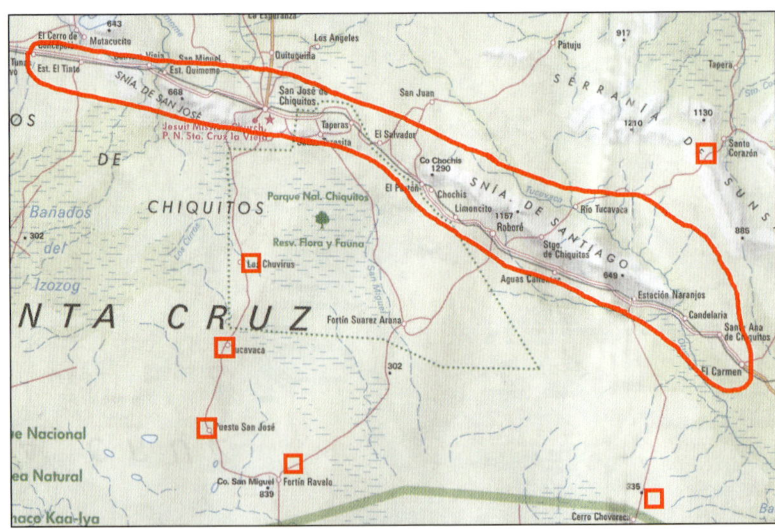

Map 67 Distribution of *G. anisitsii damsii* in Dept. Santa Cruz,, Bolivia. The form 'multiprolifer-um' occurs further east in Brazil

Commentary on *G. damsii* subsp. *evae*

This is a name for the pink flowering forms from Bolivia and may be a redescription of *G. joossensianum.*

Distribution (Map 67)

The stated distribution is north of that of *G. anisitsii* subsp. *anisitsii* and is said to overlap it near to the Río Apa junction with the Río Paraguai. Many locations of subsp. *damsii* are reported from the Serranía de Santiago and the Serranía de San José in Bolivia. The synonyms appear to have localised distributions and are shown as red squares on the map.

Till & Amerhauser (2004a) claim that the distribution is explained by floods which can occasionally be extreme, washing away the forest and distributing the plants down river.

Conservation Status

Least Concern. There are many recent reported localities from the known distribution area.

History

Described by Schumann from plants sent by Anisits from northern Paraguay, there have long been doubts about whether this is different from *G. anisitsii* or just a selection from a variable population (see notes on that taxon).

The first description and illustration in *Gesamtbescheibung der Kakteen, Nachträge* (1903) appear to be rather different from plants usually seen in cultivation under this name which probably originate from Bolivia.

Erich Dams (1904) wrote about the plant named after him, commenting on the thin flower tube which he says was mainly the reason for Schumann erecting it as a new species. He goes on to say that it was not possible to pollinate the flowers on his plants to produce good seed. Perhaps all his plants were propagations of the same clone.

Fig.530 *G. anisitsii damsii, Metzing* 98 from south of San José de Chiquitos, Prov. Santa Cruz, Bolivia. It is similar to the plant named *G. anisitsii* subsp. *volkeri.*

Fig.531 *G. anisitsii damsii* (multiproliferum), *Horst-Uebelmann* 557 from Porto Murtinho, Mato Grosso, Brazil

Later, the plant was illustrated with a fine plate in *Blühende Kakteen* Lief. 21, Tafel 83 (1906). I have a set of *Blühende Kakteen* in which there are some pencil annotations. On the text page of Tafel 83 *Echinocactus damsii* is written *Gymnocalycium damsii*. Occurs in Prov. Chiquitos Dep. Sta Cruz Bolivia. Few plants were sent by me to Europe in 1963. 21-VI-63 M.C.' which I take to be written by Martín Cárdenas. There is a report (Till & Amerhauser 2004a) that Hammerschmid supplied plants of this to Cárdenas in 1962.

Berger (1929) was of the opinion that *E. damsii* was a hybrid. Moser (1969a) tells us that he received a consignment of plants from his friend, A. M. Friedrich, which he had collected near Concepción. Among the contents were several plants that he says were typical of *G. damsii*, of different sizes and ages, uniform and corresponding accurately with the various published pictures. He concludes that *G. anisitsii* grows with *G. damsii* in the immediate neighbourhood of Concepción.

Backeberg invalidly published four poorly differentiated varieties of *Gymnocalycium damsii* in *Descriptiones Cactacearum Novarum* 3: p.6 (1963), all from near the Serranía de Santiago in Dept. Santa Cruz, Bolivia. All these were validated as varieties of either *G. damsii* or *G. anisitsii* by Till & Amerhauser (2004a) as part of a complicated scheme of names listed on page 560 of their article.

Commentary

I have found it impossible to decide the right taxonomic treatment for *G. damsii*. When I made it a subspecies of *G. anisitsii* (Charles 2005), I thought that was a reasonable approach, but more recent publications suggest that the situation is more complicated, for instance, an unsubstantiated report that they can grow together.

Till & Amerhauser (2004a) maintain that the two are different species, but give no clear way to tell them apart. Their differentiation by the flower tubes does not explain their attributions of various populations, judging by their own pictures. The many published subspecies and varieties are also not clearly differentiated, so I was tempted to treat all the populations as belonging to one variable species, *G. anisitsii*. However, in the interest of stability, I have decided to retain the subspecies rank for *G. damsii* in this book.

It is also clear that this species is related to *G. stenopleurum*. The picture (Abb. 19) on page 553 of Till & Amerhauser (2004a) shows the plant they call *G. damsii* var. *torulosum*. It looks remarkably like *G. stenopleurum*. My comments are based on the assumption that the illustrations in *Gymnocalycium* are accurately captioned. However, Abb. 18, also on page 553, is captioned "subsp. *evae* var *torulosum* mit rosa Blüte von San José. Foto: H. Till". The very same image later appeared on page 649 (Till & Amerhauser 2005), this time captioned " *G. friedrichii* Nachzuchten. Foto: H. Till" with the text telling us that the plant is the scarce, small, pink-flowering type-form of *G. friedrichii* [= *G. stenopleurum*] found by Till on the approaches to the Serranía de Charagua, which is 350km from San José!

I am left with the suspicion that *E anisitsii* and *E. damsii* were both collected in the vicinity of Concepción and represent extreme examples of a diverse population in that area. The plants we call *damsii* today, which come from a disjunct habitat area in Bolivia, may be a different taxon, but not *damsii* and so are without a name, or should it be *joossensianum*?

Cultivation

See the notes for *G. anisitsii* but with the added suggestion that the small-bodied caespitose forms often cultivated under this name are prone to dehydration if left dry for too long. Better-looking plants can be produced by removing at least some of the offsets while they are still small. It has been suggested by Till & Amerhauser (2003b) that the excessive offsetting of subsp. *multiproliferum* is caused by disease.

Gymnocalycium eurypleurum Plesnik ex Ritter

First Description (Plesnik)

Kaktusy 8(3): pp.70-72 (1972)

Gymnocalycium eurypleurum F. Ritter ex Plesnik sp. n. (nom. inval.). Validated by Ritter in *Kakteen in Südamerika* 1: pp.268, 340 (1979)

Body green, a little glossy, not offsetting, body 7cm diameter, with 8 ribs, continuous, straight, very low, rounded and wide. Areoles are 1.5-2cm apart, elliptical with scanty yellowish-white felt, soon becoming grey, and later fall off entirely. A shallow sharp groove between the areoles produces two opposed triangular facets. Radial spines, 5-6, 1 pointing straight down, the others pointing sideways in pairs, central spines, 0-1, all are acicular and slightly curved, 2.5-3cm long, at first the spines are brown, later horn coloured with a small dark tip, later projecting.

Fig.532 *G. eurypleurum, Bercht* 2215 from Cruce 4 de Mayo, Paraguay

Buds green, scales greenish with a light carmine tip, resolving into the outer petals. Petals faintly violet with a somewhat darker violet mid-stripe, tinted brown at the tip, lanceolate. Flower tube, c. 1cm long, light green, scales lighter with a lighter edge and a carmine tip. Inner petals faintly violet-pink, sometimes with a pink mid-stripe and a darker tip, lanceolate, there are two series of inner petals. The flower is 4.5-5cm diameter. Filaments white, anthers creamy. Stamens very densely packed, inclined towards the stigma, which they do not conceal. Style white, stigma creamy white, with 12 lobes.

Fruit globular, c. 2cm diameter, carmine with scales the same colour, flesh carmine. Seeds brown, 0.5mm diameter, appear to be the subgenus *Muscosemineum* Schütz.

Habitat: Paraguay, in the territory of the savage Moro Indian tribe.

Fig.533 Habitat of *G. eurypleurum, Bercht* 3036, west of Cerro León, Paraguay

Photo: Ludwig Bercht

Photo: Ludwig Bercht

Fig.534 *G. eurypleurum, Bercht* 3036, west of Cerro León, Paraguay

Discovered by Friedrich Ritter, seeds denoted as FR 1178.

Type: *Ritter* 1178, 1963, Paraguay, base of Cerro León (U).

Etymology

Derived from two Greek words, 'eurys' meaning broad and 'pleuron', a rib, hence broad-ribbed.

Map 68 Distribution of *G. eurypleurum* in north-west, Paraguay

Base Map © Nelles Verlag GmbH

Photo: Volker Schädlich

Fig.535 *G. eurypleurum, Schädlich* 264, south of El Palmar, Paraguay

Distribution (Map 68)

According to Schädlich (2002), the distribution area is not only the neighbourhood of the type locality, the Cerro León, Paraguay, but extends some way from the hill and to near the Bolivian border to the north.

Conservation Status

Data deficient. There are general comments in the literature about land being cleared for agriculture in this region, but no specific references to the effect this is having on *G. eurypleurum*. However, there are several reported habitat localities so it is unlikely to be under immediate threat.

History

This distinct species was discovered by Friedrich Ritter in 1963 when he travelled from Asunción, Paraguay, the 400km to Filadelfia and then on to the Cerro León, the first cactus explorer to go there. The region is part of the Chaco, flat land dominated

Photo: Volker Schädlich

Fig.536 *G. eurypleurum, Schädlich* 264, south of El Palmar, Paraguay

Fig.537 *G. eurypleurum, Piltz* 434 from Nueva Asunción, south of Cerro León, Boquerón, Paraguay

Fig.538 *G. eurypleurum, Bercht* 2215 from Cruce 4 de Mayo, Paraguay

by thick thorn bush, making exploration difficult. The isolated hill, the Cerro León, was in the territory of the Moro Indians, known to be dangerous at the time.

He found the plants with flowers and fruits under bushes at the base of the hill. None of the plants he collected reached Europe, but seeds were offered in 1964 by Ritter's sister, Frau Winter, under the collection number FR 1178. It was on the same occasion that he discovered *G. paediophilum* and *G. stenopleurum*.

Some of the seeds were bought by the Czech collector Ferdinand Plesnik who flowered them and set fruit after four years. He made the first description in 1972 but the name was invalid because he failed to deposit a type. It was Ritter, in Volume 1 of his *Kakteen in Südamerika* (1979), who validated the name by placing a type specimen in the herbarium at Utrecht.

Commentary

After its discovery, a considerable time passed before it was rediscovered because of the dangers of visiting the area. The first record I can find is of the trip made by Metzing, Bercht and Piltz in 1988,

Fig.539 *G. eurypleurum, Metzing* 20 from south of Cerro León, Boquerón, Paraguay

after which seed became available again and we could add this fine plant to our collections. Metzing (1997) comments that it is closely related to *G. damsii*. He also points out how similar the plants in habitat are to those in cultivation.

Cultivation

This easily grown species is very distinct and should not be confused with any other. The fresh-green body, attractive golden spines and pretty white or pale-lilac flowers make it one of the best species for culture. It rarely offsets and grows to 15cm in diameter when it makes a spectacular specimen. At first it was rare in culture, but now seed and seedlings are freely available.

Illustrations

The Chileans 17(55): p.26 (1997) of *Metzing* 20 in habitat.

Gymnocalycium marsoneri Fric ex Y. Ito subsp. *marsoneri*

First Description (Y. Ito)

Gymnocalycium Marsoneri Fric ex Kreuzinger nom. nud. in *Verzeichnis amerikanischer und andere Sukkulenten mit Revision der Systematik der Kakteen*: p.14 (1935). Validly published by Y.Ito in *Explanatory diagram of Austroechinocactinae*: p.175, 293 (1957).

Gymnocalycium Marsoneri Y. Ito, nov. sp.

Simple, depressed-globular, greyish-green or dull green. Ribs c.15, somewhat rounded, low, divided into large nipple shaped tubercles. Areoles oblong with whitish-brown felt. Radial spines c. 7, 2-3cm long, at first whitish-brown at the base with white tips, later becoming darker. Flowers bell to funnel shaped, 3-3.5cm long, 3-4.5cm wide, opaque whitish-yellow to opaque white.

Probably *G. Marsoneri* Fric nomen nudum 1934.

Etymology

Named for Oreste Marsoner who was an Argentinian cactus collector.

Distribution (Map 69)

The typical subspecies of *G. marsoneri* is the most southern form of the species, now known to occur to the south of Salta City. It is unclear to me

Fig.540 *G. marsoneri marsoneri*, GC434.03 from Campo Quijano, Prov. Salta, Argentina

Fig.541 *G. marsoneri marsoneri*, GC434.03 in habitat at Campo Quijano, Prov. Salta, Argentina

Fig.542 *G. marsoneri marsoneri*, GC434.03 in habitat at Campo Quijano, Prov. Salta, Argentina

Map 69 Distribution of *G. marsoneri marsoneri* in Prov. Salta and Tucumán, Argentina

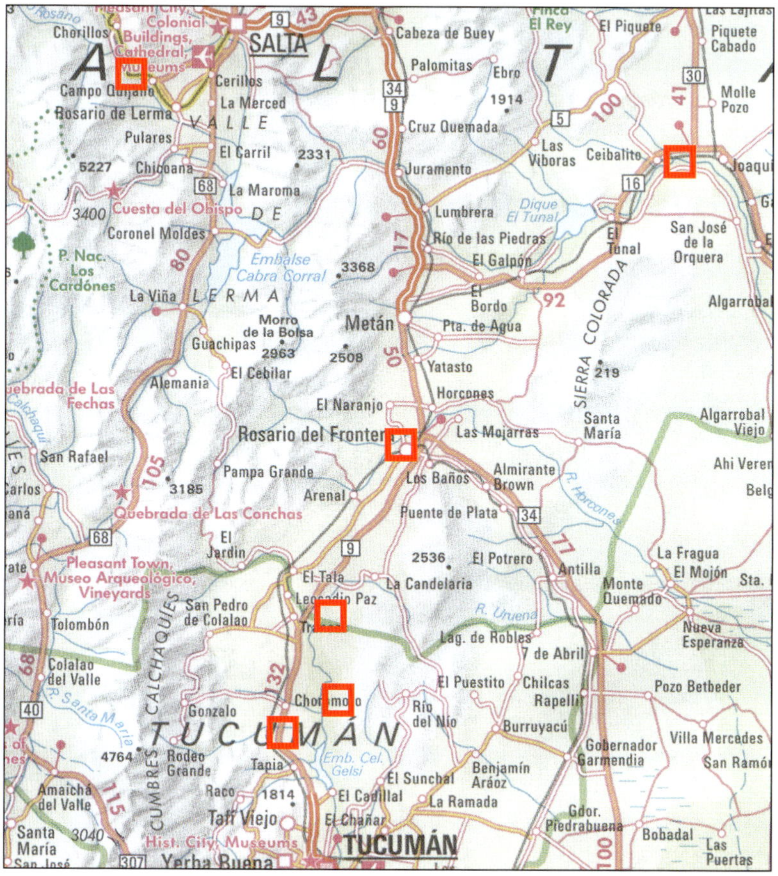

Fig.543 *G. marsoneri marsoneri,* GC434.03 in habitat at Campo Quijano, Prov. Salta, Argentina

who determined that this is the same plant that Ito described, since he designated only Argentina as its habitat. The most southerly locality I have seen recorded was by Neuhuber (2001b) in Prov. Tucumán, south of Trancas, where it is said to grow with *G. schickendantzii* subsp. *delaetii*. Till & Amerhauser (2003) report localities near Vipos and San Vicente also in northern Tucumán, although Meregalli (pers. com.) thinks these could really belong to *G. schickendantzii* subsp. *delaetii*.

I found it growing with *Echinopsis ancistrophora* and *Echinopsis albispinosa* on waste ground near the town centre of Campo Quijano. The climate in this area is warm and rainfall is plentiful in the late spring and summer. In consequence, where there is sufficient soil depth, natural woodland occurs.

Conservation Status

Near Threatened. There are only a few recorded localities in the region which is subject to a rising population and hence, an increase in human activity, such as building and agriculture.

History

The name appeared with a brief German description in the Kreuzinger catalogue: *Verzeichnis amerikanischer und andere Sukkulenten mit Revision der Systematik der Kakteen*: p.14 (1935) with an attribution to Fric (1934), presumably a reference to just the name in a catalogue.

Fig.544 *G. marsoneri marsoneri,* GC434.03 from Campo Quijano, Prov. Salta, Argentina

Fig.545 *G. marsoneri marsoneri,* GC434.03 from Campo Quijano, Prov. Salta, Argentina

Fig.546 *G. marsoneri marsoneri,* GC434.03 from Campo Quijano, Prov. Salta, Argentina

Commentary

The closest relative of *G. marsoneri* is the wide-spread *G. schickendantzii,* which it replaces north of its distribution range in Argentina. The name *G. marsoneri* is often found mis-applied to other species in culture.

Cultivation

The typical subspecies of *G. marsoneri* is not often seen in collections, the northern subspecies *megatae* being much more common. It makes a large flat plant up to 15cm in diameter that flowers from areoles around the shoulder. It has been reported to be difficult to grow but, so long as adequate water is given in warm summer weather, it should grow into an impressive specimen.

Gymnocalycium marsoneri subsp. *matoense* (Buining & Brederoo) Braun & Esteves

First Description (Buining & Brederoo)

KuaS 26(12): pp.265-268 (1975)

Gymnocalycium matoense Buining & Brederoo

Plant flattened-globular, sometimes elongated, up to 15cm diameter, 7cm high or higher, from dark green, dark brownish-green to chocolate green, roots branching, crown sunken, unarmed. Ribs up to 21, fairly sharp, 2-2.5cm wide at the base, separated into small warts that protrude from between and usually 6-7mm above the areoles, acute, 1.5-2cm high. Areoles oval, 8mm long, 4mm wide, at first with a greenish felt, later nearly bald, 1.5cm apart, sitting at the base of the wart.

Flower buds a little sunken in the upper part of the areole at the base of the wart. Spines develop later from the older of the unarmed areoles, first yellowish, later light grey, radial spines 9, of which one points downwards, c. 15mm long, 3 pairs pointing sideways, comb-like lying on the plant, 13mm long, usually another pair above having the same characteristics, 11mm long; on older areoles 1 central spin vertically away standing, 17-19mm long.

Flower when closed 38mm long, 14mm wide, narrow funnel shaped. Pericarp stalk-like, conical pointed, 15mm long, up to 8mm wide, covered with nail-like, heart shaped fleshy scales, with smooth edges 2-3mm long, 3-4mm wide, without hairs, growing into elongated, tapering, round at the top, fleshy, transitional petals, 9mm long, 5mm wide, white, the point notched. Ovary oval, elongated heart shape, 11mm long, 4mm wide, ovules grow from the wall, tree-like, branched in bundles of 10-12.

Fig.547 *G. marsoneri matoense, Horst-Uebelmann* 452 from near to Porto Murtinho, Mato Grosso, Brazil

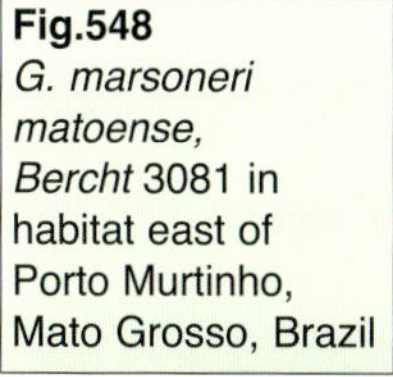

Fig.548 *G. marsoneri matoense, Bercht* 3081 in habitat east of Porto Murtinho, Mato Grosso, Brazil

Photo: Ludwig Bercht

Fig.549 *G. marsoneri matoense, Braun* 248 south-east of Porto Murtinho, Mato Grosso, Brazil

Nectar chamber conical, 4.5 long, 3.5 wide: Primary stamens standing in a closed circle on the edge of the nectar chamber that is not closed, 5mm long, curved, pointing towards the style, the receptacle wall is behind the thickened base of the primary stamens, secondary stamens in 3-4 rings, the highest ring being just below the top of the receptacle, the higher stamens 8mm long, very thin, white, anthers 2-2.5mm long, white. Style 8mm long,1.5mm thick, white, the 8 stigma lobe 2.5mm long, with papillae.

Fruit a juicy berry, oval, light blue with a few grey-blue, lighter at the edges, nail-like scales, on top a small lid with a curved light brown edge, carrying the flower remains, flesh dark red. Seed helmet shaped, 0.9-1mm long, 0.8-0.9mm wide, testa matt ochre, with little round warts of uneven size, smaller along the hilum edge, hilum elongated-oval. Micropyle and funicle lying deep in the hilum, seed threads growing from the testa wall, light ochre colour. Embryo egg-shaped, without perisperm, cotyledon not conspicuous.

Habitat; near Porto Murtinho, Mato Grosso, Brazil, c. 140m., on flat stony soil, often found growing in Caatinga with *Discocactus silicicola* Buin. & Bred. *Frailea melitae* Buin. & Bred., as well as Bromeliads and small bushes. A.F.H. Buining and L. Horst were at the habitat on the 16th Sept. 1974. It was the first time that a *Gymnocalycium* had been discovered in the Mato Grosso.

Type: *L.Horst* & *W.Uebelmann* 452, 16.9.1974, Brazil, Mato Grosso, near to Porto Murtinho (U).

Etymology

Occurs in the Brazilian state of Mato Grosso do Sul

Other Name

Gymnocalycium matoense Buining & Brederoo in KuaS (26(12): pp.265-268 (1975)

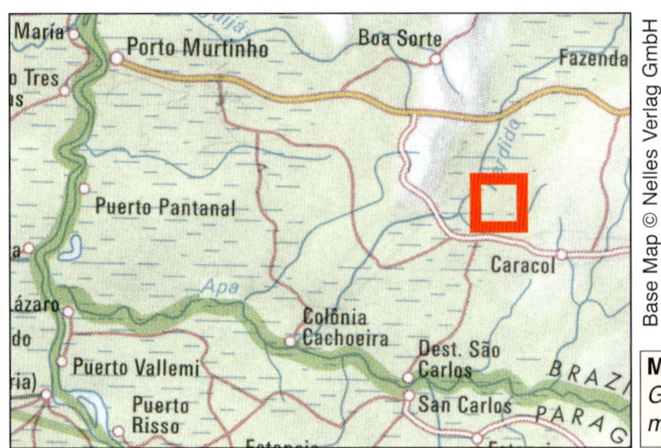

Map 70 Location of *G. marsoneri matoense* in Brazil

Distribution (Map 70)

This taxon is known from a few rocky localities in the forest just in Brazil, to the east of the Río Paraguai which here is the border with Paraguay.

Conservation Status

Data Deficient. There are a few reported habitats for this taxon but little information on the threats to its survival. It is known only from the vicinity of the type locality but this is a poorly explored region. It grows on rocky patches in the forest and may be vulnerable to human activity.

History

G. matoense was described by Buining and Brederoo in KuaS (1975), claiming it to be the first *Gymnocalycium* to be described from Mato Grosso, Brazil. It represents the most easterly known population of this species. Braun and Esteves (1995) then made it a subspecies of *G. marsoneri*.

Commentary

It was unfortunate that we were forced by the rules of nomenclature to use the name subsp. *matoense* for the Bolivian, Paraguayan as well as the Brazilian populations of *G. marsoneri* in the *New Cactus Lexicon* (2006) where they were treated as a single subspecies.

The plants described by Buining & Brederoo (1975) as *G. matoense* have a distinctive appearance, but the name had to be used because it is the oldest published for these plants at the rank of subspecies, if the Bolivian and Paraguayan populations are considered to be the same subspecies.

Because of the clear differences, I have here recognised the Bolivian and Paraguayan plants as a separate subspecies *megatae*. It is actually more different from subsp. *matoense* than it is from the type subspecies from Argentina.

Cultivation

This is an example of one of the more difficult Gymnocalyciums to grow. It is used to a shady environment with high temperatures in the growing season when the rains come. It is quite rare in culture, even though it was described some time ago.

Gymnocalycium marsoneri subsp. *megatae* (Y. Ito) G. Charles

First Description (Y. Ito)

Explanatory diagram of Austroechinocactinae: p.292, 172 (1957)

Gymnocalycium megatae Y. Ito, nov. sp.

Flattened-discoid, c. 6mm high, 15-20cm diameter, greenish-grey. Ribs 9-13, acute angled. Spines radial, up to 5, slender, c. 2cm long, brownish when young, blackish-brown later. Flowers white, funnel-form, c. 5cm long and diameter, tube naked, narrow, blue. Paraguay.

New Combination

Gymnocalycium marsoneri subsp. ***megatae*** (Y. Ito) G. Charles **comb. et stat. nov. Basionym:** *Gymnocalycium megatae* Y. Ito in *Explanatory diagram of Austroechinocactinae*: 292, 1957.

Synonyms

G. onychacanthum Y. Ito in *Explanatory diagram of Austroechinocactinae*: pp.292, 173 (1957)

First description of *Gymnocalycium onychacanthum* Y. Ito, nov. sp.

Discoid, 15-20cm broad, glaucous-green; ribs resolved into hatchet-like humps; spines awl-shaped, up to 9, c. 3cm long, comb-like, dark grey; flowers bell-shaped to funnelform, c. 2.5cm long, c. 3cm wide, dingy white, tube naked, narrow, short, deep sky blue. Uruguay.

G. tudae Y. Ito in *Explanatory diagram of Austroechinocactinae*: pp.292, 172 (1957)

First description of *Gymnocalycium tudae* Y. Ito, nov. sp.

Flattened-globular, c. 6cm tall, 17cm diameter. Velvety glaucus green. Ribs 9-14, tubercles axe like. Spines, radials, up to 7, hardly curved, base white to deep yellow, dark brown at the tip. Flowers white, funnel shaped, 4cm tall and diameter, white, tube naked, thickened, straw colour. Paraguay.

G. eytianum Cárdenas in KuaS 9(2): pp.25-26 (1958)

Fig.550 Habitat of *G. marsoneri megatae, Schädlich* 234, south-west of Toledo, Boquerón, Paraguay

Photo: Volker Schädlich

Fig.551 *G. marsoneri megatae, Kiesling* s.n. from west of Estación La Patria, Boquerón, Paraguay

Commentary on *G. eytianum*

This plant was compared by Cárdenas (1958) with *G. pflanzii* and treated as a variety of that species by Donald (1971). However, the picture on page 25 clearly shows *G. marsoneri*. Both plants grow near the town of Eyti and the description has elements of both, so making this name valueless.

G. hamatum Ritter in *Kakteen in Südamerika* 2: pp.663-664 (1980)

First description of *Gymnocalycium hamatum* Ritter spec. nov.

Body grey-green, simple, flowering size 6-16cm diameter, in habitat up to 8cm high with a sunken spineless crown. Ribs, 9-16, from 10-15mm high, 15-25mm wide, horizontal cross-cuts over the areoles, below these the rounded chin that more or less protrudes, the humps are rounded, not axe-shaped, the grooves separating the ribs are sinuous. Areoles on the humps are 5-8mm long, 3mm wide, with whitish felt, 10-18mm apart. Spines, only radials, awl-shaped, slightly flattened, pointing sideways, light to mid brown, (5-)7-9, curved towards the body, more strongly curved at the ends (hooked), usually the bottom pair are hooked, all are 12 to less than 20mm long.

Flowers fairly wide, 5cm long, opening for only a few hours for a few days, one after the other, odourless. Ovary 15mm long, 9mm diameter, skittle-shaped, olive-green with ruby-brown, white edged rounded scales, c. 2mm long, 3-4mm wide. Nectar chamber tubular, 3mm long, well covered by the basal filaments; tube cup-shaped at the top, 12mm long, inside white, outside brownish carrying large brownish blunt scales with pink tips. Filaments white, over the

Photo: Volker Schädlich

Fig.552 *G. marsoneri megatae* (holdii), *Schädlich* 34 from near Est. El Tinto, Dept. Santa Cruz, Bolivia

Fig.553 The habitat of *G. marsoneri megatae* (hamatum), GC863.04 south of Palos Blancos, Dept. Tarija, Bolivia

Photo: Ludwig Bercht

Fig.554 *G. marsoneri megatae, Bercht* 3428 in habitat east of Boyuibe, summer

Photo: Ludwig Bercht

Fig.555 *G. marsoneri megatae, Bercht* 3023 in habitat east of Boyuibe. Same habitat as Fig.554 but in the dry season.

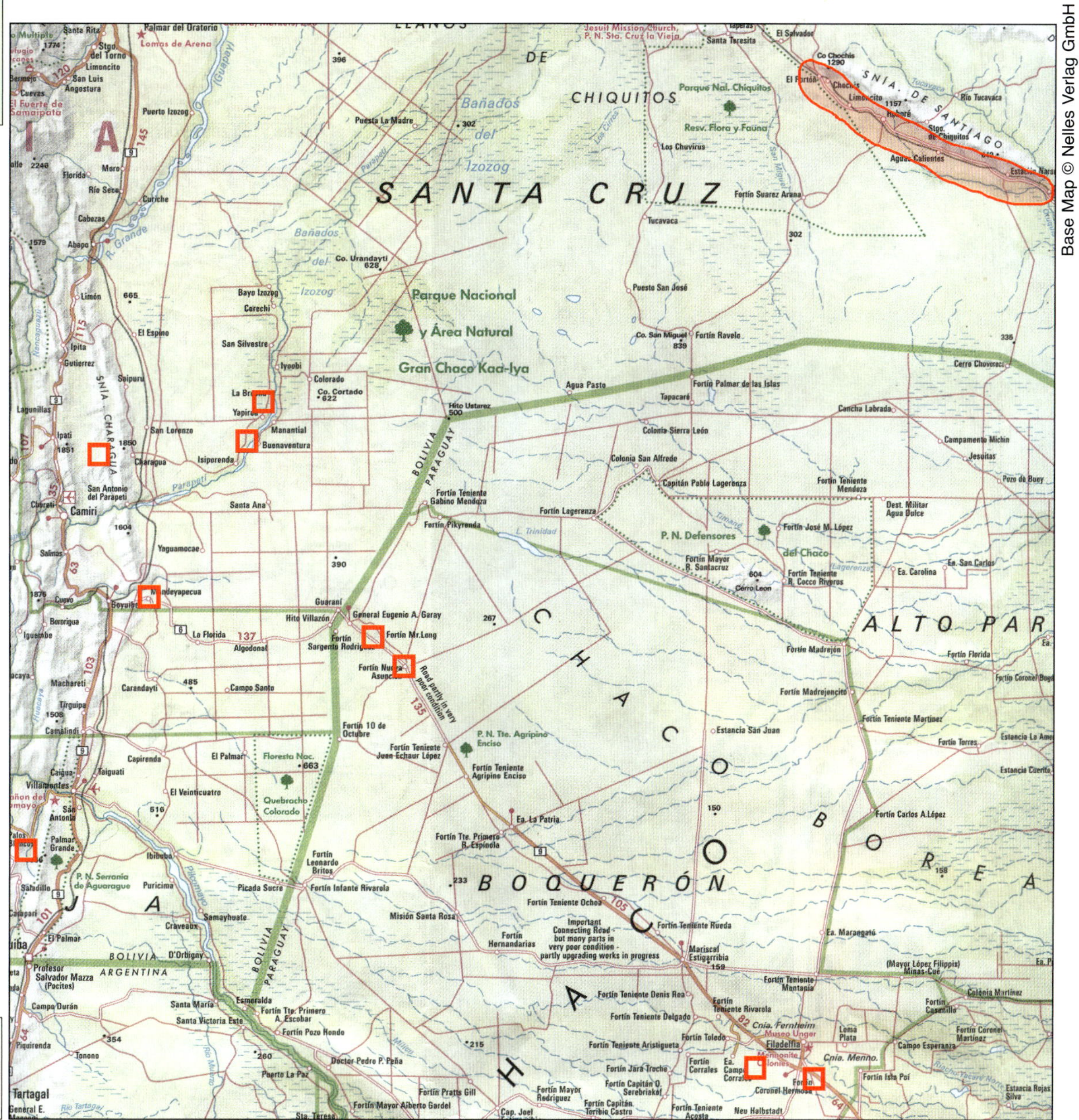

Base Map © Nelles Verlag GmbH

Map 71 Locations of *G. marsoneri megatae* in Bolivia and Paraguay

Photo: Volker Schädlich

Photo: Volker Schädlich

Photo: Volker Schädlich

Figs.556-558 (above) *G. marsoneri megatae* (pseudo-malacocarpus), *Schädlich* 54 in habitat west of El Carmen, Dept. Santa Cruz, Bolivia. The type locality of *G. anisitsii* var. *pseudomalacocarpus.*

Photo: Volker Schädlich

Fig.559 *G. marsoneri megatae* (holdii), *Schädlich* 34 from near Est. El Tinto, Dept. Santa Cruz, Bolivia

Photo: Volker Schädlich

Photo: Volker Schädlich

Figs.560 & 561 (above) *G. marsoneri megatae* (pseudomalacocacarpus), *Schädlich* 51 from near Santa Ana de Chiquitos, Dept. Santa Cruz, Bolivia

Photo: Volker Schädlich

Fig.562 *G. marsoneri megatae* (pseudo-malacocarpus), *Schädlich* 325 in habitat east of Limoncito, Dept. Santa Cruz, Bolivia

Fig.563 *G. marsoneri megatae* (hamatum), GC863.04 south of Palos Blancos, Dept.Tarija, Bolivia

nectar chamber is a thick enclosing ring of 4-5mm long filaments that lean against the style, above these is a gap of c. 2mm, then the insertions increase in density up to the seam, all curved towards the centre, increasing in length, up to 10mm at the top. Anthers grey black, long oval, pollen cream. Style with the stigma lobes in the centre of the anthers, c. 18mm long without the stigma lobes, white; c.12 cream coloured stigma lobes, radiating, curved upwards, 4-6mm long. Petals, innermost 12mm long, 2-3mm wide, white, blunt at the top, narrow at the bottom, middle petals increasing in length to 22mm, 5-6mm wide, white with a pale pink central stripe, the outer petals with a greenish-red tip and a greenish central stripe, spread out and curved upwards.

Fruit reddish with a blue bloom, spindle-shaped, 25-30mm long, 15-20mm diameter, flesh juicy, pink. Seeds nearly globular, light brown, 0.8-1.0mm diameter, with many fine humps running into each other, the grooves forming a fine network, hilum basal, long oval white.

Type Habitat: Bolivia, Tarija, Prov. Gran Chaco, Palos Blancos. Discovered by me, 1958. Related to G. tudae. No. FR819 (areoles and seeds only at U)

Gymnocalycium anisitsii subsp. *holdii* Amerhauser in *Gymnocalycium* 16(3): pp.531-532 (2003)

Distribution (Map 71 on page 243)

This taxon occurs over a large region of western and north-western Paraguay as well as in eastern Bolivia. It grows in open, flat or gently sloping sandy patches, often under trees which bear leaves mainly in summer. The 'pseudo-malacocacrpus' form always grows on rocky surfaces according to Schädlich (pers. com.)

Conservation Status

Least Concern. There are many reported habitats for this taxon but little information on the threats to its survival. It grows in flat land, probably suitable for cultivation so may be effected by human activity in the future.

History

The oldest names at species rank for this taxon were published by Ito in 1957. They were *G. megatae*, *G. tudae* and *G. onychacanthum*. The descriptions fall within the variability of this taxon so they are synonyms. The first listed, *G. megatae*, has usually been used and is chosen here as the subspecies' name.

Photo: Volker Schädlich

Fig.564 *G. marsoneri megatae* (pseudo-malacocarpus), *Schädlich* 54 from west of El Carmen, Dept. Santa Cruz, Bolivia

Photo: Volker Schädlich

Fig.565 *G. marsoneri megatae* (holdii), *Schädlich* 34 from near Est. El Tinto, Dept. Santa Cruz, Bolivia

Fig.566 *G. marsoneri megatae* (pseudo-malacocarpus), *Lau* 365 from Taperas, Dept. Santa Cruz, Bolivia

Fig.567 *G. marsoneri megatae,* GC852.02 in habitat near Eyti, Dept. Santa Cruz, Bolivia

Fig.568 *G. marsoneri megatae, Metzing* 106 from Boyuibe, Dept. Santa Cruz, Bolivia

Fig.569 *G. marsoneri megatae, Kiesling* s.n. from west of Est. La Patria, Boquerón, Paraguay

They were first discovered around the period of the Chaco War (1933-35) by Adolfo Friedrich who was a war reporter for the Paraguayan army. He collected cacti for Prof. Hassler, director of the Botanical Institute in Asunción. Moser (1971) explains how the collected plants were bought by Oreste Marsoner on the instructions of Harry Blossfeld, the nurseryman from Potsdam, Germany.

However, most of the plants went direct to the USA and Japan, which explains how they came to be described by Yoshio Ito. Moser also offers an explanation for the strange statement by Ito in his description of *G. onychacanthum* that it came from Uruguay. Apparently, Marsoner made no notes about the finding places told to him by Friedrich, that likely resulted in the error.

Fig.570 *G. marsoneri megatae,* GC852.02 in habitat near Eyti, Dept. Santa Cruz, Bolivia

Commentary

This subspecies is well known as *G. megatae* or *G. tudae* in collections so, if you consider them to be a different species from *G. marsoneri,* then *G. megatae* or *G. tudae* would be the right name to use. An interesting observation by Moser (1971) is that the flowers have a strong scent of lavender. He also tells us that the seeds are small and he found a single fruit with 4,900 seeds! I wonder.

The plants look very different in the dry season when they shrink and become brown, blending with the ground in which they grow. After the rains in late spring, the trees around grow their leaves and shade the plants which consequently become greener and flower.

I have included here the plant which was invalidly described by Backeberg as *G. pseudo-malacocarpus*, rather than under *G. anisitsii* where it was validated as a variety by Till & Amerhauser (2003b).

Cultivation

It is certainly the case that these are among the more difficult Gymnocalyciums to grow. They are used to a shady environment with with high temperatures in the growing season. The summer is hot in their Chaco habitat when the rainfall can be heavy and the trees grow a dense canopy of leaves.

If you succeed, the reward can be an impressive, large, flat brown or olive-green body with a ring of flowers from the shoulders of the plant. Such specimens are rarely seen, but a well-grown individual is sure to get noticed at a show and do well against most other species. It is a spectacular sight when large and in full growth.

Illustration

The National Cactus and Succulent Journal 26(1): pp.4-5 (1971)

Gymnocalycium mihanovichii (Fric & Gürke) Britton & Rose

First Description (Fric & Gürke)

Echinocactus Mihanovichii Fric & Gürke in *Monatsschrift für Kakteenkunde* 15(9): pp. 142-143 (1905)

Body wider than high, 2-3cm high, 4-6cm broad, grey-green, sometimes tinted red, the crown flat and a little sunken, not covered with humps, very few spines. Ribs, 8, wide, in cross-section wide triangular, blunt with very flat indents between the areoles, from the areoles, grow slanted, easily recognisable (because of their light colour) cross ribs, the vertical ribs are very sharply defined towards the crown, towards the base they are very flat.

Areoles 10-12mm apart, fairly round, 1-2mm diameter, covered with dirty white wool-felt. Radial spines, 5, sometimes 6, very often some drop off with age, usually 2 pairs remain, pointing diagonally upwards and diagonally downwards, one pointing straight down, and the 6th, (often missing) pointing straight upwards, at first, all straight, later lying close to the body and curved, in youth grey-yellow

Fig.571 *G. mihanovichii, Bercht* 2195 from north of Filadelfia, Paraguay

Fig.572 The habitat of *G. mihanovichii, Bercht* 3059, east of Agua Dulce, Paraguay

Photo: Ludwig Bercht

Photo: Volker Schädlich

Fig.573 *G. mihanovichii, Schädlich* 26, in habitat east of Filadelfia, Boquerón, Paraguay

Photo: Volker Schädlich

Fig.575 Habitat of *G. mihanovichii, Schädlich* 231, at Km 426 Ruta Trans Chaco, Boquerón, Paraguay

with red-brown tips, eventually becoming grey, the longest 1cm long, central spines missing.

Flowers appear near the crown from behind bundles of spines, 4-4.5cm long. Buds are a reddish colour. Flowers are small funnel shaped, the tube in which the ovary does not bulge, is cylindrical, widening towards the top, 2cm long, 7-8mm diameter at the top, mostly curved, shiny light green, becoming emerald green, completely bald, covered in scales set in a spiral order. Scales c. 16-18, semi circular, 3mm wide, 1-2mm high, slightly pointed, green at the base, running into pink with a transparent edge, the colour of the 40-50 petals is generally light olive green, 12-14 outer petals are wide spatulate, yellowish-green, reddish at the tips, c. 6-10mm long, 5-6mm wide, the edges turned outwards. The 12-14 central petals are longer and narrower, 12-15mm long, 4-5mm wide, also yellowish-green, the tips more olive coloured with a red sheen. The innermost petals (14-16) are again shorter than the centrals, 8-10mm long, 2-3mm wide, lanceolate, greener than the others, with no red sheen.

The anthers are found in two circles, the lower consists of c. 20 stamens, originating at 3-4mm from the base of the carpel, the stamen 4mm long slanting towards and pressed onto the style; the upper circle has many anthers, 6-10mm long, they do not reach to the top of the inner petals but bend over the style covering the lobes, all filaments are yellowish, almost white; anthers light yellow, longish, nearly 1mm long. Ovary 11-13mm long, fills nearly 75% of the flower tube. Style 8-9mm long (excluding lobes), very strong, light green, the 5 lobes are yellowish, 4-5mm long, they split nearly to the base into 2 branches.

This species belongs to the sub-family Hybocactus, and specially in relationship to *E. denudatus* Link & Otto of which the body possesses similarities, one would mistake it in a non-flowering condition as a form of this variable species, yet the spines are not pressed so strongly to the body as on the other species.

In a flowering condition *E. Mihanovichii* differs from *E. denudatus* and the related *E. Damsii* K. Schum. and *E. Quehlianus* Haage jr. by way of its olive-green flowers.

Etymology

Named for Nicolás Mihanovich, a Croatian businessman who lived in Argentina and supported the activities of the Czech cactus specialist Alberto Fric.

Photo: Volker Schädlich

Fig.574 *G. mihanovichii, Schädlich* 280, in habitat north-east of Madrijon, Alto Paraguay, Paraguay

Photo: Ludwig Bercht

Fig.576 *G. mihanovichii, Bercht* 3059 in habitat east of Agua Dulce, Paraguay

Map 72
The distribution of *G. mihanovichii* in Paraguay. There are also recorded localites further south in Prov. Formosa, Argentina

Distribution (Map 72)

The habitat range of this species stretches from the Chaco Austral in Prov. Chaco, Argentina north into the chaco region of Paraguay to the Chaco Boreal. It grows in flat sandy ground under the protection of bushes and palms.

Fig.577 *G. mihanovichii* (stenogonum), *Piltz* 242 from Chaco Austral, Prov. Formosa, Argentina

Fig.578 *G. mihanovichii* (filadelfiense), *Piltz* 430 from Mariscal Estigarribia, Boquerón, Paraguay

Conservation Status

Near Threatened. The plant is widespread with many reported habitat localities but the region is being increasingly used for agriculture.

History

First found by Fric in 1903, plants were sent to the nurseryman De Laet in Belgium. He considered it to be a form of *G. denudatum* but, of course, there were only a few *Gymnocalycium* species known to him then. It was formally described by Gürke in 1905, but the locality of the discovery was not given at the time. It was thirty years later that Fric revealed that he found the plant near Porto Casado, a port on the River Paraguay, a locality which Schütz (1979) maintains was false.

The var. *stenogonum* was found by Fric in 1926 and described by him and Pazout (1948). It was said to come from Toro Alarachii in the Chaco Austral, Prov. Chaco, Argentina.

Werner Hoffmann from Heidelberg found a form of this species near the Mennonite settlement of Filadelfia some 160km to the west of the type locality. It was subsequently named invalidly by Pazout (1965) as var. *filadelfiense*, but it is likely that this is just a redescription of the type form. Ritter had also found the plant and gave it the invalid name *G. mihanovichii* var. *chlorostrictum*.

Commentary

This plant has been the subject of intense interest among early students of the genus, particularly from Czech enthusiasts. There has been much confusion between this species and the similar-looking *G. stenopleurum*, with which it sometimes grows sympatrically. At first, all the variants were described as varieties of *G. mihanovichii*, but now many of these forms are considered to belong to *G. stenopleurum*.

Cultivation

A well grown plant of *G. mihanovichii* will always do well at shows because it presents some difficulty in cultivation. In summer, it prefers to be protected from strong sunshine, enjoying more humidity than most, and a well drained acid compost with a steady supply of moisture. Then in winter, it appreciates a minimum of 10°C with a little moisture to prevent shrivelling.

Its distinctive appearance and curious green flowers will always make this plant popular, although large specimens are rare. Plants of the Argentinian var. *stenogonum* in collections often originate from seed collected by Piltz under his field number P 242.

Illustration

Monatsschrift für Kakteenkunde 29: p.67 (1919)

Fig.579 *G. schickendantzii schickendantzii*, GC951.04 in habitat near the road to Agua de Ramón, Prov. Córdoba, Argentina

First Description (F.A.C.Weber)

Echinocactus Schickendantzii F.A.C.Weber in Bois, *Dictionnaire d'Horticulture* p.470 (1896)

Catamarca. – Close to *Echinocactus saglionis*, no central spines; six radial spines, recurved, flattened angular, coarse on upper face. Flowers hardly different.

A more complete description was given by Karl Schumann in *Gesamtbescheibung der Kakteen* (1899)

Body simple, semi-globular, depressed, round on top, depressed in the crown, sparsely covered with woolly felt, also covered with spines that are interlaced, dark green, c. 10cm diam. Ribs 7, separated by sharp vertical grooves, c. 1cm high, straight, blunt, these are separated by sharp cross cuts, forming plump humps, 5 sided at the base. Areoles 10-12mm apart, small elliptical, 5-6mm diameter, covered with short woolly felt. Radial spines clearly in two rows, comb-like, 6-7, sometimes the first pair, sometimes the second pair are the longest, up to 3cm long, the smaller pair are round, the larger pairs at the top and sides are flattened, sometimes with a groove on the top side, these are strongly curved outwards, rough, red into horn colour. Central spines 0.

The flowers, according to Weber, are not much different from *Echinocactus Saglionis*.

Fig.580 *G. schickendantzii schickendantzii*, GC995.04 in habitat south of Quilino, Prov. Córdoba, Argentina

Fig.581 *G. schickendantzii schickendantzii*, GC958.03 in habitat at Los Colorados, Prov. La Rioja, Argentina

Fig.582 *G. schickendantzii schickendantzii*, GC988.03 growing epiphytically on a tree trunk south of San Martín, Prov. Catamarca, Argentina

Geographical habitat: Near Catamarca in the state of Tucumán, Argentina. Schickendantz. [In fact, Catamarca is not and was never in Tucumán, presumably a mistake caused by Schickendantz sending plants from Tucumán City]

Epitype designated by Neuhuber (2004): Argentina, Province Catamarca, south of Catamarca City near Coneta, 540m, 18 Dec. 1991, leg. G. Neuhuber GN91-456 (WU, dry prep. no.447)

Etymology

Named for Frederick Schickendantz (1837-1896), a German chemist who worked in Argentina and established the Lillo Herbarium in Tucumán.

Synonyms

G. michoga Fric ex Y. Ito *Explanatory Diagram of Austroechinocactinae*: p.175, 292-293 (1957). The first appearance of the name was in a Fric catalogue of 1926 followed by a brief description in Kakteenjäger: p.8 (c.1929)

Fig.583 *G. schickendantzii schickendantzii*, GC987.03 in habitat south of San Martín Prov. Catamarca, Argentina

First description of *Gymnocalycium michoga* Y. Ito, nov. sp.

Simple, globular; grey-green, a little paler in the crown, often light brownish-purple. Ribs, up to 11, lost in sharp edged, large wart-like humps. Areoles longish, c. 1.5cm apart, with blackish-green spots on the humps. Spines needle-like, c. 7 radial spines, up to 2cm long, at first dark brown, later grey. Flowers bell to funnel shape, 3-5cm long, 3-4cm diameter. Inner petals white with a pale reddish-green central stripe, outer petals pale to dark green; flower tube long, naked, rich green. Argentina.

Fig.584 *G. schickendantzii schickendantzii*, GC978.03 in habitat south of Villa Sanagasta, Prov. La Rioja, Argentina

Commentary on G. michoga

It is not known where the Fric plant was collected, although the origin is given as 'St Jago' [could this be San Tiago, Santiago?] in his English 1928 price list Cacti, *The Coming Fashion*. Other cacti listed there with this origin suggest that this is a place in Argentina. The brief description in *Kakteenjäger* simply says it has a colourful body and is like *G. mihanovichii* but with stronger spines, which Backeberg (1959) took to mean a Paraguayan origin. The illustration and description by Ito is clearly of *G. schickendantzii* so belongs here if he described the same taxon as Fric had intended.

Gymnocalycium schickendantzii subsp. *bergeri* Neuhuber in *Gymnocalycium* 14(4): p.420 (2001)

Fig.585 *G. schickendantzii schickendantzii*, GC984.04 in habitat south of Concepción, Prov. Catamarca, Argentina

Commentary on *G. schickendantzii* subsp. *bergeri*

Neuhuber (2001b) published this name for an apparently isolated population of *G. schickendantzii* which occurs in woodland to the north of the type subspecies distribution and east of the subsp. *delaetii*. Its habitat is east of Rosário de la Frontera.

Gymnocalycium schickendantzii subsp. *periferalium* Neuhuber in *Gymnocalycium* 17(2): p.564 (2004)

Fig.588 *G. schickendantzii schickendantzii*, GC937.03 in habitat near Quilmes, Prov. San Luis, Argentina

Fig.586 *G. schickendantzii schickendantzii*, GC18.03 in habitat at Chepes Viejo growing with *G. castellanosii* (bozsingianum), Prov. La Rioja, Argentina

Commentary on *G. schickendantzii* subsp. *periferalium*

A name applied by Neuhuber to a minor variant with flowers produced from areoles lower down the sides. It occurs in the north-west of Prov. Córdoba. In the same publication, he validated var. *pectinatum,* first mentioned as a name by Lambert in his field list.

Distribution (Map 73)

This is the most widely distributed and most southern species in the subgenus *Muscosemineum* as well as being one of the most commonly encountered Gymnocalyciums in habitat. It is unusual in the genus for its liking of flat sandy areas away from rocks and steep slopes, but often in the shade of shrubs. It will often be found in the same general

Fig.589 *G. schickendantzii schickendantzii*, GC216.03 in habitat at Los Colorados, Prov. La Rioja, Argentina

Fig.587 *G. schickendantzii schickendantzii*, GC218.02 in habitat near Marayes, Prov. San Juan, Argentina

Fig.590 *G. schickendantzii schickendantzii, Piltz* 17i from Cruj del Eje, Prov. Córdoba, Argentina

Fig.591 *G. schickendantzii schickendantzii, Piltz* 17s from Chepes Viejo, Prov. La Rioja, Argentina

area as other species of the genus but only *G. bodenbenderianum* shares its habitat preference.

Its range stretches from Prov. Salta in the north to as far south as Prov. San Luis and just into Prov. Mendoza, all in Argentina. Neuhuber (2004) speculates on the origins of the distribution we see today and describes the minor variants found in the various populations.

Conservation Status

Least Concern. It must be one the most plentiful of Gymnocalyciums, often locally common and with a wide distribution, often colonising land where other species do not grow.

History

The first brief description by Weber appeared in the *Dictionary of Horticulture* (1896) of D. Bois. Schumann, in his *Gesamtbescheibung der Kakteen* (1898) tells us that he got a dried plant from Dr. Weber and publishes a more complete description. In his Nachträge (1903), Schumann includes a picture captioned *E. schickendantzii* provided by the nurseryman De Laet which had first appeared in MfK 11(12): p.187 (1901) captioned as *Echinocactus De Laetii*. He said that he had compared the two species and concluded that they were the same, claiming that he was mislead by Weber's comment that the flowers of *E. schickendantzii* were like *E. saglionis* which he correctly says they are not.

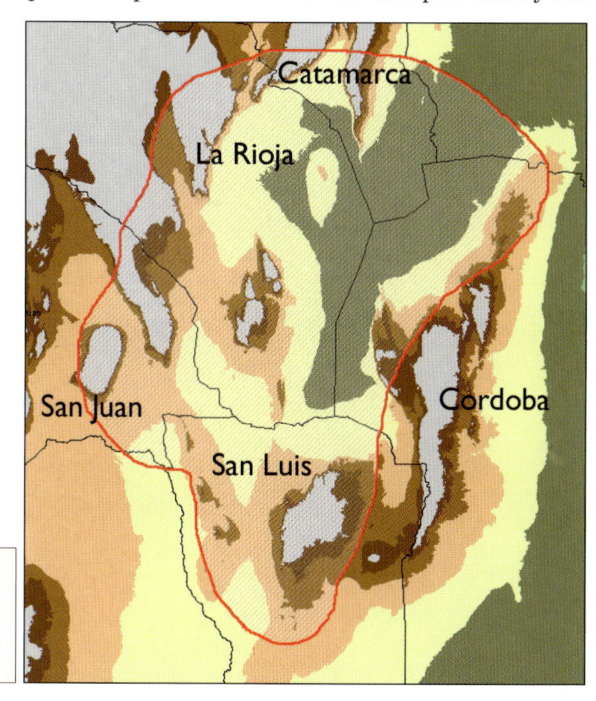

Map 73 The extensive distribution of *G. schickendantzii schickendantzii* in Argentina

Werdermann, in the notes to Plate 15 of his *Blühende Sukkulenten* (1931), comments on the confusion with *Echinocactus delaetii* which he claims to be a different species, but with many transitional forms between it and *E. schickendantzii*. He explains that Stümer sent him material of both species from the same habitat.

Commentary

This taxon is really distinct and should not be confused with any other. It is often found at habitat localities where other *Gymnocalycium* species grow, although usually in a different habitat niche. It likes the flat sandy plains, so grows in places where cactus enthusiasts would probably not have any reason to stop. It can also sometimes be found in open patches in woodland.

In Prov. La Rioja, it often grows in the same kind of habitat as *G. bodenbenderianum*, sandy patches near to the bases of bushes. It presumably finds an environment there which helps the seed to germinate and then the bush protects it from trampling by animals.

Cultivation

Although it responds to the same cultural conditions as other Gymnocalyciums, I have some experience of it losing its roots. For such a large-growing plant, you would expect it to develop quickly but for me, that is not the case, and my old plants are still not large. It is uncommon to see big specimens of this species in cultivation.

Piltz has offered seed from various localities with his field number P 17 followed by various letters, each corresponding to a different finding place. There is an attractive densely-spined form which one sees in cultivation called *G. pungens*, a

Fig.592 *G. schickendantzii schickendantzii, Neuhuber* 86 13-17, from San Agustín del Valle Fertil, Prov. San Juan, Argentina

Fig.594 *G. schickendantzii schickendantzii*, GC21.03 from south-east of Los Colorados, Prov. La Rioja, Argentina

Fig.593 *G. schickendantzii schickendantzii, Piltz* 17 from Salinas Grandes, Prov. Catamarca, Argentina

name which was invaldly described by Fleischer (1962)

Illustrations

Britton & Rose *The Cactaceae* III: p. 165 Figs. 179 and 181. Plate XIX.2 (1922)

Werdermann *Blühende Sukkulenten* Mappe 4 Tafel 15 (1931)

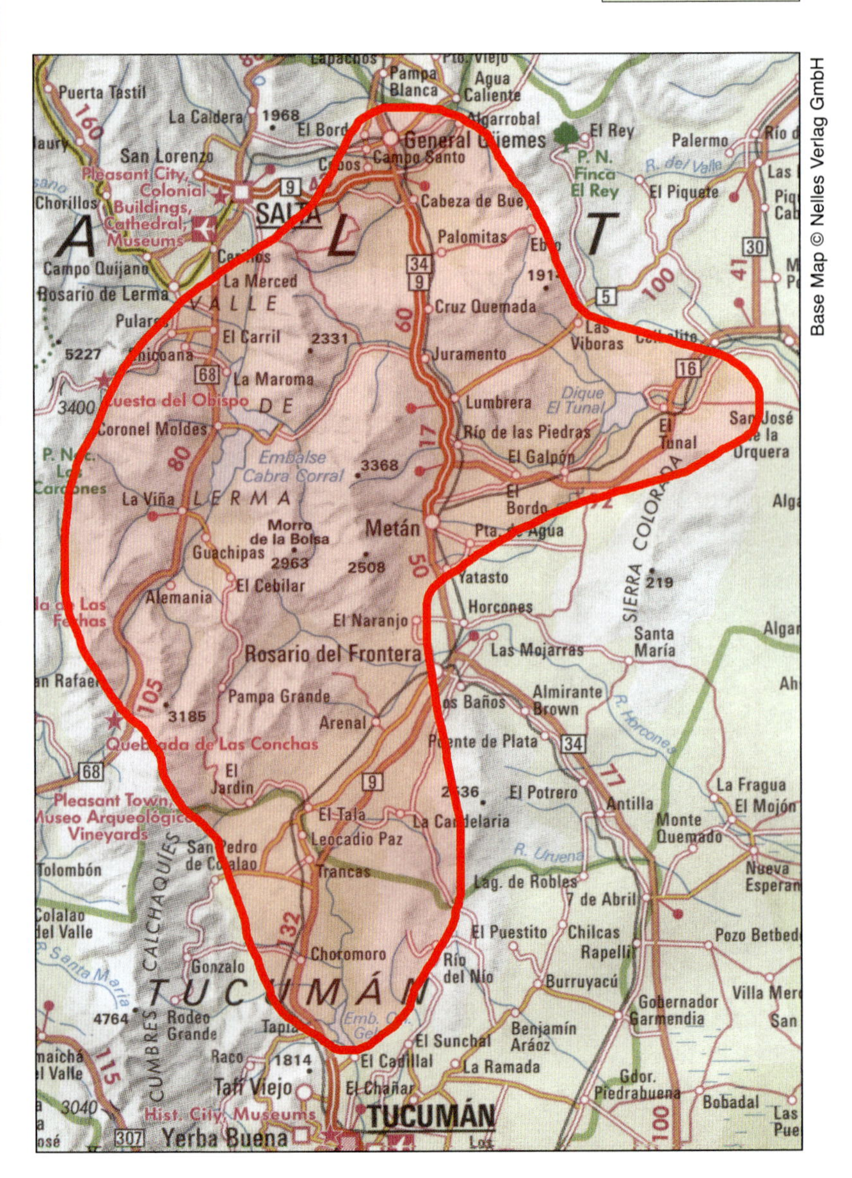

Gymnocalycium schickendantzii subsp. *delaetii* (K. Schumann) Charles

Fig.595
G. schickendantzii delaetii, Piltz 55 from Quebrada de Cafayate, Prov. Salta, Argentina

Map 74
Distribution of *G. schickendantzii delaetii* in Prov. Salta & Tucumán, Argentina
⟵

Fig.596
G. schickendantzii delaetii, GC65.04 in habitat north of Alemania, Prov. Salta, Argentina

First Description (Schumann)

Echinocactus de Laetii eine neue Art in *Monatsschrift für Kakteenkunde* 11(12): pp.186-187 (1901)

Body semi-globular, depressed to short columnar; crown rounded and a little sunken, without wool-felt, covered with twisted spines, in new growth light matt green, warts in 8 and 13 regular rows, 4 sided, rounded, with the areoles sitting on the short sloping tips, Areoles 6mm long, with white wool-felt, soon becoming grey, then quickly becoming bald. Spines usually 7, compressed, curved, sharp pointed, on the new growth dark, later reddish, then grey, matt, from the top, the second pair are the longest, up to 3cm long.

Flowers appear on the top of the plant near the crown. Ovary is thin cylindrical, olive-green, with small scales in the form of a half ellipse, the scales olive green with pink edges, the whole flower is 5cm long, diameter smaller, the lower seldom fully open.

Outside petals are trapeziform to wide spatulate, nail-like at the base, blunt at the tips, inner petals oblong, rounded with fine points, pinkish-red. Stamens reach almost to the top of the flower tube, inserted (as in *Echinopsis*) in 2 sections, the lower are inserted in a circle on the edge of the base where the beaker shape begins, the rest are separated, inserted further up the tube, the stamens are white, anthers grey-yellow, the almost strong white style has 12 yellow lobes.

Geographic distribution: Argentina. This species probably originates from Tucumán; it was imported with *Echinocactus Saglionis* Cels. 1901.

Etymology

Named for Franz De Laet (1866-1928), a Belgian succulent specialist and nurseryman from Contich.

Other Names

Gymnocalycium delaetii (K. Schumann) Hosseus in *Revista del Centro Estudiantes de Farmacia, Córdoba* 2(6): p. 22 (1926)

Synonym

G. antherostele Ritter in *Kakteen in Südamerika* 2: pp.475-476 (1980)

First description of *Gymnocalycium antherostele* Ritter sp. nov.

Body only slightly raised, Single, 8-12cm thick, dark green. Ribs 13, c. 15mm high, strongly notched, becoming thicker at the areole, thin between the areoles, the separating groves are snake-like, humps rounded, without chins. Areoles oval, 4-6mm long, c. 2cm apart, on top of the humps. Spines, 5, rarely 7, 1 pair below, two pairs (sometimes 3 pairs) pointing sideways, all 10-15mm long, matt brown-yellow to brown-red, the ends darker, thick awl-shaped, all slightly porrect and recurved toward the body, the fine points are rarely straight, slightly curved, sometimes hooked, no central spines. Flowers white, curved at the top 5cm long, odourless, opening only for a short time during the afternoon, not very wide. Ovary 18mm long, 9mm diameter, 3mm at the bottom, obconical, curved at the bottom, grey-green, with very small pale red, hardly white edged, 2-4mm wide 1.5-2mm long, rounded scales, at a distance of 4-8mm. Nectar chamber tubular, grey, 3mm long, c. 0.6mm wide round the style, style being enclosed with a ring of stamens.

Flower tube almost beaker-shaped, 12mm long, 12mm wide at the top, inside grey, outside grey-green, the lower scales the same colour as the ovary, the higher scales are larger and more green. Stamens over the nectar chamber, one basal ring, c. 5mm long, white, hair-like, close together and leaning in towards the style, the upper part has very few stamens all leaning against the wall, the top 4mm is thickly covered with stamens, the anthers pointing inwards, meeting over the top of the style, together forming a pillar, therefore the pollen falls onto the style (self-fertile), the filaments are light carmine with white ends, up to 5mm long, at the tube seam stands a denser ring of filaments, from 12mm long and much thicker, their anthers forming the top of the pillar. Anthers grey-green, pollen grey. Style pale grey-green, 1.5mm thick, 14mm long, of which 4mm is the 12 cream-white radiating lobes. Petals 20mm long, 6-8mm wide, pale pink, the lower half very narrow and straight, 1.5mm wide, at top blunt with very small points, a few inner petals, only c. 15mm long, 2.3mm wide, outer petals more

Fig.597 *G. schickendantzii delaetii, Horst-Uebelmann* 1642 from El Carril, Prov. Salta, Argentina

reddish with a pale green central stripe. Fruit usually curved, green, 25-55mm long, 15-18mm diameter at the bottom, tapering towards the top, 10-12mm thick, scales similar to those on the ovary, very little fruit flesh, slimy. Seeds brown, 1.0mm long, 0.7mm wide, 0.4mm thick, thickly covered with small humps, hilum white, long and thin, without the spongy growth.

Type: General Moldes, Prov. Salta. I discovered this plant in January, 1959. FR 963.

Commentary on *G. antherostele*

Neuhuber (2001b) points out that the correct locality for this is Coronel Moldes. He goes on to resurrect Ritter's invalid name *G. antherosacos* as a form of *G. delaetii*, which he accepts as a separate species from *G. schickendantzii*. Its type locality is south-west of Alemania, within the distribution of *G. schickendantzii* subsp. *delaetii*.

Distribution (Map 74 on page 254)

This subspecies has a much smaller and more northerly distribution than the type subspecies. It can be found to the south of Salta City, usually growing under bushes, often in the company of other *Gymnocalycium* species.

Conservation Status

Least Concern. The plant appears to be plentiful over its habitat range without any immediate threats to its survival.

History

The first description of *Echinocactus de Laetii* in MfK (1901) was accompanied by a good photograph from de Laet. Soon after, Schumann concluded that his new taxon was the same as *E.schickendantzii,* so when he reproduced the same photograph in his *Nachträge* (1903), he captioned it as such. He said that he had not realised before because he was misled by Weber's comment about the flowers being like *E. saglionis*, which actually they are not.

Schelle (1907) listed both species separately and published Fig.124 on page 191 showing a large number of imported plants of *E. schickendantzii*, presumably at the De Laet nursery, since the same picture appeared in their catalogue captioned as *E. De Laetii*.

Fig.598 *G. schickendantzii delaetii, Piltz* 55 from Quebrada de Cafayate, Prov. Salta, Argentina

Britton & Rose (1922) also considered this taxon to be a synonym of *G. schickendantzii*.

Backeberg made *G. delaetii* a variety of *G. schickendantzii* in *Kaktus-ABC*: p.296 (1936).

Commentary

This taxon is similar to the type subspecies except for a more applanate body and broader ribs divided by transverse furrows. It is the northern subspecies but has a much smaller range than the type subspecies.

Cultivation

Although it is not difficult to grow, one rarely sees large specimens of this attractive plant. It remains solitary and is rather flatter than subsp. *schickendantzii*. The flowers, which are produced over a long period, grow from areoles on the shoulder of the plant.

Illustrations

Monatsschrift für Kakteenkunde 11(12): p.187 (1901) The same image republished in Schumann *Gesamtbeschreibung der Kakteen Nachträge*: p.123 Fig.29 (1903) captioned as *Echinocactus schicken-dantzii*.

Backeberg *Die Cactaceae* III: p.1778 Abb.1709

New Cactus Lexicon Atlas (2006) Illn. 286.1

Gymnocalycium stenopleurum Ritter

First Description (Ritter)

Kakteen in Südamerika 1: pp.265-266, 339-340 (1979)

Gymnocalycium stenopleurum Ritter spec. nov.

Body simple, 6-12cm diameter at flowering age, the same height or greater. Epidermis dark green, granular with a few (or none) green spots, (*G. mihanovichii* is not grainy, but has a completely smooth epidermis). Ribs, 8-11, later 10-14 with 2-3cm high, extraordinary thin matt edges, thickened at the areole and elevated, cross-ribs the same colour the surface also has irregular wrinkles. Areoles 2.5-5mm diameter, round, with a dirty white felt on young areoles, becoming blackish, later bald, 7-17mm apart.

Spines, needle fine, awl-shaped, light to dark brown, mostly slightly curved, radial spines 3-6 c. 7-40mm long, pointing diagonally upwards and outwards, central spines usually missing, sometimes 1, away-standing, usually a little longer and stronger, the spines do not fall off and are not brittle.

Flowers, from the crown, only warmth will open them for a few hours each day, 5-7cm long, no scent. Ovary, c. 30mm long, 6-8mm thick, indian club shape, green to bluish green, scales as little wider than long, whitish, often with a purple edge. Nectar chamber, c. 3mm long, very narrow, pale, enclosed by a ring of anthers that lean in onto the style. Tube 7-10mm long, c. 10mm wide at the top, inside pale, outside greenish-red with large white edged scales. Stamens white, those of the basal ring hardly becoming thick, (on *G. mihanovichii*, thicker and woolly) with the anthers under the stigma lobes, the upper insertions from 3-5mm higher, the topmost anthers curving inwards, with the anthers over the stigma, anthers small, blackish, pollen pale yellow. Style whitish, c. 13mm long, of which 4-5mm consists of the 13-15 radiating lobes between the anthers. Petals, inner petals upright when open, 3-5mm wide, the outer petals well spread out and well curved back, inner petals white, 10-15mm long, 4-5mm wide, oblanceolate, middle petals a little larger, white, the ends often greenish-brown, outer petals 18-20mm long, 6-7mm wide, more green with reddish-brown tips and white edges, straight and short pointed.

Photo: Volker Schädlich

Fig.600 *G. stenopleurum, Schädlich* 249 from the Cerro León, Alto Paraguay, Paraguay

Fruit c.4cm long, 1cm thick, grey-green, fruit flesh white. Seeds, brownish, nearly globular, 0.8mm diameter, with small, mainly tapering humps, hilum basal, long, narrow, white.

Type habitat, Cerro León. Department Boquerón. N.W. Paraguay, discovered by myself, September, 1963. FR 1176.

Type: *F. Ritter* 1176, Paraguay, Cerro León (U)

Photo: Volker Schädlich

Fig.599 The Cerro León, Alto Paraguay,Paraguay The type locality of *G stenopleurum*

Photo: Volker Schädlich

Fig.601 *G. stenopleurum, Schädlich* 22a in habitat at the Cerro León, Alto Paraguay, Paraguay

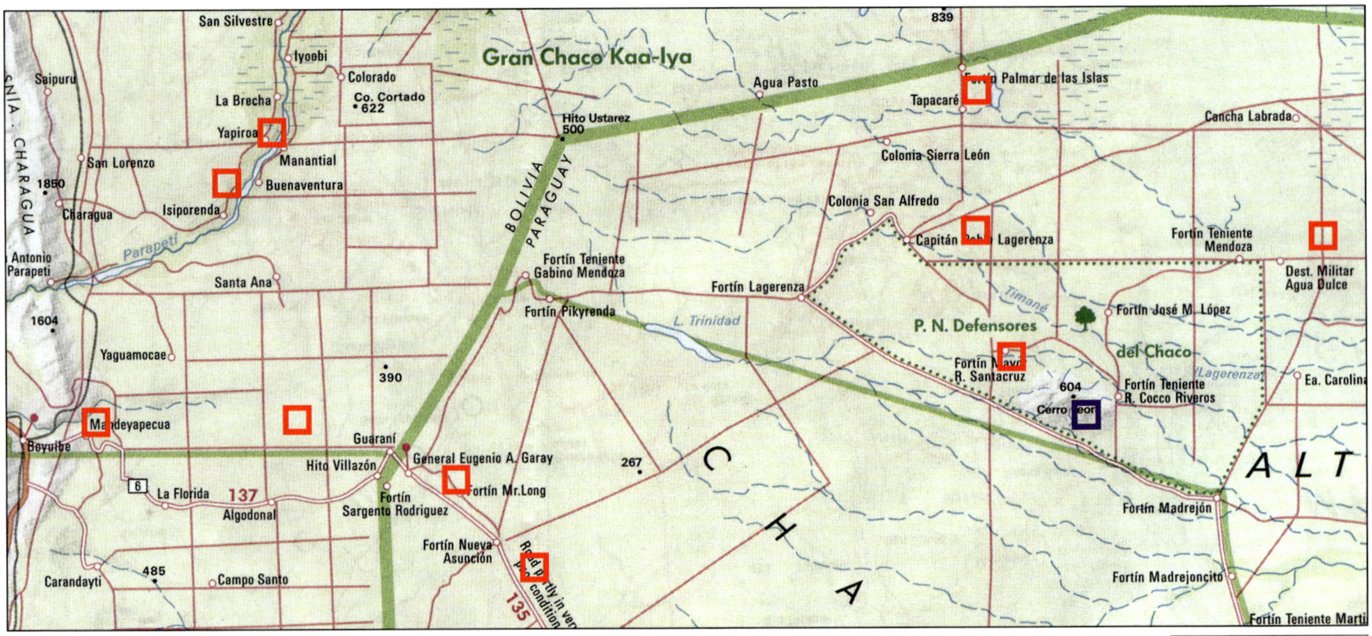

Map 75 Locations of *G. stenopleurum*, the type location is indicated with the blue square

Etymology

From the Greek words 'stenos' meaning narrow and 'pleuron' meaning rib.

Synonym

G. mihanovichii var. *friedrichii* Werdermann in *Blühende Kakteen und andere sukkulente Pflanzen* Mappe 29, Plate 113 (1936)

Photo: Volker Schädlich

Fig.602 *G. stenopleurum* (moserianum), *Schädlich* 239 in habitat south-west Nueva Asunción, Boquerón, Paraguay

Photo: Volker Schädlich

Fig.603 *G. stenopleurum* (albiflorum), *Schädlich* 244 in habitat west of Garay in Bolivia, Prov. Santa Cruz

First description of *Gymnocalycium mihanovichii* (Fric et Gürke) Britton et Rose var. *Friedrichii* Werdermann

Differs from the type being taller in proportion and narrower, increasingly thickened near the areoles, areoles themselves sunken, having more spines, which are occasionally deciduous or sometimes completely absent. Flowers deep pink, petals frequently recurved before anthesis.

Commentary on *G. mihanovichii* var. *friedrichii*

Till & Amerhauser (2005) designated a neotype: HA 94-976 = HT 2463 from Isiporenda, Bolivia, near the border with Paraguay, 270m (WU)

Pazout realised that this taxon is a different species from *G. mihanovichii* but his attempt to raise the name to specific status in *Friciana* 4(23): pp.3-19 (1964) is considered to be invalid, as is the later publication at species rank by Schütz (1979) in the same journal. As a result, *G. stenopleurum* is the oldest name for this taxon at specific rank.

Gymnocalycium anisitsii subsp. *volkeri* Amerhauser, subsp. nov. in *Gymnocalycium* 17(1): pp. 559-560 (2004)

Distribution (Map 75)

Including its synonyms, the distribution range extends from the Chaco Boreal of northern Paraguay into the adjacent Bolivian chaco. It lies to the north and west of the similar species *G.*

Photo: Volker Schädlich

Fig.604 *G. mihanovichii*, *Schädlich* 267 (left) growing with *G. stenopleurum* (friedrichii), *Schädlich* 266 north-east of Lagerenza, Alto Paraguay, Paraguay

Fig.605 *G. stenopleurum* (moserianum), *Metzing* 24 from Americo Picco, Paraguay

Fig.606 *G. stenopleurum* (moserianum), *Bercht* 2211 from Nueva Asunción, Boquerón, Paraguay

mihanovichii but the distributions overlap. The two species can occur side by side, an example being LB2219 and LB2220 growing near Capitán Pablo Lagerenza in northern Paraguay.

Conservation Status

Least Concern. There is no published information on the conservation status of this taxon but its wide distribution in a little-used region would suggest that there is no immediate threat.

History

This taxon was among those collected by Friedrich during his time as a war correspondent in Paraguay during the Chaco War (1933-35). He grew the plants in his garden at Asunción where they were seen by Harry Blossfeld and Oreste Marsoner when they visited Friedrich in 1935. Blossfeld was a nurseryman and brought plants and seeds back to Europe.

The formal description of *G. mihanovichii* var. *friedrichii* by Werdermann followed in 1936. The type locality was not stated at the time, but Moser ((1985) in his book about Friedrich called *Kakteen. Adolfo Maria Friedrich und sein schönes Paraguay*) tells us that it is by Huirapitindi, a place just to the east of Boyuibe, Bolivia.

There followed a number of publications of varieties which were just cultivated selections of the progeny of this early collection. However, there is a form that is known from the Paraguayan Chaco Boreal and has recently been recollected. It was found by Friedrich in 1964 and named invalidly as *G. friedrichii* var *moserianum* by Pazout (1966). It is said to differ from *G. friedrichii* by its larger stems and only three spines per areole. It was described from a habitat near Yrendague (today called General Garay on the Paraguay/ Bolivia border) just 50km from the type. Schütz (1979) says that the plants described as *G. mihanovichii* var. *albiflorum* by Pazout (1963) also came from this place.

Realising that *G. mihanovichii* var. *friedrichii* was a separate species, Pazout (1964) tried to publish the name at specific rank but, according to Metzing et al (1995), his change was invalid. Likewise, a later attempt by Schütz (1979) to do the same thing was also invalid. So, the valid publication of this taxon at specific rank was first achieved by Ritter when he published *G. stenopleurum* in volume 1 of *Kakteen in Südamerika* (1979). He had found the plants in 1963, but no collected plants reached Europe at the time, only seeds which were sold by his sister, Frau Winter.

Commentary

G. mihanovichii var. friedrichii is not accepted as a synonym of *G. stenopleurum* by all students of the genus, although the stated differences are slight and their habitats are close. Both have the rough epidermis which separates them from *G. mihanovichii*. Schädlich (2005b), who does not accept that they are synonyms, gives us more information about plants he saw at the type locality. He reports that the plants there can grow to 300mm tall, the flower varies from white to pink and ripe fruits are actually red. Bercht (pers. com.) says that *G. stenopleurum* usually grows on rocks whereas *G. friedrichii* prefers sand.

Cultivation

This taxon is probably most familiar to cultivators as '*G. friedrichii*' or one of its forms. It can have pretty pink flowers, even on small plants. The type form of *G. stenopleurum* is less common in collections, with its pale green, striped body and white flowers. It is notoriously tricky to grow, liking a warm environment (min. 10°C) protected from strong sunshine and with a little water to prevent excess shrivelling.

Variegated selections of this taxon are among the best selling cacti in commerce. More information on these can be found in Gordon Rowley's book 'Teratopia' (2006). First of all there was the red 'Hibotan', but now many other colour forms are propagated by grafting on to a strong stock, usually *Hylocereus* (see page 280). These colour variants have insufficient chlorophyll to survive on their own roots, so must be grafted on a stock left above ground. The *Hylocereus* stock is cold sensitive so such grafts must be kept warm in winter.

Names of Uncertain Application

Some botanists like to utilise old names for *Gymnocalycium* taxa even if their application is only a guess. Others prefer to dismiss these confused old names and use a later, preferably typified, name. There are examples of both approaches in this book depending on the history of usage and the acceptance of a neotype where available.

The following are here treated as *nomina confusa* and rejected as unattributable: Page

Gymnocalycium hybopleurum (K. Schumann) Backeberg

First Description (Schumann)

Echinocactus multiflorus var. *hybopleurus* K. Schumann in *Gesamtbescheibung der Kakteen*: p.405 (1898)

[After the entry for *E. multiflorus*]

Var. *hybopleura* [sic] (-rus) K. Sch. Body flattened, plump humps separated by deep cross-cuts, fewer spines. Flowers white with red flecks at the base.

Amplified by Backeberg when he transferred it to *Gymnocalycium*:

Gymnocalycium hybopleurum Backbg. in *Kaktus-ABC*: pp.289-90 (1936)

Gymnocalycium hybopleurum Backbg. - *Echinocactus multiflorus* hyb. K. Sch. 1896 – wide, flat, dull grey-green, sunken in the centre of the crown. c. 13 broad ribs, half round with weak transverse grooves between the areoles forming ± sharp sided chin-like humps above the grooves. Areoles elongated, with grey-brown felt. Spines whitish-grey, 3-4 pointing to each side, 1 pointing down, no central spine, all spines flexible, interwoven together, basket-like, up to 3cm long. Flowers white to greenish, with a pink throat, bell to funnel shaped, c. 4cm long. Fruit green. – N. -Argentina.

This species willingly sets side shoots, forming cushions. This species is not, as K. Schumann believed, attached to *Gymnocalycium multiflorum*. The green part of the plant is very flat. - Var. euchlorum Backbg., is grey-green, with fewer and much shorter spines.

From the Greek 'hybos' meaning tubercle or hump and 'pleuron' for rib., referring to the tuberculate ribs.

Commentary

First described by Schumann as a variety of *Echinocactus multiflorus*, which is usually considered as a synonym of *Gymnocalycium monvillei* although it was said to come from Paraguay. Till & Till (1994) believe that the var. *hybopleurus* was also from Paraguay and hence Backeberg's interpretation of the name as a plant from Córdoba was an error. Backeberg invalidly described a number of varieties, among them var. *ferox* and var. *ferocior* have particularly strong spines.

Slaba (1984) agreed with Backeberg that the plants he thought were varieties of Schumann's *hybopleurus* were a good species, so he endeavoured to describe them as *Gymnocalycium ferox* but he quoted the invalid *G. hybopleurum* var. *ferox* as its basionym, so his publication was invalid as well. Interestingly, Slaba illustrates the seeds (on page 82) of his proposed species and comments on its different appearance. It has prominent tubercles on the surface, the reason why I made it a subspecies of *G. castellanosii* which has a similar seed testa. [see *G. castellanosii* subsp. *ferocius* on page 140]

Koop (1975) illustrates and describes a number of plants from Argentina that he considers to be Backeberg's interpretation of *G. hybopleurum*. These are not from Prov. Córdoba but from Prov. Catamarca, specifically from Andalgalá and Hualfín. This is probably why plants from there were known by this name for so long. However, this interpretation cannot be accepted because it is based on a mis-application of Schumann's name. So, for this reason, Till & Till (1995) described the plants from Prov. Catamarca, Argentina as *G. catamarcense* and for which the earlier name *G. pugionacanthum* is used in this book.

Gymnocalycium leptanthum (Spegazzini) Spegazzini

First Description (Spegazzini)

Anales del Museo Nacional de Buenos Aires XI (Ser.3a;IV): p.504 (1905)

Echinocactus platensis var. *leptantha*

Body size, colour and ribs, as in the proceeding species, Spines mostly 7, strong, (7-10mm long), lying flat, straight or recurved, Flowers erect, long, thin, (60-65mm. long), petals white. 1/3 longer.

Later in *Nuevas Notas Cacológicas* (1925): p.138

Gymnocalycium leptanthum Speg.

Habitat: In the dry stony hills near Cosquin. Prov. Córdoba.

Observations: After a meticulous examination of many notes and drawings of the flowers of this species, I am convinced that this is an independent species quite separate from *Gymnocalycium platense* (Speg.) Speg.

The flowers have a different appearance from those on *G. platense*. Compare the picture of this species (Br. & R. Fig. 176) with the typical *G. platense* (Br. & R. Fig. 177), where you see the ovary extremely elongated (pedicel like) the style and stigma are the same colour, white, reaching only half way up the filaments of the uppermost stamens.

Commentary

This is another of Spegazzini's names which has been associated with a number of different taxa known in habitat today. His initial description of the plant as variety of his *Echinocactus platensis* can probably be explained by the fact that there were very few taxa of *Gymnocalycium* known to him, so the two would appear to be similar to him, even though we might now consider them as different species, as he did later.

Fig.607 Spegazzini's illustration of *Gymnocalycium leptanthum* not published until1925

Fig.608 Fig.176 of *Gymnocalycium platensis leptanthus* from *The Cactaceae* (1922) Britton and Rose, determined by Spegazzini.

Spegazzini refers to Fig.176 in Britton & Rose *The Cactaceae* Vol.III (1922) which they say he determined as *G. leptanthum,* even though it looks rather different from the plant pictured in Spegazzini (1925d). The curious shape of the flower tubes was probably caused by the plant being poorly lit so that the buds would grow towards the light.

No location is given in the first description but later, Spegazzini (1925d) tells us that it comes from dry stony hills near Cosquin in Prov. Córdoba. In this region today we can find plants here considered to belong to the widespread *G. calochlorum*. However, these plants do not look much like the old published pictures of *G. leptanthum*.

Meregalli (pers. com.) thinks *G. leptanthum* could have been the plant recently described as *G. erolesii*. It certainly has long, thin flower tubes reminiscent of the old pictures and description of *G. leptanthum*.

Till & Amerhauser (2007) put forward the case for this name being closely related to their concept of *G. quehlianum* which is not accepted here. They made the combination *G. quehlianum* subsp. *leptanthum* and treated *G. calochlorum* as a variety of it. This is a further example of the regrettable displacement of a well-known name without any persuasive evidence. They go on to make *G. terweemeanum* a further variety, a name here also considered to be unattributable (see page. 267).

Gymnocalycium mucidum Oehme

First Description (Oehme)

KuaS 1(12): pp.197-198 (1937)

Gymnocalycium mucidum sp. n. Oeh.

From Hanns Oehme

Wide globular, reddish grey-green with a grey bloom, crown sunken, light green in new growth, crown with spines and some felt. Ribs 12, with chin-like pointed humps. Areoles c. 10-15mm apart, longish, with a little wool-felt. Radial spines 6-8, the lowest pointing downwards, 4-6 spines pointing sideways, the top spine erect, all round, somewhat curved, 1-2cm long, grey, horn coloured in new growth. Flowers coming from the crown, in the white wool-felt immediately behind the spines, the wool-felt later becoming grey. Flowers c. 4-4.5cm long, wider when fully expanded, up to 4.5-5cm. Ovary naked, light mossy brown, with a dark bloom that can be wiped off, a few staggered scales with delicate pink edges, becoming progressively longer.

Outer petals a delicate yellowish-pink with a dark mid-stripe, also with a blue bloom, inner petals a delicate pinkish-yellow, the largest petals 1cm wide, from c. 2cm long, Throat a delicate madder red. Style 14mm long, delicate green (almost white), with 9 delicate ivory stigma lobes. Filaments dark ochre, with a yellowish pink bloom, many anthers overhanging the style, those below the stigma lobes form a garland around the style. The plant is conspicuous by way of its greyish body tones, dark blue ovary and its multi coloured stamens.

I found this plant as a single example in a large quantity of imported plants from South America at Mr Fortenbacher's in Dusseldorf about 5 years ago (with *Gymnocalycium gibbosum* (Haw.) Pfeiff). It originates from Argentina; I name this plant *mucidum* after its mouldy-grey body colour. The example at hand is 8cm diameter, although very shrunken it produces many flowers. This plant is placed near *Gymnocalycium sutterianum* (Schick.) Berg. Seed unknown.

Lectotype: Illustration published with the first description. Designated by Piltz. J.; Metzing, D. & Schweitzer, B. in *Gymnos* (11)22: p.56 (1994).

From the Latin 'mucidus' meaning mouldy, referring to the 'mouldy' grey body colour.

Commentary

Gymnocalycium mucidum was described in 1937 by Oehme from a single individual without knowledge of its provenance. He got the plant from the nursery of Fortenbacher from an importation of plants from South America. No type was placed in a herbarium and Oehme's collection was destroyed in the war.

Fig.609 The illustration of *Gymnocalycium mucidum* which accompanied the first description in KuaS (1937)

A cactus collector called Wilhelm Simon visited Fortenbacher in 1948 and found three plants which resembled *G. mucidum*.

Backeberg, in his *Die Cactaceae* (1959) expressed the opinion that *G. mucidum* is related to *G. mazanense*, a name treated here as a synonym of *G. hossei*. He formed this opinion because of its yellowish-pink filaments.

Most recent treatments regard this taxon as the earliest name for the plant widely known as *G. glaucum*. According to Piltz et al. (1994), there is complete agreement between its description and that of *G. glaucum* and the picture published with the first description is a good likeness for *G. glaucum*.

This opinion was refuted by H. Till (2000) who made a case for *G. mucidum* actually being a relation of *G. capillaense* and so belonging to subgenus *Gymnocalycium*. Oehme, himself, actually suggested this relationship in his first description of *G. mucidum*. Till illustrated such plants found in Córdoba which he claims to be a perfect match for the description of *G. mucidum*. Later Till (2003a) made the combination *G. capill(a)ense* var *mucidum* in support of his view.

The latest candidate for the name has been suggest by Papsch (pers. com.) to be the recently described *G. prochazkianum*, presumably because of its glaucous body.

Although I personally have a leaning towards the case of *G. mucidum* being the oldest name for *G. glaucum*, I do not think the evidence is strong enough to justify displacing the well-typified and familiar name *G. glaucum*. The added confusion caused by Till's alternative opinion further supports the view taken by The *New Cactus Lexicon* editorial team who decided to reject *G. mucidum* as being of uncertain attribution.

Gymnocalycium parvulum (Spegazzini) Spegazzini

First Description (Spegazzini)

Anales Mus. Nac. Buenos Aires II(Ser.3a;IV) *Cactacearum Platensium Tentamen*: p.505 (1905)

Echinocactus platensis var. *parvula*

Body elongated globose, small to very small, (10-30mm diameter), and high. Dirty greyish-green. Ribs often 13, straight , with flat humps. Spines 5-7, stiff, strong, flexible, adpressed. (2-4mm long), greyish-white, all radiating. Flowers erect, larger than the plant, (45-60mm long), flower tube thin, petals white, moderately long.

Then in *Nuevas Notas Cactológicas* 1925 p.141

Gymnocalycium parvulum Speg. = *Echinocactus platensis* Speg. v. *parvula* Speg., *Cact. Plat. Tent.* No. 84d.

Habitat; in the dry, very rocky hills of the Sierra de San Luis

Having carefully reviewed my notes, and descriptions of this species, I am convinced that it is not a variety of *Gymnocalycium platense* Speg. even though it is associated with, and closely resembles the external characteristics of *Gymnocalycium platense*. It is a good distinct species because of the following characteristics of the flowers:- It differs by having uniform and clearly distinct scales, the style terminating in 10-12 stigma lobes and being much shorter. Also it differs in the way the stigma lobes project beyond the lower series of anthers, but only reach the filaments of the outer stamens.

Commentary

There are no published pictures which help us to decide which plant known to us today was described by Spegazzini. The name has not been treated consistently by authorities since its description.

Till (1994a) provides a detailed explanation of his opinion. He says that Spegazzini's locality is not the familiar Sierra San Luis in Prov. San Luis but in fact refers to a sierra near a place called San Luis. He identifies this as the town of Villa San Luis which is in Prov. Córdoba between Villa Cura Brochero and Panaholma. From this, he deduces that *G. parvulum* is an older name for *G. proliferum*, a synonym of *G. calochlorum*. He describes a similar plant from further west as a new variety *amoenum*.

Lambert (2002) later raised *amoenum* to specific rank because he believed that he had found the the 'true' *G. parvulum* that was a different plant from further north. His picture of *G. amoenum* on page 227 of his article looks a lot more like *G. taningaense* than Till's original pictures of his var *amoenum*. Both taxa grow in that part of Prov. Córdoba.

The latest publication to take Till's application of the name even further is that of Berger (2008) who makes var. *amoenum* a subspecies and describes a further subspecies *huettneri* for plants from the north east of Prov. Córdoba. All these plants are considered to be forms of the variable and widespread *G. calochlorum*.

In the light of the inconsistent application of the name, and no conclusive argument to use it in place of a younger typified name, *G. parvulum* is best rejected as of uncertain application.

Gymnocalycium platense (Spegazzini) Britton & Rose

First description (Spegazzini)

Estudio de la Flora de la Sierra de la Ventana. (1896) p.28

Echinocactus platensis Speg. (n. sp.)

This species is common in the Curá-malal and in the other Sierras and Pampas, it looks similar to the plants that grow in the Sierra de Córdoba, only different by the fewer and sharp spines. The body is almost globose, the lower part elongated in the shape of a cone (6-10cm diameter, 8-10cm high) the top part is convex, the sunken centre has no spines, colour dark green to ash-grey. Usually 14 rows of tubercles (ribs), which are separated by pronounce furrows are broad and blunt, the tubercles somewhat exaggerated, (15mm diameter, 5-7mm high), hemispherical, almost 5 sided, flat on the top edge, sharp angular below with a small tooth. Areoles elliptical, sit in the centre of the tubercles, somewhat sunken, almost hairless, typically armed with 7 spines (frequently only 5 spines), irregular, almost lying flat on the surface of the plant, the 2 topmost are the shortest (5mm long), 4 laterals (10mm long), the bottom one the longest (15mm long), white, base purple, not swollen at the base, thin, sharp pointed, somewhat curved. Flowers coming from the areoles near to the shoulder of the plant, tubular bell-shaped, externally dark green with transparent-purple scales, hairless, petals white.

Habitat: Found growing in the fissures in the rocks on the hills of the valley.

Later in *Cactacearum Platensium Tentamen*: p.504 (1905)

Habitat: Generally in the area of the Sierra pampeanas (Ventana, Curámalál, Olavarría etc.) and near Córdoba.

Fig.610 Fig.177 of *G. platense* from *The Cactaceae* (1922) Britton and Rose, contributed by Spegazzini

Fig.611 the illustration of *Echinocactus platensis* from the Graessner catalogue of approx. 1909

Commentary

This is perhaps the most inconsistently applied *Gymnocalycium* name of all. It has been confused with many other species since its description. The first description states clearly that it is common in the Sierra de Cura Malal which is in the south of Prov. Buenos Aires. The only *Gymnocalycium* growing there is *G. reductum* subsp. *reductum* so I would conclude that this name is a synonym of that taxon.

Localities supporting this identification are repeated and enlarged upon in subsequent publications. Fig.177 on page 164 of Britton & Rose *The Cactaceae* Vol. III (1922) is confirmed by Spegazzini (1925) as a good representation of his *G. platense* but the buds on this plant are not like those of *G. reductum* as explained by Papsch (2001a), being more like those of *G. schroederianum*.

Fric visited the Sierra de la Ventana in 1926 looking for *G. platense* but he reported being unable to find it but rather he found a clumping *Gymnocalycium* related to *G. gibbosum* to which he gave the provisional name *G. caespitosum*.

Kiesling (1982) tried to show that there were two taxa in this area which he called *G. platense* var. *platense* and *G. platense* var. *ventanicola* but in fact there is only one variable species here for which the oldest name is *G. reductum*.

Papsch (2001a) also considered the question of *G. platense* in depth while trying to establish *G. hyptiacanthum* as the oldest name for the plants from the Sierra Bayas. He concluded that the type of *G. platense* must have come from near Olivarria and that *G. schroederianum* subsp. *bayensis* is a redescription of the same taxon. So, bearing in mind all this confusion, it is best to regard *G. platense* as unattributable.

G. rauschii H. Till and W. Till

First description (H. Till & W. Till)

Succulenta 69(2): pp.27-31 (1990)

Plants simple, strongly offsetting from the top part of the areoles, with c. 12 strong roots; body flattened, semi-globular, 50-70(-90)mm diameter, 20-30(-40)mm high, crown only slightly sunken, usually with spines; epidermis dark green to grey-green, matt.

Ribs 10-12(-14), straight, separated by long sharp furrows, flattened at the base, 15-18mm wide, towards the crown distinctly angular, with chin-like humps formed by distinct cross-cuts.

Areoles 10-12mm apart, sitting on the humps, occasionally sunken, (on older plants the areoles on the upper part of the body often project over the chin-like humps), mostly oval to longish, 5 x 2-3mm, young areoles with brownish-white felt that soon disappears.

Spines (7-)9 per areole, arranged in (3-)4 pairs (in older forms the top pair may be missing) 12-17mm long comb-like, radiating sideways and downwards: the central pair radiating horizontally, the lower pair radiating diagonally downward, the uppermost pair always the shortest horizontal or sloping upwards, a shorter, usually only 7mm long, points straight down, all spines are usually straight, seldom slightly curved or twisted ± stalk-like, stiff, awl shaped, pointing towards the body, these however are not lying flat, in new growth yellowish, becoming red-brown, later, as the spines become rough-scaly, they appear to be grey in colour, occasionally the base of the spines remain yellow.

Flowers coming from the crown usually all at the same time, funnel shape, 25-35mm long, 23-30mm in diameter, light pink ± odourless, with a tendency to single sex flowers. Pericarp slightly conical, thick walled, 8 x 8mm, dark green, with many egg-shaped spatulate reddish-brown white-edged scales, as the higher ones are longer they become the outer petals. Receptacle ± bell shaped, thick walled, without a nectar chamber, outer petals spatulate, rounded at the tip, c. 18 x 7mm yellowish-green at the base, pale pink towards the top with a wide brownish-green central stripe. Inner petals lanceolate, pointed, 21-23 x 4-5mm, whitish-pink with a dark pink central stripe, white at the base.

Stamens mainly curved to the wall of the perianth tube, not in two separate circles, filaments thin, 9mm long, whitish to pale green, anthers oval, 1.2 x 0.7mm, whitish to cream coloured, pollen whitish, on female plants the anthers are stunted, style ± free standing, not as tall as the stamens, stigma c. 9-10mm, long, at the base c.1.4mm, diameter, above c. 1.1mm diameter, yellowish-white, with 7-9 stigma lobes, stigma whitish, on the male plants the style is aborted; ovary in section is heart to urn shaped 6 x 3-4mm, inside wall whitish; with many ovules filling the ovary. Fruit dry when seeds are ripe, small, oval, 7 x 5mm, still with the small thick flower remains, few seeds, 5-7, seeds blunt elongated-egg shape, black (brown), matt, bare, 1.5-1.56 x 1.5-1.56mm, testa cells small, curved wart-like, with a net like thickened border, hilum micropyle area rhomboid, with a distinct reduced funicle, thinner, light cream colour cellular tissue. (Ovatisemineum Schütz).

Fig.612 *G. rauschii, Rausch* 350A, a propagation of one of the original clones.

Type: *W. Rausch* 350 p.p. = HT 408 in coll. H.Till. Uruguay, Dept. Tacuarembo, near Ansina.

Commentary

This plant looks much like *G. hyptiacanthum uruguayense*, a *Macrosemineum* plant also from Uruguay, but *G. rauschii* has seeds more like those of subgenus *Gymnocalycium*.

Rausch brought back four small plants from his visit to Uruguay in 1967 but it is not known if he found them in the wild. He gave his number *Rausch* 350 to all Gymnocalyciums from Uruguay. Although we are told a locality, the plant has not been found there since.

The suspicion is that these plants are hybrids produced in cultivation from *G. hyptiacanthum* and a member of the *Gymnocalycium* seed group such as *G. schroederianum*. The name is best abandoned and treated as unattributable.

Gymnocalycium stuckertii (Spegazzini) Britton & Rose

First Description (Spegazzini)

Cactacearum Platensium Tentamen (1905) p.502

Echinocactus stuckertii Speg. (n. sp.)

Diagnosis: Hybocactus, flattened-globular, dark green, with a small depression in the centre of the crown. Ribs 9-11, wide blunt with serrated, wide at the base humps. Areoles elliptical. Spines 7-9, all radials, 6-8 pointing sideways, the lowest pointing straight down, round in section, all curved back. ash grey, dark brown at the tip. Flowers medium size, coming from the side of the crown, outside dark green, tinted blue, almost covered with scales; petals spatulate, whitish to pink; stigma whitish, 12 lobes.

Habitat: On the dry hills, Prov. San Luis, Córdoba, Tucumán and Salta.

Observations: Body 60-65mm diameter, 35-40mm high, crown slightly sunken, with humps, almost without spines and felt. Ribs on top sharp, later flattened, with 3-5 humps. Areoles elongated, 7-8mm long, 4-5mm diameter, 10-15mm apart. Spines rigid, rough, 10-24mm long. Flowers odourless, often solitary, 40mm diameter. Scales semi-circular with lilac-white edges; petals somewhat fleshy. Stamens and pollen yellow, style greenish.

Fig.613 Fig.180 of *Gymnocalycium stuckertii* from *The Cactaceae* (1922) Britton and Rose. The picture was supplied by Spegazzini and resembles *G. monvillei.*

Commentary

Named by Spegazzini for Theodore Stuckert, a Swiss pharmacist who lived in Córdoba and helped him with his cactus studies.

The identification of the name with plants known today has been inconsistent. This has, in part, been caused by the illustration reproduced in Britton & Rose [*The Cactaceae* Vol. III, Fig.180, page 165 (1922)] which, although said to be provided by Spegazzini, does not match the original description and looks like *G. monvillei*.

Its stated widespread distribution and flowers from the side of the crown could only be *G. schickendantzii* which is the opinion of Dölz (1957) judging by his unattributed illustration. Kiesling associates the name with *G. sutterianum* which he believes came from Prov. San Luis, the plant that here is called *G. fischeri*.

This name is a catalogue of confusion and is best ignored forever. It is here treated as a *nomen confusum*.

Gymnocalycium terweemeanum (Teucq ex. Duursma) Borgmann & Piltz

First Description (Duursma)

***Echinocactus platensis* v. *Terweemeanus* Teucq in *Succulenta* 12(7): pp. 137-139 (1930)**

Here we give the scientific diagnosis of a novelty, imported by the firm De Laet in 1908, which is named after our member Mr. Arnold Ter Weeme of Neede:-

Body simple, flattened globular, up to 7cm diameter, shrinking at the base. No offsets forming, slightly sunken in the crown. Colour dark green, with a coppery tint. Ribs 8-9, flat, formed by vertical furrows, which are separated into tubercles by shallow horizontal cross-cuts. Areoles on the tubercles, at first woolly, later naked. Radial spines 5-7, up to 1cm long, thin, sometimes slightly curved, grey, brownish at the base. Centrals lacking, or sometimes one, then darker and shorter than the radials.

Flowers from the youngest areoles, 5cm long and broad, thin funnel shape. Flower tube 18mm long, with mostly 6 small rounded scales, pink, white edged. Outer petals 6mm long, gradually becoming longer towards the centre, there, up to 25mm long, 6mm wide, lanceolate with a rounded tip, light to dark pink, deep pink to red at the base. Filaments pink at the base, white at the top, anthers yellow. Style thick and short, the same colour as the filaments, with 10 whitish stigma lobes, hardly reaching the stigma. Fruit oval, 15-30mm, broad, green, with few white edged, 5mm wide scales. Seeds 1mm broad, black, globular with a flat hilum.

Habitat: Argentina near Mendoza.

Other Names

Echinocactus platensis v. *Terweemeanus* Teucq in *Succulenta* 12(7): pp. 137-139 (1930)

G. quehlianum subsp. *leptanthum* var. *terweemeanum* (Teucq ex. Duursma) H. Till and Amerhauser

Commentary

Pictures of this plant appeared in the De Laet catalogue of 1930, the same year as the above description was published by Duursma, the then editor of *Succulenta*. The name was attributed to Arille Teucq who took over the running of the De Laet nursery after the death of Franz De Laet in 1928 until it closed in 1930. Its stated habitat of Mendoza can be disregarded since at that time origins were often confused, due to plants being collected on the way to or from the principal destination of an expedition.

Although this name was validly published, it was overlooked by authors of cactus books until Borgmann & Piltz (1997) brought it to our attention and elevated it to specific rank. Associating it with a population known today is a challenge and there are differing opinions. It clearly belongs to the *Gymnocalycium* seed group so Borgmann & Piltz (1997) consider that the most likely candidates are *G. berchtii* or the undescribed *G. lepidum* n. n. *Piltz* 192 from southern Córdoba or even *G. altagraciense* nom. prov.

Fig.614 The illustration of *Echinocactus platensis* var. *Ter Weemeanus* A.Tq. from the DeLaet catalogue (1930)

Till & Strigl (1998) conclude that its correct identification is *G. leptanthum* Spegazzini but this is another name that is difficult to attribute. Neuhuber (1999) suggests that plants of *Neuhuber* 88-81/203 from La Toma, Prov. San Luis, are a good match for *G. terweemeanum* and his illustration (p.299) appears to support this view.

Most recently, Till & Amerhauser (2007) have made *G.. leptanthum* a subspecies of *G. quehlianum* (sensu Till) and *G. terweemeanum* a variety of that. This speculation only adds to the confusion and supports the treatment of this name as unattributable.

Glossary and Abbreviations

Anther Part of the stamen which produces the pollen.

Applanate Flattened. Often referring to a plant body which is broader than high

Appressed Flattened against the stem.

Areole A highly modified short shoot unique to cacti, from which flowers and offsets arise. It is usually a felty cushion with spines or hairs growing out of it.

BCSJ Journal of the British Cactus and Succulent Society.

Caespitose Growing a clump of stems from a common base.

Calyx The outer series of perianth segments comprising a number of scale-like tepals which protect the flower.

CAM Crassulacean Acid Metabolism. A method of assimilation adopted by cacti to reduce water loss by allowing the stomata to open only at night.

Central Spine One of the innermost spines of the areole.

CITES Convention on International Trade in Endangered Species.

Corolla The inner whirl of the perianth comprising the petals of a flower.

Cristate A form of abnormal growth when the growing point becomes linear and the resulting growth is in the shape of a fan. Known also as fasciation or a crest.

CSI Cactaceae Systematic Initiatives. Bulletin of the International Cactaceae Systematics Group published since 1996.

CSJ(GB) Journal of the Cactus and Succulent Society of Great Britain.

CSJ(US) Journal of the Cactus and Succulent Society of America.

Cultivar A propagated strain of a plant developed or selected in cultivation.

Dehiscent The natural opening or splitting of a fruit when ripe.

Dioecious Each flower has functional male or female parts only, but may include non-functional parts of the opposite sex. Each individual only bears flowers of one functional sex.

Endemic Having a limited or localised distribution.

Epidermis The outermost protective layer of cells of the stem or leaf.

Epiphytic A plant which is growing on another plant using it for support but not obtaining any nutrient from it.

Filament The elongated stalk of a stamen which carries the anther at its tip.

Funicle The thread-like structure that connects the ovule to the carpel.

Geophyte With much of the plant below ground, often with a thickened root or caudex.

Growing-point The cells at the tip of a shoot or root which are capable of division to produce new tissue.

Halophyte A plant able to live in a saline habitat

Hilum The scar-like part of the seed where it was attached to the funicle.

HMR The hilum-micropylar region comprising the hilum and micropyle of a seed.

Holotype The preserved specimen designated by the author as the nomenclatural type of a name.

Infraspecific Ranks of classification below the level of species

Isotype Duplicates of the type collection which have not been designated as the holotype.

KuaS: Kakteen und andere Sukkulenten, journal of the German Cactus Society (D.K.G.).

ICBN The International Code of Botanical Nomenclature

Lectotype The specimen selected from the original collection where a holotype does not now exist or was not originally designated.

MfK Monatsschrift für Kakteenkunde. The first German journal which became the official publication of the German Cactus Society. Published from 1891/2 until 1922

Micropyle The small opening in a seed left by the pollen tube after fertilisation, and through which water can enter prior to germination.

Neotype The type selected when the original type, or material relating to the original description, no longer exists.

NCSJ Journal of the National Cactus and Succulent Society (UK).

Nom. confus. A name of uncertain or variable application.

Nom. illeg. A illegitimate name because it does not comply with the rules of the ICBN.

Nom. inval. An invalid name.

Nom. nud. An undescribed name which is therefore invalid.

Offset A side shoot which can be detached for propagation.

Ovary The chamber at the base of the style that contains the ovules which, after fertilisation, will become the seeds.

Papillose Covered in minute projections of epidermal cells.

Pectinate In the shape of a comb, radiating from along a straight line.

Perianth The envelope of a flower consisting of segments which constitute the calyx and corolla.

Pericarp The wall of the fruit which is derived from the wall of the ovary.

Pericarpel The wall of the ovary.

Radial Spine One of the outermost of the spines of an areole which are usually radiating or appressed.

Rank The levels of classification of a plant accepted by the ICBN. e.g. Genus, species.

Receptacle Referring to the outer part of the floral tube in cacti, otherwise termed hypanthium.

s.n. sine numero. Without number, usually referring to a collection number.

Scion The upper part of a graft. The piece of the desired species which is grafted onto a stock of a strong-growing species, usually from another genus.

Self-fertile Able to produce fertile seed having been pollinated with its own pollen.

Self-sterile An individual plant unable to produce viable seed or any seed from pollination using its own pollen.

Solitary An undivided stem, singular.

Stamen The pollen-bearing male part of the flower.

Steppe Grassland plain with only scattered vegetation, such as the pampa.

Stigma The female part of the flower at the tip of the style which receives the pollen.

Stigma lobes The divisions of the stigma, often papillate or sticky to trap the pollen.

Stock The strong-growing base of a graft onto which the desired plant (scion) is placed.

Stomata Minute pores in the epidermis through which gases can pass.

Style The stalk-like carpel projection which bears the stigma and through which the pollen tube grows to reach the ovary.

Sympatric Occurring together or at least having overlapping distributions.

Synonym A plant name later included in a broadened concept of a species (or other taxanomic rank) with an older name.

Taxon A general term for any taxonomic category, e.g. species, genus etc.)

Taxonomy The study of the classification of organisms.

Tepals The segments of the corolla of a flower, often coloured to attract the pollinator.

Testa The outer protective layer of a seed, the seed coat.

Tricolpate Having three apertures, used in the description of pollen.

Tubercles Conical outgrowths from the body of a plant.

Turgid Filled with water.

Type The permanent record of the identity of a name. Normally a preserved specimen of parts of a plant.

Type Locality The place where the type was collected

Type Species The species selected as the type of a higher rank, such as a genus.

Type Specimen The permanently preserved specimen representing the type or anchor of a name.

Valid Name One published in accordance with the rules of the International Code of Botanical Nomenclature.

Further Reading

Anderson, E. F. (2001) The Cactus Family. Timber Press Inc. Portland, Oregon.
Backeberg, C. (1958-1962) The Cactaceae. Handbuch der Kakteenkunde Vols. 1-6. Gustav Fischer, Jena
Backeberg, C. & Knuth, R. (1936) Kaktus-ABC, Nordisk Forlag, Kobenhavn
Backeberg, C. (19780 Cactus Lexicon. Blandford Press, Poole, Dorset.
Berger, A. (1929) Kakteen, Eugen Ulmer, Stuttgart
Britton, N. L. & Rose, J. N. (1919-1923) The Cactaceae. Carnegie Institution, Washingtom
Buxbaum, F. (1958) Cactus Culture Based on Biology
Buxbaum, F. (1968) in Krainz Die Kakteen
Charles, G. (2003) Cacti and Succulents. An illustrated guide to the plants and their cultivation
Eggli, U. (2005) Das Grosse Kakteen-Lexikon
Eggli, U. & Newton, L.E. (2004) Etymoogical Dictionary of Succulent Plant Names
Förster, C. F. (1846) Handbuch der Cacteenkunde. Leipzig
Ginns, R. (editor) (1966?) Gymnocalyciums. The Succulent Plant Institute.
Haage, W. (1981) Kakteen von A bis Z. Neumann Verlag, Leipzig, Radebeul
Hosseus, C.C. (1939) Notas Sobre Cactaceas Argentinas
Hunt, D. (1999) CITES Cactaceae checklist, 2nd edition. R.B.G., Kew & IOS
Ito, Y. (1957) Explanatory Diagram of Austroechinocactinae. Japan Cactus Laboratory, Tokyo.
Kiesling, R. (1984) Recompilación, en edición facsimilar, de todos los trabajos o referencias sobre Cactaceas Publicados por el Dr. Carlos Spegazzini. Librosur Ediciones, Buenos Aires.
Kreuzinger, K. (1935) Verzeichnis amerikanischer und anderer Sukkulenten mit Revision der Systematik der Kakteen
Kulhánek, T. & Odehnal, J. (2006) Variabilität von Gymnocalycium spegazzinii. Kaktusy special 3.
Labouret, J. (1853) Monographie de la Famille Des Cactées. Paris
Lambert, J. (1993) Cactus d'Argentine. Concordia, Roeselare.
Meregalli, M. (1985) Il genere Gymnocalycium Pfeiffer. Piante Grasse 5(1): pp.5-63
Moser, G. (1985) Kakteen. Adolfo Maria Friedrich und sein schönes Paraguay. Kufstein.
Osten, C. (1941) Notas Sobre Cactáceas. Anales del Museo de Historia Natural de Montevideo 2(5) No.1
Papsch, W. (2008) Il Genere Gymnocalycium. Piante Grasse Speciale.
Pilbeam, J. (1995) Gymnocalycium. A Collector's guide. A. A. Balkema Publishers, Rotterdam/Brookfield
Pin, B.P. & Simon, J. (2004) Guía ilustrada de los Cactus del Paraguay.
Putnam, E. W. (1978) Gymnocalyciums. National Cactus and Succulent Society Handbook No. 5.
Ritter, F. (1979-1980) Kakteen in Südamerika Volumes 1 & 2. Self published.
Rowley, G (2006) Teratopia. The World of Cristate and Variegated Succulents. Cactus & Co.
Schelle, E. (1907) Handbuch der Kakteenkultur. Eugen Ulmer. Stuttgart.
Schumann, K. (1897-1899) Gesamtbescheibung der Kakteen. N. Neumann, Neudamm, Germany.
Schumann, K. (1903) Nachträg 1898-1902. N. Neumann, Neudamm, Germany.
Schütz, B. (1986) Monografie rodu Gymnocalycium. Vydal Klub Kakusaru Astrophytum, Brno.

Journals

British Cactus & Succulent Journal	Journal of the BCSS since 1983 (Titled *Cactus World* since 2006)
Cactus (Paris)	Journal of the French 'Association des Amateurs de Cactées et Plantes Grasses', published from 1946 to 1967, some issues are rare
Cactus Adventures	Journal of the French Society ARIDES, English edition published since 1996
Cactus & Co	Open sale journal published in Italy since 1997 (Italian & English)
Cactus & Succulent Journal of Great Britain	Journal of the C.& S. Society of Great Britain from 1932 to 1982 (German)
Cactus & Succulent Journal (US)	Journal of the C.& S. Societ of America since 1929
Chileans	Published by The Chileans, study group of South American Cacti since 1966
Gymnos	Published by Arbeitsgruppe Gymnocalycium from 1984 to 1997 (German)
Gymnocalycium	Published by Arbeitsgruppe Gymnocalycium Österreich since1988
Kakteen und andere Sukkulenten	Journal of the German Society (DKG), published under various names since 1891, early issues are very rare (German)
Kaktusy	Journal of the Czech Society, published since 1965 (Czech)
National C & S Journal (UK)	Journal of the National Cactus & Succulent Society from 1946 to 1982
Piante Grasse	Journal of the Italian Society AIAS, published since1981 (Italian)
Succulenta	Journal of the Dutch Society, published since 1919 (Dutch)

References

Amerhauser, H. (1998) *Eine neue Subspecies von G. pflanzii aus dem Tal des Río Paichu.* Gymnocalycium 11(4): pp.265-268

Amerhauser, H. (1999) *Gymnocalycium chacoense, eine bemerkenswerte neue Art aus Südbolivien* Gymnocalycium 12(4): pp.301-304

Amerhauser, H. & H. Till (2006) *Ergebnisse langjähriger Pflanzenstudien und Beobachtungen an den heimatlichen Standorten. Teil 2: G.castellanosii subsp. bozsingianum.* Gymnocalycium 19(3): pp.671-676.

Amerhauser, H. (2008) *Die beiden Taxa G.horstii Buining und G.bueneckeri Swales.* Gymnocalycium 21(3): pp.791-802

Arakaki, M.; Soltis, D.E. & Speranza, S. (2007) New Chromosome counts and evidence of polyploidy in Haageocereus and related genera in tribe Trichocereeae and other tribes of Cactaceae

Arechavaleta, J. (1905a) *Echinocactus uruguayensis Arech. nov. sp.* Flora Uruguaya - Anales del Museo Nacional de Montevideo V: pp.218-220

Arechavaleta, J. (1905b) *Echinocactus melanocarpus Arech. nov. sp.* Flora Uruguaya - Anales del Museo Nacional de Montevideo V: p.220

Backeberg, C. & Knuth, F. (1936) *Gymnocalycium albispinum Bckbg. sp. n.* Kaktus-ABC: pp.285, 416

Backeberg, C. & Knuth, F. (1936) *Gymnocalycium Castellanosii Bckbg. sp. n.* Kaktus-ABC: pp.287, 416-417

Backeberg, C. & Knuth, F. (1936) *Gymnocalycium grandiflorum Bckbg. sp. n.* Kaktus-ABC: pp.289, 417

Backeberg, C. & Knuth, F. (1936) *Gymnocalycium Ochoterenai Bckbg. sp. n.* Kaktus-ABC: pp.293, 417

Backeberg, C. (1932) *Echinocactus (Gymnocalycium Br. u. R.) prolifer Bckbg. n. sp.* Der Kakteen-Freund 1(12): pp.132-133

Backeberg, C. (1934a) *Gymnocalycium nigriareolatum Backeberg.* Blätter für Kakteenforschung, Liefg. 5, Bl. 74-1

Backeberg, C. (1934b) *Gymnocalycium oenanthemum Backeberg.* Blätter für Kakteenforschung Liefg. 9: Bl.74-4

Backeberg, C. (1966a) *Gymnocalycium hybopleurum var. ferocior Backeberg* (nom. inval.) Kakteenlexicon: p.168

Backeberg, C. (1966b) *Gymnocalycium pugionacanthum Backeberg* (nom. inval). Kakteenlexicon: pp.172-173

Beckert, K. (2006) *Gymnocalycium armatum Ritter - 2001 wiederentdeckt.* Gymnocalycium 19(1): pp.651-654

Bercht, C.A.L. (1985) *Gymnocalycium hybopleurum - Wat is dat?* Succulenta 64(7-8): pp.159-162

Bercht, C.A.L. (1994) *Gymnocalycium baldianum - seine Synonyme und eine neue Varietät.* Gymnos 11(21): pp.1-9

Berger, A. (1929) *Echinocactus Bodenbenderianus Hosseus (1928). – Gymnocalycium Berger.* Berger's Kakteen: pp.221-222

Berger, F. (2003) *Gymnocalycium fischeri, eine weit verbreitete Art aus der argentinischen Provinz San Luis.* Gymnocalycium 16(4): pp.533-540

Berger, F. (2005) *Eine interessante Sippe aus der nördlichen Sierra de Sañogasta, Prov. La Rioja, Argentinien: Gymnocalycium ritterianum Subsp. acentracanthum.* Gymnocalycium 18(1): pp.599-602

Berger, F. (2006) *Eine beachtenswerte Sippe aus der Sierra de Los Colorados, Prov. La Rioja, Argentinien. Teil 4: Gymnocalycium coloradense spec. nov.* Gymnocalycium 19(4): pp.691-696

Berger, F. (2008) *Charakterisierung Verbreitung und geografische Differenzierung von Gymnocalycium parvulum.* Gymnocalycium 21(2): pp.761-766

Bödeker, F. (1918) *Echinocactus Joossensianus Böd. spec. nov.* MfK 28: pp.40-44

Bödeker, F. (1930) *Echinocactus Andreae Böd sp. n.* Monatsschrift der DKG 2(10) pp.210-212

Bödeker, F. (1932) *Echinocactus calochlorus Böd sp n.* Monatsschrift der DKG 4(11): pp.260-262

Bödeker, F. (1934) *Echinocactus (Gymnocalycium) rhodantherus Böd., sp.n.* Kakteenkunde 2(1): pp.13-14

Borgmann, W. & Piltz, J. (1997) *G. terweemeanum (Teucq ex. Duursma) Borgmann & Piltz.* Gymnos 13(25/26): pp.37-50

Borth, H. & Koop, H. (1976) *Gymnocalycium carminanthum Borth & Koop spec. nov.* KuaS 27(4): pp.73-79

Braun, P.J. & Esteves, E. (1995) *Nieuwe Combinaties en Namen voor Cactussen uit Brazilie, Bolivia en Paraguay.* Succulenta 74(3): pp.130-135

Braun, P.J. & Hofacker, A. (2002) *Gymnocalycium horstii subsp. bueneckeri comb. et stat. nov.* Schumannia 3 p.188

Britton, N. & Rose, J. (1922) *Gymnocalycium spegazzinii nom. nov.* The Cactaceae Vol. III

Buining, A .(1968) *Collecting Cacti in Uruguay.* National Cactus and Succulent Journal 23(4): p.95

Buining, A. (1970) *Gymnocalycium horstii Buining spec. nov.* KuaS 21(9) pp.162-165

Buining, A. & Brederoo, A. (1975) *Gymnocalycium matoense Buining & Brederoo.* KuaS 26(12): pp.265-268

Buxbaum, F. (1968) *Gattung Gymnocalycium.* Krainz, H. (ed.) Die Kakteen Lieferung 38-39 CVIf.

Cárdenas, M. (1958) *Die Gattung Gymnocalycium in Bolivien.* KuaS 9(2): pp.21-27`

Cárdenas, M. (1960) *Die Gattung Gymnocalycium in Bolivien.* KuaS 11(3): pp.46-47

Cárdenas, M. (1963) *Gymnocalycium chiquitanum Card. spec. nov.* Cactus (Paris) 18(78): pp.95-97

Cárdenas, M. (1966) *New Bolivian Cactaceae Part XI.* CSJ(US) 38(4): pp.141-147

Castellanos, A. (1950) *Gymnocalycium ragonesii Castell. sp. nov. Dos especies nuevas de Cactaceas Argentinas.* Lilloa 23: pp.5-13

Castellanos, A. & Lelong (1939) *Gymnocalycium immemoratum Castellanos et Lelong nov. sp.* Lilloa 4: pp.195-196

References

Cels, J. (1847) *Echinocactus saglionis Cels.* Portefeuille des Horticulteurs: 180

Charles, G. (2005) *Gymnocalycium.* Cactaceae Systematics Initiatives 20: pp.17-18

Charles, G. & Meregalli, M.(2008) *Our latest thoughts on Gymnocalycium.* Cactaceae Systematics Initiatives 24: pp.21-27

Dams, E. (1904) *Echinocactus Damsii K. Sch.* Monatsschrift für Kakteenkunde 14(5): pp. 76-77

Diers, L. (2008) *Over Chromosomen bij Cactussen - ter Herinnering aan Albert Buining* (I). Succulenta 87(2): pp.83-92

Dölz, B. (1938) *Gymnocalycium sanguiniflorum Werd., ein Synonym.* KuaS 2(3): pp.79-82

Dölz, B. (1943) *Gymnocalycium deeszianum Dölz. spec. nov.* Kakteenkunde 11(3): pp.54-55

Dölz, B. (1957) *Zur Klärung einiger Gymnocalycium-Arten.* Sukkulentenkunde VI: pp.28-32

Donald, J. (1971) *Gymnocalyciums from Alfred Lau (Part. 1).* NCSJ 26(4): pp.96-100.

Donald, J. (1972) *Gymnocalyciums from Alfred Lau (Part. 2).* NCSJ 27(1): pp.6-9.

Duursma, G.D. (1930) *Echinocactus platensis var. Terweemeanus Teucq.* Succulenta 12(7): pp.137-139

Eggli, U. (1987) *A type specimen register of cactaceae in Swiss herbaria.* Trop. Subtrp. Pflanzenwelt 59: pp.1-124.

Eggli, U. & Metzing, D. (1992) *Proposal to conserve the orthography of 5408a Gymnocalycium Pfeiffer ex Mittler (Cactaceae)* Taxon 41(1); pp.141-142

Eggli, U. & Leuenberger, B. (2005) *The Cárdenas type specimens of Cactaceae names in the herbarium of the Instituto Miguel Lillo, Tucumán, Argentina (LIL).* Willdenowia 35: pp.179-192

Elsner, J. (1970) *An Attractive Hybrid Gymnocalycium.* NCSJ 25(2): p.52

Ferrari, O. (1985) *Gymnocalycium kieslingii a new species from La Rioja, Argentina* CSJ(US) 57(4): pp.244-246 (1985)

Fleischer, Z. (1962) *Gymnocalycium pungens Fleischer, sp. n.* Friciana 1(7): pp.1 & 8

Frank, G. (1963) *Gymnocalycium horridispinum Frank* (nom. inval). KuaS 14(1): pp.8-10

Frank, G. (1964) *Schwierigkeiten der Artdiagnose bei variablen Formen Erläutert an Importpflanzen von Gymnocalycium spegazzinii.* KuaS 15(6): pp.116-118

Frank, G. (1967) *Gymnocalycium damsii - Formen aus Ostbolivien.* KuaS 17(8): pp.155-158

Frank, G. (1970) *Der Formenkreis Gymnocalycium capillaense-sigelianum-sutterianum.* KuaS 21(8): pp.142-145

Frank, G. (1976-77) *The Genus Gymnocalycium. Parts I -VI.* CSJ(US) (58): pp.215-218; 265-267 CSJ(US) (59) pp.18-21; 66-70; 128-132; 149-151

Fric, A.V. & Gürke, M.(1905) *Echinocactus Mihanovichii Fric & Gürke.* Monatsschrift für Kakteenkunde 15(9): pp.142-143

Fric, A.V. & Pazout, F.(1948) *Gymnocalycium mihanovichii var stenogonum.* Kaktusarske Listy (3): pp.17-19

Gerloff, N. (1993) *Am Heimatstandort von Gymnocalycium denudatum Pfeiffer ex Mittler.* Gymnos 10(19): pp.11-19

Gerloff, N. (1997) *Gymnocalycium - Arten in Brasilien.* Gymnos 13(25/26): pp.18-24

Gerloff, N. (1998) *The cacti of Western Rio Grande do Sul.* Cactus & Co. 2(3): pp.5-15

Gerloff, N. (1999) *Where Gymnocalycium horstii Buining has its roots.* Cactus & Co. 3(4): pp.179-186

Gerloff, N. & Metzing, D.(2002) *Zur Verbreitung der Gattung Gymnocalycium in Südbrasilien.* KuaS 53(2): pp.29-39

Gürke, M. (1905) *Echinocactus saglionis Cels.* Tafel 58 in Blühende Kakteen Band 5

Gürke, M. (1906a) *First description of Echinocactus Kurtzianus* MfK 16: p.55

Gürke, M. (1906b) *Echinocactus Mostii Gürke n.sp.* Monatsschrift für Kakteenkunde 16(1): pp.11-12

Jeggle, W. (1973) *Gymnocalycium striglianum Jeggle spec. nov.* KuaS 24(12): pp.267-268

Haage, W. (1971) *What does the true Gymnocalycium hossei really look like?* NCSJ (26)4: p.104

Haage, W. (1898) *Die Varietäten des Echinocactus denudatus Lk. & Otto.* MfK 8: pp.36-37

Halda, Kupcák, Lukasik & Sladkovsky (2002) *G. fischeri* Acta musei Richnoviensis 7(2): pp.41-70

Heese, E. (1911) *Echinocactus Gürkeanus Heese nov. spec.* MfK 21(9): p.132

Herter, G. (1951) *A new Uruguayan cactus and observations on the flowering of a Uruguayan cactus.* Revista Sudamericana de Botanica 10(1): pp.1-2

Herter, G. (1954) *Flore Illustrée de l'Uruguay Cactacées* Cactus 41: pp.91-96

Hofacker, A. (1999) *Gymnocalycium denudatum aus Südbrasilien.* KuaS 50(7): pp. 166-168

Hooker (1837) *Echinocactus mackieanus* Curtis's Botanical Magazine plate 3561

Hooker (1845) *Echinocactus leeanus* Curtis's Botanical Magazine 71: plate 4184

Hosseus, C.C. (1926a) Revista Centro Estud. Farm. Cordoba 2(6): p.16

Hosseus, C.C. (1926b) Revista Centro Estud. Farm. Cordoba 2(6): p.22

Hunt, D., Taylor, N. & Charles, G. (2006) The New Cactus Lexicon. DH books.

Hutchison, P. (1957) *Icones Plantarum Succulentarum. 5. Gymnocalycium westii.* CSJ(US) XXIX(1): pp.11-14

Hutchison, P & Donald, J.(1959) *The Genus Weingartia (Gymnocalycium).* NCSJ 14(2): p.38

Hüttner, T. (2006) *Ergebnisse langjähriger Pflanzenstudien und Beobachtungen an den heimatlichen Standorten. Teil 3: Gymnocalycium castellanosii subsp. bozsingianum var. armillatum Piltz.* Gymnocalycium 19(3): p677-682

Jajó, (1934) *Gymnocalycium Valnicekianum Jajo* Kaktusar 5(7): pp.73-74

Jeggle, W. (1973) *Gymnocalycium striglianum Jeggle spec. nov.* KuaS 24(12): pp.267-268

Kessler, A. (1981) *Wohin gehört es nun, das Gymnocalycium eytianum Cárdenas?* KuaS 32(11): pp.266-268

Kiesling, R. (1980) *Gymnocalycium mesopotamicum sp. nov.* CSJ(GB) 42(2) pp.39-42

Kiesling, R. (1982) *Nota sobre Gymnocalycium platense (Speg.) Br. et Rose (Cactaceae).* Darwiniana 24: pp.437-442

Kiesling, R. (1987) *Two new subspecies of Gymnocalycium schroederianum.* CSJ(US) 59(2): pp.44-49

Kiesling, R. (1988) *Cactaceae.* Flora Patagonica 5: pp.218-243

Kiesling, R. (1999) *Cactaceae in Zuloaga, F. O. & Morrone, O., Calálogo de las Plantas Vasculares de la República Argentina 2.* Monogr. Syt. Bot. Missouri Bot. Gard. 74: pp.423-485

Kiesling, R, et al. (2000) *Gymnocalycium kroenleinii Kiesling, Rausch & Ferrari spec. nov.* KuaS 51(12): pp.315-318

Kiesling, R. et al. (2002a) *Gymnocalycium robustum R. Kiesling, O. Ferrari & D. Metzing nov. spec.* CSJ(US) 74(1): pp.4-8

Kiesling, R. et al. (2002b) *Gymnocalycium schroederianum subsp. boessii subsp. nov.* KuaS 53(9): pp.225-232

Kreuzinger, K. (1935) *Gymnocalycium Marsoneri Fric* (nom. nud). Verzeichnis amerikanischer und andere Sukkulenten mit Revision der Systematik der Kakteen: p.14

Koop, H. (1975) *Gymnocalycium hybopleurum (K. Schumann) Backeberg.* KuaS (26)10: pp.219-223

Koop, H. (1976) *Gymnocalycium borthii Koop* KuaS 27(2): pp.25-27

Lambert, J.G. (1985) *Nieuwbeschrijving Gymnocalycium erinaceum species nova* Succulenta 64(3): pp.64-66

Lambert, J.G. (1985) *Gymnocalycium multiflorum (Hook) Br. & R.* Succulenta 64(9): pp.179-182

Lambert, J.G. (1988) *Nieuwbeschrijving: Gymnocalycium acorrugatum spec. nov.* Succulenta 67(1): pp.4-7

Lambert, J.G. (2002) *Gymnocalycium parvulum Speg.:eindelijk teruggevonden!* Succulenta 81(5): pp.227-231

Lambrou, M. & Till, W. (1993) Zur Karyologie der Gattung Gymnocalycium. Gymnocalycium 6(1): pp.85-88

Lemaire, C (1838) *Echinocactus monvillii.* Cactearum Aliquot Novarum pp.14-15

Link, H.F. (1822) *Cactus reductus Link* Enumeratio plantarum horti regii botanici berolinensis altera 2:21

Link, H.F. & Otto, F. (1828) Icones plantarum rariorum (Abbildung neuer und seltener Gewächse) 2: p.17, pl.9

Lodé. J. (1995) *Nouvelles Combinaisons.* Cactus-Aventures International 27: p.29

Lorenzo, P. & Hammerschmid, J.(1965) *Gymnocalycium Gulf Oil 70.* KuaS 16(12): pp.234-236

Meregalli, M. (1985) *Il genere Gymnocalycium Pfeiffer.* Piante Grasse 5(1): p 5-63

Meregalli, M. (1986-87) *Rapporto preliminare sui Gymnocalycium appartenenti al Gruppo Mazanense Backeberg.* Piante Grasse 6(2): pp.51-57; 6(3): pp.80-88; 7(1): pp.3-9

Meregalli, M.; Neuhuber, G. & Caramiello, R. (2000) *Seed morphology in the Gymnocalycium hossei - group (Cactaceae): a useful tool for taxonomic studies.* Allionia 37: p217-232

Meregalli, M.; Metzing, D. & Kiesling, R. (2002) *Systematics of the Gymnocalycium paraguayense-fleischerianum group (Cactaceae): morphological and molecular data.* Candollea 57: pp.299-315

Meregalli, M. & Guglielmone, L. (2006) *Gymnocalycium striglianum subsp. aeneum n. subsp. A distinct taxon from the province of San Luis, Argentina.* Cactus & Co. X(2): pp.69-81

Metzing, D. (1985) *Was ist Gymnocalycium mucidum Oehme?* Gymnocalycium-Brief 2(4): pp.16-19

Metzing, D. (1988) *Eine frühe Erwähnung der Gattung Gymnocalycium.* Gymnos 5(10): pp.82-83

Metzing, D. (1991) *Was ist Gymnocaycium leeanum?* Gymnos 8(16): pp.63-65

Metzing, D.; Meregalli, M. & Kiesling, R. (1995) *An annotated checklist of the genus Gymnocalycium Pfeiffer ex Mittler (Cactaceae).* Allionia 33: pp.181-228

Metzing, D. (1997) *A Visit to Paraguay.* Chileans 17(55): pp.23-27

Metzing, D.; Meregalli, M. & Kiesling, R (1999) *Typification of Gymnocalycium quehlianum (Cactaceae).* Hickenia 3: pp.9-11

Moser, G. (1969a) *Echinocactus damsii K. Schumann (1903) Gymnocalycium Br. u. R.* KuaS 20(2): pp.38-39

Moser, G. (1969b) *"Zeigt her Euer Wissen" Gymnocalycium anisitsii (K. Sch.) Br. & R.* KuaS 20(4): pp. 71-73

Moser, G. (1970a) *How variable can a cactus species be?* NCSJ 25(1), pp.34-37

Moser, G. (1970b) *Discoveries in Paraguay since 1963 in the Gymnocalycium mihanovichii - friedrichii complex.* NCSJ 25(3), pp.77-78

Moser, G. (1971) *One species – three names?* NCSJ 26(1), pp.4-5.

Mundt, W. (1897) *Echinocactus denudatus var. Golziana, Lk. & Otto.* MfK7: p.187

Neuhuber, G. & H. Till (1993) *Das Gymnocalycium monvillei Aggregat Teil 2: .* Gymnocalycium 6(4): pp.107-112

Neuhuber, G. (1994) *Gymnocalycium andreae subspecies carolinense eine Neuheit aus der argentinischen Provinz San Luis.* Gymnocalycium 7(3): pp.127-130

Neuhuber, G. (1997) *Gymnocalycium berchtii spec. nov.* Gymnocalycium 10(3): pp.217-220

Neuhuber, G. (1999a) *Gymnocalycium monvillei, eine Richtigstellung und zwei neue Varietäten.* Gymnocalycium 12(2): pp.283-286

Neuhuber, G. (1999b) *G. poeschlii eine beachtenswerte neue Sippe aus San Luis, Argentinien.* Gymnocalycium 12(3): pp.295-300

Neuhuber, G. & W. Till (1999) *Gymnocalycium x heidiae Neuhuber. Eine neue Nothospecies und ihre Verbreitung.* Gymnocalycium 12(1): pp.275-282

Neuhuber, G. (2001a) *Gymnocalycium gaponii* Gymnocalycium 14(3): p.410

Neuhuber, G. (2001b) *Ein Neufund in der Provinz Salta: Gymnocalycium schickendantzii subsp. bergeri und die Validierung und Neubewertung von Ritter's Gymnocalycium antherosacos als G. delaetii fa. antherosacos.* Gymnocalycium 14(4): pp.417-424

Neuhuber, G. & L.Bercht (2002) *Gymnocalycium erolesii, ein interessanter Fund aus der Provinz Santa Fé.* Gymnocalycium 15(4): pp.473-478

Neuhuber, G. (2003) *Gymnocalycium bruchii 80 Jahre nach der Erstbeschreibung.* Gymnocalycium 16(2): pp.499-510

Neuhuber, G. (2004) *Die Vielgestaltigkeit und Verbreitung von Gymnocalycium schickendantzii und seinen Verwandten in Argentinien. Teil 1.* Gymnocalycium 17(2): pp.561-568

Neuhuber, G. (2005a) *Neufunde in der südlichen Sierra de Sañogasta: Gymnocalycium jochumii und Gymnocalycium jochumii var. jugum.* Gymnocalycium 18(1): pp.603-610

Neuhuber, G. (2005b) *Ein ganz besonderes Gymnocalycium aus der Provinz San Luis: Gymnocalycium nataliae.* Gymnocalycium 18(3): pp.635-638

Neuhuber, G. (2005c) *Gymnocalycium andreae subsp. carolinense: Ein Neubetrachtung und Neubewertung.* Gymnocalcium 18(4): pp.639-640

Neuhuber, G. (2006) *Ergebnisse langjähriger Pflanzenstudien und Beobachtungen an den heimatlichen Standorten. Teil 3: Gymnocalycium acorrugatum Lambert.* Gymnocalycium 19(4): pp.687-690

Neuhuber, G. (2007) *Zur geografischen Radiation des Gymnocalycium borthii und seiner Verwandten.* Gymnocalycium 20(1): pp.697-708

Oehme, H. (1937) *Gymnocalycium mucidum sp. n. Oeh.* KuaS (Berlin) 1(12:) pp.197-8

Oehme, H. (1938) *Zu Gymnoc. Sigelianum und capillaensis.* KuaS Berlin) 2(1): pp.10-12

Oehme, H. (1941) *Der Formenkreis des Gymnocalycium Bruchii (Speg,) Hoss. bzw. Gymnocalycium lafaldense Vpl. Cactaceae.* Jahrbücher der D.K.G. pp.26-30

Osten, C. (1926) *Gymnoalycium lafaldense.* Zeitschrift für Sukkulentenkunde II(9): p.146

Osten, C. (1941) *Gymnocalycium Schroederianum n. sp.* Notas Sobre Cactáceas - Anales del Museo de Historia Natural de Montevideo 2nd series, Vol V, No.1; p.60

Papsch, W. (1996a) *Zur Herkunft und Charakterisierung von Gymnocalycium gibbosum (Haworth) Pfeiffer ex Mittler.* Gymnocalycium 9(3): p181-188

Papsch, W. (1996b) *Gymnocalycium gibbosum (Haworth) Pfeiffer ex Mittler und seine Varietäten.* Gymnocalycium 9(4): pp.189-202

Papsch, W. (1997) *Die pampinen Gymnocalycien 1. Ein Gymnocalycium mit interessanter Geschichte: Gymnocalycium reductum.* Gymnocalycium 10(4) pp.223-232

Papsch, W. (1989) *Gymnocalycium stellatum im Spiegel der Literatur.* Gymnocalycium 2(3): pp.XIX-XXVIII

Papsch, W. (2000) *Die pampinen Gymnocalycien 2. Gymnocalycium reductum subsp. leeanum (Hooker) Papsch.* Gymnocalycium 13(3) pp.363-372

Papsch, W. (2001a) *Die pampinen Gymnocalycien 3. Gymnocalycium hyptiacanthum (Lemaire) Britton & Rose.* Gymnocalycium 14(1): pp.385-392

Papsch, W. (2001b) *Die pampinen Gymnocalycien . Gymnocalycium reductum Schlüssel der infraspezifischen Formen.* Gymnocalycium 14(1) pp.393-394

Papsch, W. (2002) *Gymnocalycium prochazkianum vorgestellt.* Gymnocalycium 15(1): pp.437-440

Papsch, W. (2003) *Liebe Gymnocalyciumfreunde!* Gymnocalycium 16(3): pp. IX-XII

Papsch, W., M. Wick, & G. Hold (2008) *The Polymorphism of Gymnocalycium castellanosii Backeberg 1935 emend. Piltz 1993* Cactus & Co XII 2: pp.69-104

Pazout, F. (1943) *Unser Schönstes rotblühendes Gymnocalycium und seine Geschichte.* Kakteenkunde 11(2): pp.37-40

Pazout, F. (1963) *Gymnocalycium mihanovichii var. albiflorum Pazout var. nova.* Friciana 3(17): pp.5-8

Pazout, F. (1964) *Gymnocalycia Skupiny Muscosemineae.* Friciana 4(23): pp. 3-19

Pazout, F. (1965) *Gymnocalycium mihanovichii var. filadelfiense (Backeberg) Pazout comb. nov. Nove rostliny z okruhu druhu Gymnocalycium mihanovichii.* Friciana 5(29): pp.10-11

Pazout, F. (1966) *Gymnocalycium friedrichii var. moserianum Paz. var nov. (illeg.)* Succulenta 45(6): pp.99-100

Piltz, J. (1995) *Gymnocalycium gibbosum (Haworth) Mittler im Spiegel der Literatur.* Gymnos 12(24): pp.49-65

Piltz, J. (1977) *Beitrag zur Kenntnis der Variabilität und Verbreitung von Gymnocalycium spegazzinii in Nordargentinien.* KuaS 28(9): pp.214-217

Piltz, J. (1979) *Gymnocalycium valnicekianum Jajo.* KuaS 30(3): pp.64-68

Piltz, J. (1980) *Gymnocalycium ambatoense Piltz* KuaS 31(1): pp.10-13

Piltz, J. (1987) *Gymnocalycium bruchii var brigittae Piltz var. nov.* Succulenta 66(10): pp.213-216

Piltz, J. (1990) *Gymnocalycium taningaense Piltz.* KuaS 41(2): pp.22-26

Piltz, J. (1992) *Beitrag zur Kenntnis der Gattung Gymnocalycium. Gymnocalycium castellanosii Backeberg. Teil 1.* Gymnos (9) 17: pp.17-23

Piltz, J. (1993) *Beitrag zur Kenntnis der Gattung Gymnocalycium. Gymnocalycium castellanosii Backeberg. Teil 2.* Gymnos (9) 18: pp.33-53

Piltz, J. (1994) *Gymnocalycium erinaceum var. paucisquamulosum var. nov. Ein neue Varietät aus der Sierra Chica.* Gymnos 11(21) pp. 12-17

Piltz. J.; Metzing, D. & Schweitzer, B. (1994) *Geschichte und Typisierung von Gymnocalycium mucidum Oehme.* Gymnos (11)22: pp.51-60

Plesník, F. (1972) *Gymnocalycium eurypleurum F. Ritter ex Plesnik sp. n. (nom. inval.).* Kaktusy VIII(3): pp.70-72 (1972)

Prestlé, K.H. (2001) *Gymnocalycium melanocarpum (Arechavaleta) Britton & Rose* Gymnocalycium 14(3): pp.413-416

Prestlé, K.H. (2004) *Gymnocalycium denudatum (Link & Otto) Pfeiffer ex Mittler.* Gymnocalycium 17(2): pp.569-578

Putnam, E.W. (1967) *Notes on Gymnocalycium.* N.C.S.S.J. 22(1): pp. 21-22

Quehl, H. (1899) *Neuheiten*. Monatsschrift für Kakteenkunde 9(3): pp.43-44

Rausch, W. (1989) *Gymnocalycium bruchii var. niveum var. nova*. Succulenta 68(9): pp.179-181

Rausch, W. (1970) *Gymnocalycium tillianum Rausch spec. nov.* KuaS 21(4): p.66

Rausch, W. (1972a) *Gymnocalycium uebelmannianum Rausch spec nova* Succulenta 51(4): pp.62-64

Rausch, W. (1972b) *Gymnocalycium ritterianum Rausch spec. nov.* KuaS 23(7): pp.180-181

Rausch, W. (1985) *Gymnocalycium alboareolatum Rausch spec. nov.* Succulenta 64(10): pp.213-214

Rausch, W. & Till, H. (1990) *Gymnocalycium alboareolatum Rausch.* Gymnocalycium 2(3): pp.29-31

Ritter, F. (1963) *Gymnocalycium glaucum Ritter spec. nov.* Sukkulentenkunde (Jahrbücher der S.K.G.) 7/8: pp.37-38

Ritter, F. (1964) *Gymnocalycium cardenasianum Ritter sp. nova. Diagnosen von neuen Kakteen.* Taxon XIII(4): p.144

Ritter, F. (1979) *Gymnocalycium stenopleurum Ritter spec. nov.* Kakteen in Südamerika 1: pp.265-266, 339-340

Rivera, G. (2009) *Gymnocalycium della Provincia di Córdoba.* Piante Grasse 29(1): pp.2-19

Rowley, G. (1986) *Stoye Postcards and the Haage Connection.* Excelsa 12: pp.23-47

Schädlich, V. (2002) *Die Gymnocalycien des Gran Chaco und der Savannen aus Argentinien, Bolivien und Paraguay Teil 1.* Gymnocalycium 15(4): pp.485-488

Schädlich, V. (2005a) *Die Gymnocalycien des Gran Chaco und der Savannen aus Argentinien, Bolivien und Paraguay Teil 5.* Gymnocalycium 18(3): pp.629-634

Schädlich, V. (2005b) *Die Gymnocalycien des Gran Chaco und der Savannen aus Argentinien, Bolivien und Paraguay Teil 6.* Gymnocalycium 18(4): pp.641-644

Schädlich, V. & Bercht, L. (2007) *Ein Glückstag!* Gymnocalycium 20(3): pp. IX-XI

Schick, (1923) *Echinocactus capillensis n. sp. Neue Kakteen aus der Sierra de Cordoba.* Möller's Deutsche Gärtner-Zeitung 38(26): pp.201-202, 203

Schick, (1923) *Echinocactus sigelianus n. sp. Neue Kakteen aus der Sierra de Cordoba.* Möller's Deutsche Gärtner-Zeitung 38(26): p.201

Schick, (1923) *Echinocactus sutterianus n. sp. Neue Kakteen aus der Sierra de Cordoba.* Möller's Deutsche Gärtner-Zeitung 38(26): p.201

Schick, C. (1953) *Gymnocalycium tobuschianum Schick* CSJ(GB) 15(2): p.44, illn. p.36

Schlosser, H. & B. Schütz (1982) *Die Gattung Gymnocalycium in Uruguay.* KuaS 33(4): pp.88-91

Schumann, K. (1900) *Echinocactus Anisitsii K. Sch. n.sp.* Blühende Kakteen - Iconographia Cactacearum 1: pl.4

Schumann, K. (1901) *Echinocactus de Laetii eine neue Art.* Monatsschrift für Kakteenkunde 11(12): pp.186-187

Schumann, K. (1903) *Echinocactus Damsii K. Sch. n. spec.* Gesamtbeschreibung der Kakteen, Nachträge 1898-1902: pp.119-120

Schumann, K. (1903) *Echinocactus paraguayense* Bulletin de l'Herbier Boiss. 2, 3: p.252

Schütz, B. (1947) *Gymnocalicium guanchinense dr. Schütz spec. nova..* Zpravy Ceskoslovenske Kaktusarske Spolecnosti 2: p.21

Schütz, B. (1948) *Gymnocalycium guanchinense Schütz spec. nov.* Sukkulentenkunde II: pp.32-33.

Schütz, B. (1949) *Gymnocalycium valnicekianum var pluricentralis Schütz var. nov.* Kaktusarske Listy (6): pp.41-42

Schütz, B. (1969a) *Rodu Gymnocalycium.* Friciana 7: pp.3-23

Schütz, B. (1969b) *The Genus Gymnocalycium Pfeiff.* NCSJ 24(4): pp.74-6

Schütz, B. (1977a) *Gymnocalycium paediophylum Schütz sp.n.* Kaktusy 13(5): pp.100-101

Schütz, B. (1977b) *Gymnocalycium bozsingianum Schütz sp n.* Kaktusy13(6): pp.124-126

Schütz, B. (1978) *Gymnocalycium FR430?* Kaktusy 14(5): pp.112-113

Schütz, B. (1979) *Gymnocalycium mihanovichii-friedrichii Komplex.* Friciana 8(51): pp. 5-35

Schütz, B. (1981) *Gymnocalycium altagraciense Bozsing nom. prov.* Kaktusy 17(3): pp.59-60

Schütz, B. (1985) *Gymnocalycium valnicekianum var. polycentralis.* Kaktusy 21(1): pp.13-14

Schweitzer, B. (1994) *Ergänzungen zur Erstbeschreibung von Gymnocalycium obductum.* Gymnos 11(22): pp.44-47

Simon, W. (1974) *Ist dies das Gymnocalycium mucidum Oehme?* Stachelpost 10: pp.10-11.

Slaba, R. (1984) *Gymnocalycium hybopleurum (K.Sch.) Backbg. a jemu blízce príbuzné taxony.* Kaktusy (20)4: pp.77-83 & (20)5: pp.99-104

Slaba, R. (1985) *Gymnocalycium chiquitanum Card. – cervenokvety druh vychodni Bolivie.* Kaktusy 21(2): pp.26-28

Slaba, R. (1999) *Gymnocalycium carminanthum var. montanum.* Kaktusy (35)1: pp.3-8

Sorma, V. (1999) *Gymnocalycium prochazkianum - novy zajimavy nález ceskych kaktusaru v Argentine.* Gymnofil 28(1,2): pp.2-6

Spegazzini, C. (1902) *Echinocactus gibbosus D.C. var. chubutensis.* Anales del Museo Nacional de Buenos Aires 7(Ser. 2, IV): p.285 (1902)

Spegazzini, C. (1905a) *Echinocactus baldianus Speg. n. sp.* Cactacearum Platensium Tentamen - Anales del Museo Nacional de Buenos Aires Serie 3, 11(4) pp.505-506

Spegazzini, C. (1905b) *Echinocactus stuckertii Speg. n. sp.* Anales Museo Nacional de Buenos Aires XI (Ser.3a t.IV) Cactacearum Platensium Tentamen: p.502

Spegazzini, C. (1923) *Frailea bruchii (n. sp.) Breves Notas Cactológicas.* Anales de la Sociedad Cientifica de Argentina 96: pp.73-75

Spegazzini, C. (1925a) *Gymnocalycium brachypetalum Nuevas Notas Cactológicas* Anales de la Sociedad Cientifica de Argentina 99: pp.135-137

Spegazzini, C. (1925b) *Gymnocalycium stellatum Nuevas Notas Cactologicas.* Anales de la Sociedad Cientifica de Argentina, 99: pp.142-143

Spegazzini, C. (1925c) *Gymnocalycium loricatum Nuevas Notas Cactologicas.* Anales de la Sociedad Cientifica de Argentina, 99: p.140

Spegazzini, C. (1925d) *Gymnocalycium leptanthum Nuevas Notas Cactologicas.* Anales de la Sociedad Cientifica de Argentina, 99: pp.138-140

Strigl, F. & Till, W. (1985) *Eine neue Art aus der argentinischen Provinz Buenos Aires: Gymnocalycium schatzlianus* KuaS 36(12): pp.250-253

Swales, G. (1978) *Gymnocalycium bueneckeri sp. nov.* CSJ(GB) 40(4) pp.97-100

Till, H. (1967) *Ein schönes Gymnocalycium aus Nord-Argentinien Gymnocalycium bayrianum spec. nov.* KuaS 18(12): pp.222-224

Till, H. (1972) *Gymnocalycium baldianum (Speg.) Speg. und seine Formen.* KuaS 23(9); pp.238-240

Till, H. & Schatzl, S. (1973) *Gymnocalycium pugionacanthum Backeberg. das dolchartig bedornte Gymnocalycium.* KuaS 24(10): pp.230-233

Till, H. & Schatzl, S. (1979) *Gymnocalycium achirasense H. Till & Schatzl.* KuaS 30(2): pp.25-28

Till, H. & Schatzl, S. (1981) *Gymnocalycium schuetzianum Till & Schatzl.* KuaS 32(10): pp.234-236

Till, H. (1987) *Validierung einiger ungültig veröffentlichter Taxa von Gymnocalycium Pfeiffer* KuaS 38(8): p.191

Till, H. (1984) *Zur Identität und Verbreitung von Gymnocalycium parvulum (Spegazzini) Spegazzini.* Gymnocalycium 7(2): pp. 121-126

Till, H. & Till, W. (1987) *Gymnocalycium mazanense - ein "nomen dubium"?* Gymnos (4)8: pp.79-85.

Till, H. (1988a) *Gymnocalycium striglianum Jeggle ex Till.* Gymnocalycium 1(2): pp.3-6

Till, H. (1988b) *Gymnocalycium horridispinum Frank ex Till.* Gymnocalycium 1(3): pp.7-8

Till, H. & Till, W. (1988) *Gymnocalycium pflanzii subsp. argentinense H. Till & W. Till.* KuaS 39(12): pp.273-277

Till, H. (1989) *Gymnocalycium pugionacanthum Backeberg ex Till.* Gymnocalycium 2(3): pp.19-20

Till, H. & Neuhuber, G. (1989) *Gymnocalycium bayrianum Till* Gymnocalycium 2(4): pp.21-24

Till, H. (1990a) *Gymnocalycium kieslingii Ferrari.* Gymnocalycium 3(1): pp.25-28

Till, H. (1990b) *Ein schöne, aber oft verkannte Art: Gymnocalycium monvillei.* Gymnocalycium 3(3): pp.33-40

Till, H. & Till, W. (1990) *Een interessante nieuwe soort uit Uruguay: G. rauschii H. Till and W. Till.* Succulenta 69(2): pp.27-31

Till, H. (1991) *Gymnocalycium uebelmannianum Rausch.* Gymnocalycium 4(1): pp.43-45

Till, H. (1992) *Taxonomie und Lebensweise des Gymnocalycium andreae und seiner Varietäten.* Gymnocalycium 5(2): pp.61-66

Till, H. & Till, W. (1992) *Gymnocalycium neuhuberi - eine bemerkenswerte neue Kakteenart aus dem westlichen Argentinien.* Gymnocalycium 5(1): pp.59-60

Till, H. & Neuhuber, G. (1992a) *Die Arten des Gymnocalycium bodenbenderianum Aggregates.* Gymnocalycium 4(5): pp.71-76

Till, H. & Neuhuber, G. (1992b) *Die Arten des Gymnocalycium bodenbenderianum Aggregates.* Gymnocalycium 5(3): pp.67-71

Till, H. (1993) *Zur Identität von Gymnocalycium quehlianum.* Gymnocalycium 6(2): pp.89-98

Till, H. & Neuhuber, G. (1993) *Das Gymnocalycium monvillei Aggregat Teil 1: .* Gymnocalycium 6(3): pp.99-106

Till, H. (1994a) *Zur Identität und Verbreitung von Gymnocalycium parvulum (Spegazzini) Spegazzini.* Gymnocalycium 7(2): pp.121-126

Till, H. (1994b) *Gymnocalycium amerhauseri -eine neue Art aus der argentinischen Provinz Cordoba.* Gymnocalycium 7(3) pp.131-134

Till, H. & Till, W. (1994) *Gymnocalycium hybopleurum. 1 Teil: Zur Klärung des Schumann'schen Namens und zu seiner interpretation durch Backeberg.* Gymnocalycium 7(4): pp.135-140

Till, H. (1995) *Ein bemerkenswerter Neufund aus der Gattung Gymnocalycium in der argentinischen Provinz Salta. G. rosae H. Till.* Gymnocalycium 8(3): pp.159-162

Till, H. & Till, W. (1995) *Gymnocalycium hybopleurum. 2 Teil: Neubenennung der unter diesem Namen bekannten argentinischen Pflanzen.* Gymnocalycium 8(1): pp.141-146

Till, H. & Till, W. (1996) *Gymnocalycium stellatum Spegazzini: Geschichte, Formen und die taxonomische Stellung der Art im System. 3 Teil.* Gymnocalycium 9(2): pp.175-180

Till, H. & Till, W. (1997a) *Gymnocalycium saglionis: Verbreitung, Variabilität und Nomenklatur einer "gut" bekannten Art. Teil 1.* Gymnocalycium 10(1): pp.203-208

Till, H. & Till, W. (1997b) *Gymnocalycium saglionis: Verbreitung, Variabilität und Nomenklatur einer "gut" bekannten Art. Teil 2.* Gymnocalycium 10(2): pp.209-216

Till, H. (1998) *Zur Identität von G. nigriareolatum und seinen Formen.* Gymnocalycium 11(3): pp.251-258

Till, H. & Neuhuber, G. (1998) *Gymnocalycium baldianum seine Geschichte und seine Synonyme.* Gymnocalycium 11(1): pp.233-240

Till, H. & F. Strigl (1998) *Gymnocalycium terweemeanum (Recherche).* Gymnocalycium 11(4): pp.261-264

Till, H. (1999) *Bemerkungen zu Gymnocalycium angelae Meregalli.* Gymnocalycium 12(2): p.VI

Till, H. (2000) *Recherchen zu Gymnocalycium mucidum Oehme.* Gymnocalycium 13(4): pp.373-376

Till, H.; Till, W & Amerhauser, H. (2000a) *Revision von Gymnocalycium subgen. Pirisemineum. Teil 1: Gymnocalycium pflanzii.* Gymnocalycium 13(1): pp. 343-350

Till, H.; Till, W. & Amerhauser, H. (2000b) *Revision von Gymnocalycium subgen. Pirisemineum. Teil 2: Gymnocalycium zegarrae.* Gymnocalycium 13(2): pp. 351-358

Till, H. & Rausch, W. (2000) *Drei neue Taxa aus der Gruppe des Gymnocalycium andreae.* Gymnocalycium 13(4): pp.377-380

Till, H. (2001) *G. papschii ein interessanter Neufund vom Cerro Champaqui auf der Sierra de Comechingones.* Gymnocalycium 14(3): pp.405-408

Till, H. (2002) *Zur Klarstellung von Gymnocalycium quehlianum (Haage ex Quehl) Vaupel ex Hosseus.* Gymnocalycium 15(2): pp.441-444

Till, H. & Amerhauser, H. (2002a) *Revision des Gymnocalycium mostii-Aggregates. Teil I: Gymnocalycium mostii und seine Formen.* Gymnocalycium 15(1): pp.425-436

Till, H. & Amerhauser, H. (2002b) *Revision des Gymnocalycium mostii-Aggregates. Teil II: Gymnocalycium valnicekianum und seine Formen.* Gymnocalycium 15(2): pp.445-452

Till, H. (2003a) *Die Verwandtschaft von Gymnocalycium capillense.* Gymnocalycium 16(1): pp.489-498

Till, H. (2003b) *Gymnocalycium walteri, eine neue Zwergart aus dem Norden der argentinischen Provinz Córdoba.* Gymnocalycium 16(4): pp. 541-544

Till, H. & Amerhauser, H. (2003a) *Die Gymnocalicien des Gran Chaco und der Savannen aus Argentinien, Bolivien und Paraguay, 2.* Gymnocalycium 16(2): pp.511-516

Till, H. & Amerhauser, H. (2003b) *Die Gymnocalicien des Gran Chaco und der Savannen aus Argentinien, Bolivien und Paraguay, 3.* Gymnocalycium 16(3): pp.523-532

Till, H. (2004a) *Bemerkungen zu Gymnocalycium rhodantherum und den mit ihm verwandten (Taxa. 1.) Gymnocalycium rhodantherum.* Gymnocalycium 17(3): pp.579-584

Till, H. (2004b) *Bemerkungen zu Gymnocalycium rhodantherum und den mit ihm verwandten (Taxa. 2.) Zur identität von Gymnocalycium guanchinense.* Gymnocalycium 17(3): pp.585-588

Till, H. & Till, W. (2004) *Bemerkungen zu Gymnocalycium rhodantherum und den mit ihm verwandten (Taxa. 3.) Gymnocalycium ritterianum.* Gymnocalycium 17(4): pp.589-594

Till, H. & Amerhauser, H. (2004a) *Die Gymnocalicien des Gran Chaco und der Savannen aus Argentinien, Bolivien und Paraguay, 4.* Gymnocalycium 17(1): pp.545-560

Till, H. & Amerhauser, H. (2004b) *Bemerkungen zu Gymnocalycium rhodantherum und den mit ihm verwandten (Taxa. 4.) Gymnocalycium weissianum.* Gymnocalycium 17(4): pp.595-598

Till, H. (2005a) *Bemerkungen zu Gymnocalycium rhodantherum und den mit ihm verwandten (Taxa. 5.) Gymnocalycium alboareolatum.* Gymnocalycium 18(1): pp.611-614

Till, H. (2005b) *Beitrag zur Klärung des rosa blühenden Gymnocalycium denudatum und seine Zugehörigkeit zu Gymnocalycium megalothelon.* Gymnocalycium 18(2): pp.617-626

Till, H. & Amerhauser, H. (2005) *Die Gymnocalicien des Gran Chaco und der Savannen aus Argentinien, Bolivien und Paraguay, Teil 7.* Gymnocalycium 18(4): pp.645-650

Till, H. (2006) *Ergebnisse langjähriger Pflanzenstudien und Beobachtungen an den heimatlichen Standorten. Teil 1: G.castellanosii s.str.* Gymnocalycium 19(2): pp.663-670.

Till, H. & Amerhauser, H. (2007) *Gymnocalycium leptanthum und Gymnocalycium frankianum Raush n.n. Zwei Taxa aus dem Quehlianum Aggregat.* Gymnocalycium 20(3): pp.725-732

Till, H. & Till, W. (2007) *Ist Echinocactus ourselianus Cels ex Salm-Dyck wirklich ein Synonym von Gymnocalycium multiflorum (Hook) Britton & Rose?* Gymnocalycium 20(2): pp.715-720

Till, H. & Amerhauser, H. (2008) *Ein fastvergessenes Taxon: Gymnocalycium sutterianum (Schick) Hosseus.* Gymnocalycium 21(3): pp.783-790

Till, H.; Amerhauser, H. & Till, W. 2008) *Neuordnung der Gattung Gymnocalycium.* Gymnocalycium Sonderausgabe 2008 pp.815-838

Vala, L (2003) *Gymnocalycium fleischerianum: ist die Einreihung in de Verwandtschaft von Gymnocalycium denudatum richtig?* Gymnocalycium 16(3): pp.519-522

VanVliet, D.J. (1971) *Gymnocalycium ragonesei Castellanos.* Succulenta 50(4): pp.67-72

Vaupel, F. (1923) *Echinocactus pflanzii Vpl. spec. nov.* Zeitschrift für Sukkulentenkunde 1(8): p.83

Vaupel, F. (1924) *Gymnocalycium lafaldense Vpl. Spec. Nov.* Zeitschrift für Sukkulentenkunde 1(14): p.192

Weber, F.A.C. (1896) *Echinocactus Schickendantzii F.A.C.Weber.* Bois, Dictionnaire d'Horticulture:p.470

Werdermann, E. (1930) *Über Echinocactus loricatus Speg.* Monatsschift der Deutchen Kakteen Gesellschaft 2(8): pp.170-172

Werdermann, E. (1932) *Echinocactus sanguiniflorus Werd. nov. spec.* Feddes Repert. 30: pp.56-57

Werdermann, E. (1936) *G. mihanovichii var. friedrichii Werdermann.* Blühende Kakteen und andere sukkulente Pflanzen Mappe 29, Plate 113

Gymnocalycium Chromosome Counts

Contributed by Gordon Rowley

The first and only general survey of the genus karyologically is that by Lambrou & Till (1993), to which Diers (2008) added nine counts. All these are mitotic counts; meiosis has yet to be studied. As Lambrou and Till cautiously point out: "The amount of samples, however, is still too small for phylogenetic conclusions."

Gymnocalycium matches all other genera of Cactaceae in having a base number of 11 approximately similar-sized chromosomes, and in showing a low level of polyploidy. There are no aneuploids. The score so far, equating the published names, where possible, to the species accepted here, is 27 diploid species with 2n = 22, plus three, or possibly four tetraploids with 2n = 4x = 44. In addition, there is *G. proliferum* (here treated as a synonym of *G. calochlorum*) found to be both tetraploid with 2n = 4x = 44 as well as hexaploid with 2n = 6x = 66; the highest count so far in the genus.

Elsewhere in Cactaceae isolated counts go up to decaploid and dodecaploid (= 220 and 264 respectively). We have conflicting accounts for *G. denudatum*, reported as diploid by Diers but tetraploid by Lambrou & Till. There have been other reports of different ploidy levels within a single species of the Cactaceae (Arakaki et. al 2007) and this case could be the result of hybridisation.

The Chromosome Counts (x = 11 throughout)
† Lambrou & Till (1993) * Diers (2008)

DIPLOID with 2n = 22

*andreae
†*anisitsii subsp. damsii
*baldianum
†capillaense
†castellanosii
†*chiquitanum
†erinaceum
†gibbosun subsp. borthii
†glaucum subsp. ferrarii
*marsoneri subsp. matoense
†mesopotamicum
*mihanovichii
†mostii
†neuhuberi
†ochoterenae
†nigriareolatum
†*pflanzii subsp. pflanzii
†pflanzii subsp. zegarrae
†rhodantherum (acorrugatum)
†quehlianum (obductum)
*saglionis
†spegazzinii
†stenopleurum
†taningaense
†uebelmannianum

DIPLOID and TETRAPLOID with 2n = *22 and †44

denudatum

TETRAPLOID with 2n = 44

†bruchii
†calochlorum
†paraguayense

TETRAPLOID and HEXAPLOID with 2n = †44 and †66

calochlorum (proliferum)

Cristate and Variegated Plants

There is still no explanation of exactly what causes plants to develop cristation. The growing point becomes a line, so that the plant develops like a ribbon (a crest) rather than a sphere or cylinder. Sometimes it happens in a young seedling so that the plant will be cristate from the beginning but a plant can turn cristate at any time, such as the one in Fig 618. In habitat, a cristate plant can sometimes be found in a population of normal ones. These oddities are often larger than their companions, perhaps because the elongated growing point results in faster growth. Occasionally, a number of cristates can be found in the same location giving rise to speculation about their occurrence being a result of their genes, or even a virus spread by insects. Interestingly, cristation can affect just one head of a clustering plant or sometimes all of them. Propagation is usually achieved by grafting part of a crest onto a strong stock such as *Trichocereus*.

Photo: Geoff Bailey

Fig.615 A large cristate head of *G. saglionis* growing in an outdoor collection in Chilecito, a town on the east side of the Sierra Famatina, Prov. La Rioja, Argentina

Fig.616 A cristate of *G. spegazzinii,* GC52.02 in the dry season, north of Cachi, Prov. Salta, Argentina

Fig.617 A remarkable cristate of *G. rhodantherum,* GC961.03 on road to Guanchin, Prov. La Rioja, Argentina

Fig.618 A cristate of *G. hyptiacanthum* that happened spontaneously in cultivation.

Fig.620 This large cristate of *G. gibbosum* is photographed in the private collection of Southfields Nursery, Bourne, UK

Figs.621-624 These four pictures show commercial production of variegated clones of *G. stenopleurum* (friedrichii), grafted onto *Hylocereus* stock (Pugh's Cacti)

Fig.619 Variegation occasionally occurs naturally as on this cultivated seedling of *G. bayrianum*, *Neuhuber* 46-90

Ever since they were first propagated, brightly coloured variegated clones of *Gymnocalycium stenopleurum* (friedrichii) have been very popular in the retail trade. People seem to believe that the coloured part of the plant is a flower, perhaps fulfilling the desire for long-lasting colour in a house plant which cactus flowers cannot satisfy.

There are many differently-coloured clones in the trade, but the reds are still the most popular. You either like them or you don't! Because the plants have so little chlorophyll, they have to be grafted onto a green stock which will photosynthsise and provide nutrients to the scion. The most popular stock is *Hylocereus* which grows very quickly in the Far East where most of the propagation is done. The plants are then packed in boxes, usually with the roots cut off, and shipped to markets around the world.

More information about cristates and variegated plants can be found in Gordon Rowley's entertaining book on the subject, *Teratopia*.

GC Gymnocalycium Field Numbers

When visiting cactus habitats, it is usual for cactophiles to allocate numbers to their finds. There is no consistent format or meaning to these numbers, but they are a convenient way to trace the finding place of a plant with such a number. The best practice is for each habitat of a species to be given a separate number, but some historically important field numbers, such as those of HU (Horst-Uebelmann) and FR (Ritter) apply to a perceived species rather than a single locality. In these cases, the same number has been applied to plants from different localities. In the case of my numbers, the first part is the locality then each species has been given a different number after the dot. There are a number of other people using this system but many just use consecutive numbers, or only give numbers to species where seeds or plants were collected.

I considered publishing more of these field number lists in this book but, for *Gymnocalycium*, there are very many and they display great variation in nomenclature. They soon become out of date because of new additions or new identifications of the plants. The best way to find the data for a number is to use the internet resources: The Field Number Database at http://ralph.cs.cf.ac.uk/cacti/fieldno.html which can be accessed via the Cactus and Succulent Plant Mall at http://www.cactus-mall.com

For *Gymnocalycium*, some commonly seen acronyms on plant labels and in seed lists are:

Name	Acronym
Amerhauser, H.	HA
Bercht, L.	LB
Borth	BO
Charles, G.	GC
Gerloff, N.	GF
Hamester, M.G.	MGH
Herm, K.	H
Hillman, R.	RH
Hofacker, A.	AH
Hole, G.	HG
Horst-Uebelmann	HU
Kühhas	FK
Janeba, Z.	ZJ
Knize, K.	KK
Lambert, J.	JL
Lau, A.	L
Machado, M. (Brazil)	MM
Meregalli, M.	MM
Metzing, D.	M
Muhr, D.	B
Neuhuber, G.	GN
Papsch, W.	WP
Piltz, J.	P
Preston-Mafham, K.	PM
Rausch, W.	R or WR
Ritter, F.	FR
Schlosser, H.	Schl
Slaba, V.	VS
Städlich, V.	VoS
Standort (German = Habitat)	STO
Stockinger, F.	FS
VanVliet, D.	DVV
Winberg, M.	MN

Argentina

No.	Species/subspecies	Location
1.01	mostii mostii	S of La Granja
2.01	calochlorum	E La Falda
2.02	mostii mostii	E La Falda
2.03	monvillei	E La Falda
3.01	capillaense	Capilla del Monte
3.02	quehlianum	Capilla del Monte
4.01	mostii valnicekianum	W Capilla del Monte
5.02	mostii valnicekianum (bicolor)	Cruz del Eje
6.02	quehlianum	N La Higuera
8.01	calochlorum	N Villa Cura Brochera
10.01	ochoterenae (vatteri)	Las Rojas
11.02	ochoterenae	SE Quines
12.02	gibbosum borthii	Luján
12.03	schickendantzii	Luján
12.04	ochoterenae	Luján
17.02	schickendantzii	W Marayes
18.03	schickendantzii	Chepes Viejo
18.04	castellanosii (bozsingianum)	Chepes Viejo
18.10	saglionis	Chepes Viejo
18.11	schickendantzii	Chepes Viejo
19.02	castellanosii (bozsingianum)	NW Chepes Viejo
19.03	schickendantzii	NW Chepes Viejo
20.01	schickendantzii	S Patquia
21.02	bodenbenderianum	SE Los Colorados
21.03	schickendantzii	SE Los Colorados
22.01	bodenbenderianum	Los Colorados
22.02	rhodantherum (coloradense)	Los Colorados
22.03	schickendantzii	Los Colorados
23.02	rhodantherum	N Chilecito
23.03	saglionis	N Chilecito
23.04	schickendantzii	N Chilecito
24.02	bodenbenderianum (piltziorum)	W Schaqui
24.03	schickendantzii	W Schaqui
26.04	glaucum ferrarii	Villa Mazán
27.02	hossei	Cuesta de Cebile
28.02	bodenbenderianum (occultum)	E Catamarca City
29.02	nigriareolatum	W Cuesta El Portezuela
29.07	baldianum	Cuesta El Portezuela
30.02	nigriareolatum	E Catamarca City
30.03	saglionis	E Catamarca City
31.01	baldianum	El Rodeo
31.02	oenanthemum	El Rodeo
32.01	nigriareolatum?	NW Catamarca City
32.02	bodenbenderianum (occultum)	NW Catamarca City
32.03	saglionis	NW Catamarca City
33.01	baldianum	La Merced
33.02	nigriareolatum?	La Merced
34.02	nigriareolatum	Pomancillo
35.02	baldianum	Singuil
35.03	oenanthemum	Singuil
36.01	baldianum	E Andalgalá
37.02	pugionacanthum	N Andalgalá
40.02	spegazzinii	Amaicha del Valle
40.07	spegazzinii	SE Amaicha del Valle
43.03	saglionis	S Cafayate
44.01	spegazzinii	Las Ventanas
44.05	spegazzinii	NE Cafayate

No.	Species/subspecies	Location
45.02	schickendantzii delaetii	S Alemania
45.03	saglionis	S Alemania
45.04	spegazzinii	S Alemania
51.02	spegazzinii	S Cachi
52.02	spagazzinii	N La Poma
55.01	spegazzinii	Chorrillos
56.01	spegazzinii	S Puerta Tastil
61.01	saglionis (tilcarense)	S Tilcara
65.04	schickendantzii delaetii	N Alemania
65.05	saglionis	N Alemania
66.02	schickendantzii	N Patquia
67.01	mostii valnicelianum (bicolor)	E Cruz del Eje
67.02	mostii valnicelianum (bicolor)	N Capilla del Monte
67.04	calochlorum	N La Falda
67.05	monvillei	N La Falda
68.01	calochlorum	W Alta Gracia
68.02	mostii mostii	W Alta Gracia
160.01	spegazzinii	N Chorrillos
162.02	spegazzinii	Quebrada del Toro
164.03	spegazzinii	Santa Rosa de Tastil
165.02	spegazzinii	N Las Cuevas
170.02	saglionis (tlcarense)	E Purmamarca
185.02	spegazzinii	N. Amblayo
200.04	bodenbenderianum	E Jáchal
201.03	bodenbenderianum	E Jáchal
202.02	rhodantherum	E Villa Union
203.05	rhodantherum	Piedra Pintado
204.01	rhodantherum	Cuesta Miranda
205.01	rhodantherum	Guanchin
206.03	rhodantherum	Angulos
208.01	rhodantherum	N Chilecito
208.02	saglionis	N Chilecito
210.02	glaucum glaucum	San Blas
210.03	saglionis	San Blas
213.01	pugionacanthum (schmidianum)	Qu. Zapata
214.01	saglionis	N Chilecito
215.02	kieslingii (castaneum)	Cuesta de Huaco
215.03	saglionis	Cuesta de Huaco
216.02	bodenbenderianum	SE Los Colorados
216.03	schickendantzii	SE Los Colorados
217.02	rhodantherum (acorrugatum)	San Agustín del Valle Fértil
217.03	saglionis	San Agustín del Valle Fértil
218.02	schickendantzii	N Marayes
219.03	striglianum	Ugarteche
224.03	striglianum	N. Tunuyán
390.01	bruchii	W Alta Gracia
390.02	monvillei	W Alta Gracia
391.01	bruchii	S Observatory
391.02	monvillei	S Observatory
391.03	mostii	S Observatory
392.01	bruchii	E La Falda
392.02	monvillei	E La Falda
392.03	mostii	E La Falda
393.01	capillaense	E Sierra Chica, E La Falda
394.01	andreae	Los Gigantes
394.02	monvillei	Los Gigantes
395.02	monvillei	E Taninga
395.03	capillaense	E Taninga
396.01	calochlorum	W Taninga
396.02	taningaense	W Taninga
397.02	schickendantzii	N Marayes

No.	Species/subspecies	Location
399.04	rhodantherum (acorrugatum)	Las Chacras
399.05	schickendantzii	Las Chacras
400.02	schickendantzii	Marayes
401.02	saglionis	San Agustín del Valle Fértil
402.02	bodenbenderianum	W Patquia
402.03	saglionis	W Patquia
403.02	rhodantherum (coloradense)	E Los Colorados
403.03	saglionis	E Los Colorados
404.02	bodenbenderianum	SE of Los Colorados
404.03	schickendantzii	SE of Los Colorados
411.02	glaucum glaucum	E Tinogasta
412.03	pugionacanthum (schmidianum)	S Belen
412.04	saglionis	S Belen
413.03	pugionacanthum	Cuesta Belen
414.04	pugionacanthum	N Belen
415.02	pugionacanthum	N Villa Vil
416.03	pugionacanthum	S Hualfin
417.03	pugionacanthum	Hualfin
419.02	spegazzinii	Punta de Balasto
420.03	spegazzinii	N San Carlos
422.02	spegazzinii	Angastura
432.02	spegazzinii	Chorillos
434.03	marsoneri	Campo Quijano
435.01	saglionis (tilcarensis)	S. Tilcara
445.01	spegazzinii	NE Puerta Tastil
446.03	schickendantzii delaetii	Cobra Corral
446.04	saglionis	Cobra Corral
447.04	schickendantzii delaetii	E Cobra Corral
447.05	saglionis	E Cobra Corral
448.05	schickendantzii delaetii	E Cobra Corral
448.06	pflanzii argentinense	E Cobra Corral
448.07	saglionis	E Cobra Corral
449.02	bayrianum	Sierra de Nogolito

Brazil

No.	Species/subspecies	Location
755.02	denudatum	S. Gruta do Segredo
756.03	denudatum	S. Gruta do Segredo
758.03	denudatum	Minas do Camaquã
759.01	denudatum	Minas do Camaquã
761.01	denudatum	E Minas do Camaquã
763.02	denudatum	W Caçapava do Sul
766.03	denudatum	S Caçapava do Sul
768.02	denudatum	W Lavras
771.02	denudatum	Ibaré
775.03	denudatum	São Gabriel
777.04	denudatum	E Livramento
779.02	hyptiacanthum uruguayense	W Livramento
780.02	denudatum	W Livramento
785.02	hyptiacanthum uruguayense	N Livramento
790.04	horstii buenekeri	W São Francisco de Assis

Bolivia

No.	Species/subspecies	Location
851.05	pflanzii pflanzii	S Santa Cruz
852.02	marsoneri megatae	W Eyti
853.02	pflanzii pflanzii	Lagunillas
854.02	pflanzii pflanzii	W Lagunillas
858.05	pflanzii pflanzii	S Camiri
859.03	marsoneri megatae	S Camiri
859.04	pflanzii pflanzii	S Camiri
860.02	pflanzii pflanzii	Boyuibe
861.04	pflanzii pflanzii	W Buyuibe
863.04	marsoneri megatae	S Palos Blancos

No.	Species/subspecies	Location
864.03	pflanzii pflanzii	Tepecua
865.03	pflanzii pflanzii	W Tepecua
887.02	cardenasianum	N. Tarija
907.04	pflanzii zegarrae	N Tomina
913.05	pflanzii zegarrae	N. Nuevo Mundo
919.03	pflanzii zegarrae	E Mataral

Argentina

No.	Species/subspecies	Location
926.01	reductum leeanum	N. Tandil
927.01	schroederianum bayensis	N. Olavarrio
928.01	reductum reductum	Sierra de la Tuna
929.02	reductum reductum	Sierra de la Ventana
930.01	horridispinum achirasense	E S.del Morro
931.01	fischeri	La Petra
932.01	fischeri (poeschlii)	E Saladillo
933.01	fischeri (poeschlii)	E Saladillo
934.02	aff. fischeri	Donovan
935.01	monvillei	Suyuqui Nuevo
935.02	fischeri	Suyuqui Nuevo
936.02	fischeri (?)	Villa de la Quebrada
936.03	monvillei	Villa de la Quebrada
936.04	schickendantzii	Villa de la Quebrada
937.02	ochoterenae	Quines
937.03	schickendantzii	Quines
938.02	gibbosum borthii	Los Chañares
939.01	berchtii	E Los Chañares
940.01	mostii mostii	N Carlos Paz
940.02	quehlianum	N Carlos Paz
940.05	calochlorum	N Carlos Paz
941.01	calochlorum	S Icho Cruz
941.02	monvillei	S Icho Cruz
941.03	mostii mostii	S Icho Cruz
941.04	quehlianum	S Icho Cruz
942.01	andreae	E El Condor
942.02	monvillei	E El Condor
943.02	calochlorum	Ojo de Agua, W Nono
944.01	ochoterenae (vatteri)	N Altautina
945.01	calochlorum	S Mina Clavera
945.02	taningaense	S Mina Clavera
946.02	taningaense	W Salsacate
947.02	ochoterenae (moserianum)	NW Salsacate
948.01	capillaense	Capilla del Monte
948.02	mostii valnicekianum	Capilla del Monte
949.03	bodenbenderianum	SW Tuclame
950.03	ochoterenae	Agua de Ramón
950.04	schickendantzii	Agua de Ramón
951.02	bodenbenderianum	SW Tuclame
951.03	castellanosii ferocius	SW Tuclame
951.04	schickendantzii	SW Tuclame
952.02	castellanosii ferocius	SW Tuclame
953.02	castellanosii castellanosii	W Olta
953.03	saglionis	W Olta
954.03	castellanosii castellanosii	Olta-Malanzan
954.04	saglionis	Olta-Malanzan
955.02	castellanosii castellanosii	S Malanzan
955.03	saglionis	S Malanzan
956.02	saglionis	El Portrero
957.01	bodenbenderianum	E Los Colorados
957.02	rhodantherum (coloradense)	E Los Colorados
957.03	saglionis	E Los Colorados
957.04	schickendantzii	E Los Colorados
958.02	bodenbenderianum	Los Colorados
958.03	schickendantzii	Los Colorados
960.01	rhodantherum	W Cuesta Miranda
961.03	rhodantherum	S Guanchin
962.01	rhodantherum	Guanchin
966.03	rhodantherum	N Chilecito
966.04	schickendantzii	N Chilecito
966.05	saglionis	N Chilecito
967.02	bodenbenderianum (piltziorum)	W Sasquil
968.01	glaucum glaucum	San Blas
969.01	glaucum glaucum	Sierrra Cocacabana
974.01	hossei	S Aimogasta
975.01	kieslingii (castaneum)	Anillaco
976.02	kieslingii (castaneum)	N Huaco
976.03	schickendantzii	N Huaco
976.04	saglionis	N Huaco
977.01	albiareolatum	Villa Sanagasta
977.02	schickendantzii	Villa Sanagasta
977.03	saglionis	Villa Sanagasta
978.01	albiareolatum	S Villa Sanagasta
978.02	bodenbenderianum	S Villa Sanagasta
978.03	schickendantzii	S Villa Sanagasta
978.04	saglionis	S Villa Sanagasta
979.02	uebelmannianum	Sierra Velasco
980.02	hossei	E Anillaco
980.03	schickendantzii	E Anillaco
981.02	hossei	Rte.7&9 jnc. S Aimogasta
981.03	schickendantzii	Rte.7&9 jnc. S Aimogasta
982.01	glaucum ferrari	Villa Mazan
983.02	hossei	W Cuesta de Cebila
984.02	hossei	Cuesta Cebila
984.03	kieslingii	Cuesta Cebila
985.03	oenanthemum (ambatoense)	S Concepcion
985.04	schickendantzii	S Concepcion
986.03	bodenbenderianum	N San Martin
987.02	bodenbenderianum	S San Martin
987.03	schickendantzii	S San Martin
988.02	saglionis	Casa de Piedra
988.03	schickendantzii	Casa de Piedra
989.01	bodenbenderianum	Recreo
991.01	quehlianum	E Quilino
992.01	quehlianum	S Quilino
993.01	calochlorum	W Avellaneda
993.02	mostii valnicekianum	W Avellaneda
994.02	erinaceum	E Sauce Punco
994.03	monvillei	E Sauce Punco
995.03	quehlianum	S Quilino
995.04	schickendantzii	S Quilino
996.02	quehlianum	S Quilino
996.03	robustum	S Quilino
997.01	calochlorum	N Ischilin
997.02	mostii valnicekianum	N Ischilin
998.01	calochlorum	W Ischilin
998.02	erinaceum	W Ischilin
998.03	mostii valnicekianum	W Ischilin
999.01	calochlorum	Ojo de Agua
999.02	erinaceum	Ojo de Agua
999.03	mostii valnicekianum	Ojo de Agua
1000.01	monvillei	Ongamira
1001.01	bruchii	E La Cumbre
1001.02	monvillei	E La Cumbre
1002.01	amerhauseri	W Ascochinga
1002.02	bruchii	W Ascochinga
1002a.	capillaense	Las Tres Cascades

Index of Gymnocalycium Names

Names in **bold type** are valid accepted taxa
Names in Roman type are valid synonymised taxa
Names in *italic type* are have no valid description

A

B

C

D

E

F

G

H

I

J

K

L

M

N

O

P